Y0-BYK-285

MITOCHONDRIA

STRUCTURE AND FUNCTION

Fifth FEBS Meeting

Volume 15

GAMMA GLOBULINS Structure and Biosynthesis

Volume 16

BIOCHEMICAL ASPECTS OF ANTIMETABOLITES AND OF DRUG HYDROXYLATION

Volume 17

MITOCHONDRIA Structure and Function

Volume 18

ENZYMES AND ISOENZYMES Structure, Properties and Function

FEDERATION OF EUROPEAN BIOCHEMICAL SOCIETIES
FIFTH MEETING, PRAGUE, JULY 1968

Symposium on Mitochondria—Structure and Function, Prague, 1969.

MITOCHONDRIA

STRUCTURE AND FUNCTION

Volume 17

Edited by

L. ERNSTER

Department of Biochemistry
University of Stockholm
Stockholm, Sweden

Z. DRAHOTA

Institute of Physiology
Czechoslovak Academy of Sciences
Prague, Czechoslovakia

1969

ACADEMIC PRESS · London and New York

272682

ACADEMIC PRESS INC. (LONDON) LTD.
Berkeley Square House
Berkeley Square
London, W1X 6BA

U.S. Edition published by
ACADEMIC PRESS INC.
111 Fifth Avenue
New York, New York 10003

Copyright © 1969 by the Federation of European Biochemical Societies

SBN: 12-241250-8
Library of Congress Catalog Card Number: 71-107022

Printed in Great Britain by
Spottiswoode, Ballantyne and Co. Ltd
London and Colchester

Contents

FEBS Symposium, Volume 17, 1969, pp. 1-4

Opening Remarks

L. ERNSTER

Department of Biochemistry, University of Stockholm, Stockholm, Sweden

It is my great honour and pleasure to open this *Symposium on Mitochondria–Structure and Function.*

We are indebted to the Organizers of the Fifth FEBS Meeting for having allotted time and space to a symposium on this subject, which is so close to our hearts. In particular, I wish to thank my co-chairman, Dr. Zdanek Drahota, who has carried the major part of the burden of organizing this Symposium, which, with 31 invited and 115 contributed papers on its programme, appears to be the largest meeting on mitochondria ever held in Europe.

I regret to announce that four invited speakers, Professor J. B. Chappell, Dr. E. J. Harris, Dr. B. Kuylenstierna, and Professor A. L. Lehninger, have not been able to attend the Symposium; however, by fortunate arrangements, the papers of the three last-mentioned speakers will be read by colleagues present at the meeting.

Opening remarks of a symposium should appropriately serve as a framework of the topics to be dealt with. While preparing myself for this task one evening at my desk in Stockholm, it occurred to me that it might be interesting to make a graph representing various trends in the study of mitochondrial function over the past few years. I drew a number of lines, off-hand, in a coordinate system, with the last 7 years as the abscissa and "relative excitement" as the ordinate. Next morning I showed the graph to my colleague Bo Kuylenstierna, who, a man of precision as he is, suggested that we should check my hypothetic lines experimentally. So we did, using the relevant chapters of the past years' issues of *Annual Review of Biochemistry* and calculating, roughly, the percentage of references relating to the various topics represented by the lines in the graph. There was an amazing agreement between "hypothetic" and "experimental". Thus I felt encouraged to show you my graph (Fig. 1).

Study of the respiratory chain seems to have attracted constant interest over the past 7 years. The focus of interest, however, has shifted somewhat, from the two ends of the chain, the NADH and succinate dehydrogenases and cytochrome oxidase–which elicited quite some excitement, for example, at the International

Congress in Moscow in 1961—to its middle, notably, the interrelationships of flavoproteins, ubiquinone, and cytochrome *b*—a topic that Professor Chance will elaborate on at this Symposium.

The chemical nature of high-energy intermediates of the type A~C, X~P, etc., that is high-energy compounds in the classical sense as intermediates of the energy-conservation process of the respiratory chain, has been a major objective of both experimental and guesswork during the 1950's and early 1960's. Questions such as to whether NAD enters this compound in its oxidized or reduced form, or whether the first phosphorylated intermediate occurs at or below the level of the respiratory chain, have been the subject of lively debate at

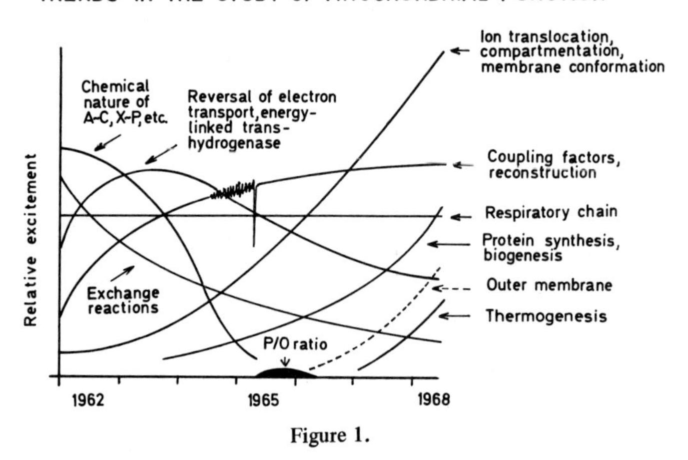

Figure 1.

many meetings. After the identification of protein-bound phosphohistidine as an intermediate of the succinyl thiokinase reaction in 1964, the number of papers in this area has decreased abruptly. Today the discussion rather concerns the very occurrence of covalent high-energy bonds in the process of respiratory chain-linked energy-conservation.

The number of published papers has likewise decreased in the field of various exchange reactions, including the P_i-H_2O, P_i-ATP, and ATP-ADP exchange. Although these reactions certainly do reflect a great deal of information of fundamental importance, their final interpretation may not be easy in the very complex systems in which they have mostly been studied so far. It may be expected, however, that we shall witness a comeback of this approach, when the various reactions can be investigated with simplified, reconstructed systems.

A very important generalization and extension of the concept of respiratory chain-linked energy transfer was achieved by the discovery of the reversibility of oxidative phosphorylation in the late 1950's. During the very fertile period that followed—and that perhaps reached its peak at the time of the First Johnson

Foundation Colloquium on Energy-Linked Functions of Mitochondria in 1963–it was first realized that ATP synthesis is but one facet of mitochondrial energy conservation, and that energy derived from the respiratory chain can be utilized for driving a number of energy-requiring processes directly, without the participation of the phosphorylating system, e.g. the energy-linked nicotinamide nucleotide transhydrogenase reaction. Although the relative number of publications on this topic has decreased somewhat during the last few years, it still occupies a major field of interest, not least from the viewpoint of the physiological regulation of the generation and maintenance of mitochondrial reducing power.

One of the major fields of research in the biochemistry of mitochondria that has evolved during this decade concerns the isolation and characterization of structural and enzymatic components involved in the process of oxidative phosphorylation and related energy-transfer reactions. The most important achievement within this field is the purification of what seems to be the terminal enzyme in ATP synthesis, the coupling factor F_1, and the chemical and morphological reconstruction of the oxidative phosphorylation system by means of this protein. We greatly regret that Professor Efraim Racker, whose group has pioneered in this brilliant achievement, has not been able to accept our invitation to attend this Symposium. As indicated by the graph in Fig. 1, there has been some disturbance in the records of this field which, however, disappeared after a sudden downward spike on a Sunday afternoon in the Spring of 1965. Since then, progress has been steady.

A small bump on the graph, hardly above the noise level, relates to high P/O ratios.

By far the most active field of research in mitochondrial function at the present time concerns problems related to membrane conformation, compartmentation, and ion translocation. Interest in these problems has arisen, on one hand, from experimental findings over the past years concerning shape and volume changes of mitochondria as well as the recognition of the inner mitochondrial membrane as the site of specific ion-translocating functions closely related to the energy-transfer system of the respiratory chain; and, on the other hand, from the chemiosmotic hypothesis of Peter Mitchell, first formulated in 1961, which has stimulated a great deal of experimentation and discussion–perhaps more than any other hypothesis has done in this field of biochemistry. These topics have been the main subject of several symposia during the past few years–among these the 1965, 1966 and 1968 Bari Symposia–and seem to account for more than 50% of all papers currently being published on mitochondria. This proportion also holds for the programme of the present Symposium, where we are glad to see some of the most prominent representatives of these lines of research. A subline of this particular field, indicated by the dotted line in the graph and representing the separation and properties of the outer mitochondrial membrane, is especially young but nevertheless already contains a stimulating amount of controversy.

Another field of exponential growth during the last few years is that of mitochondrial protein synthesis and biogenesis. To cover adequately this most exciting topic it would have been necessary to devote an entire symposium to it, as was the case at the 1967 Bari Symposium, the proceedings of which have just appeared. From the papers relating to this field on our programme, we look forward to learning the latest news concerning the mechanism by which the mitochondrial and extramitochondrial protein-synthetizing machineries of the cell interact in producing and maintaining the molecular organization of the mitochondrion.

Finally, a special problem that has elicited much interest recently concerns the biochemical mechanism of thermogenesis as it occurs in the mitochondria of brown adipose tissue. It often happens in biochemistry that problems exist for a number of years in a relatively dormant state until suddenly, probably when general knowledge in the field has reached a critical stage, they acquire a burst of interest. This apparently has happened in the case of thermogenesis, as judging from the number of papers on brown adipose tissue mitochondria that have appeared during the last few months, together with a symposium entirely devoted to this very subject that was held earlier this year in Washington, D.C., under the sponsorship of the American Oil Chemists' Association.

In terminating my opening remarks, I wish to welcome you to this Symposium, and express my hope that we shall have an exciting and profitable meeting. I may now ask Professor Chance to take the chair of the first session.

FEBS Symposium, Volume 17, 1969, pp. 5-31

Structure, Composition and Function of Mitochondrial Membranes*

L. ERNSTER and B. KUYLENSTIERNA

Department of Biochemistry, University of Stockholm, Stockholm, Sweden

The purpose of this paper is to summarize current information concerning the morphological, physical, chemical, and enzymatic properties of isolated mitochondrial inner and outer membranes, and to discuss this information in relation to the functional organization and biogenesis of mitochondria.

Methods for the separation of the outer and inner mitochondrial membranes

Although the existence of two mitochondrial membranes [1], an outer limiting membrane and a folded inner membrane giving rise to the cristae [2, 3], has been established for a long time, it is only recently that methods have been developed for the separation of the two membranes.

The first successful method for the preparation of what appears to be pure inner and outer membranes was described by Parsons *et al.* [4] in 1966 (see also refs. 5, 6). It is based on the observation [7, 8] that swelling of liver mitochondria in hypotonic phosphate buffer causes a distension and occasional rupture of the outer membrane, while the inner membrane unfolds without breaking. Centrifugation of the swollen mitochondria on a suitable gradient separates small but pure specimens of inner and outer membrane from the bulk of the unbroken mitochondria [4-6]. A somewhat similar principle was used by Schnaitman *et al.* [9, 10], who exposed mitochondria to osmotic lysis in distilled water [11] and separated the "ghosts", consisting of inner membranes, from the fragmented outer membranes by means of gradient centrifugation. Another method, introduced by Lévy *et al.* [12-14], and developed further by Schnaitman *et al.* [9, 10] and by Hoppel *et al.* [15, 16], takes advantage of a selective action of low concentrations of digitonin on the outer membrane, yielding, after suitable centrifugal separation, outer membrane fragments and relatively well-preserved inner membranes in good quantities. A procedure

* Work reported from the authors' laboratory has been supported by grants from The Swedish Cancer Society and the Swedish Medical and Natural-Science Research Councils.

elaborated by Sottocasa *et al.* [17-19] is based on a swelling of mitochondria in hypotonic phosphate, followed by a selective contraction of the inner membrane in the presence of hypertonic sucrose containing ATP and Mg^{2+}, and by gentle sonication to facilitate the detachment of the broken outer membrane. Subsequent density-gradient centrifugation separates outer and inner membranes in a high yield. Brdiczka *et al.* [20] have recently employed a procedure based on freezing of mitochondria at liquid-air temperature in a hypotonic medium to detach the outer membrane, resulting, after density-gradient centrifugation, in inner-membrane structures with a well-preserved matrix.

All the above methods have been worked out with liver mitochondria and have been based primarily on morphological criteria in identifying the two membranes. They have received wide use in the past few years in different laboratories engaged in studies of the intramitochondrial distribution of enzymes and other chemical constituents, and the results have shown remarkable agreement among the various procedures and research groups. Moreover, the results are consistent with conclusions drawn from studies with intact mitochondria.

Green and associates [21-33] have devised a number of procedures, involving disruption of the mitochondria by sonication [20, 22, 24, 25], freezing and thawing [22], organic solvents [23], detergents [25, 26], phospholipase [23, 24, 26-30], fatty acids [29, 30], or diethylstilbestrol [31-33], with the purpose of isolating outer and inner membranes from various tissues. These procedures have been based primarily on biochemical criteria for the identification of the two membranes. The conclusions reached by Green and associates regarding the intramitochondrial distribution of enzymes and other constituents differ in several important respects from those arrived at by most other laboratories. In the sections that follow we shall first summarize these more generally supported conclusions, and subsequently consider possible reasons for the discrepancies between these and the conclusions of Green and associates.

Morphological, physical and chemical characteristics

Table 1 is a survey of some morphological, physical and chemical characteristics of the inner and outer membranes of liver mitochondria. Morphologically, the two membranes are approximately equal in thickness, 50-70 Å (ref. 7), but differ of course markedly in surface area, which in the case of the inner membrane is greatly enlarged owing to the cristae. Although projecting subunits have been observed on the outer surface of the outer membrane [8, 34], they differ from the inner membrane subunits of Fernández-Morán [35, 36] in regards to size, shape and structure. After extraction of 90% of the mitochondrial phospholipid by acetone, the inner membrane retains its double-layered structure whereas the outer membrane is destroyed [37]. From

Table 1. Some morphological, physical and chemical characteristics of the inner and outer membranes of liver mitochondria

	Inner membrane	Outer membrane
Morphological		
Thickness	50 Å–70 Å	50 Å–70 Å
Shape	Folded	Distended
Surfaces	Outer surface smooth, inner surface covered with regularly spaced projecting subunits	Inner surface smooth, occasional projections on outer surface
Effect of phospholipid extraction	Double-layered structure retained	Double-layered structure destroyed
Physical		
Density	1·192-1·230	1·094-1·122
Permeability	Only uncharged molecules of MW 100-150	Most substances up to MW ~ 10,000
Osmotic behaviour	Reversible unfolding and refolding, contraction	Irreversible stretching, rupture
X-ray diffraction pattern	Fundamentally similar	
Chemical		
Phospholipid/protein (w/w)	0·27	0.82
Cardiolipin	High	Low
Phosphatidyl inositol	Low	High
Cholesterol	Low	High
Ubiquinone	Present	Absent

isopycnic centrifugation data the inner membrane appears to have a density of about 0·1 unit higher than the outer membrane [4].

X-ray diffraction patterns of the two membranes show a fundamental similarity [38, 39] despite striking differences in physical and chemical properties.

Very striking differences in physical properties between the two membranes are found in regards to osmotic behaviour and permeability. Extensive studies [40-45], based on measurements of the space occupied by various substances present in the medium in relation to the total water space of the mitochondria, and correlated with morphological observations, have led to the conclusion that the inner mitochondrial membrane possesses only a very limited permeability to most substances except uncharged molecules of a molecular weight not greater than 100-150. The majority of charged molecules of physiological importance pass through the inner membrane by way of specific translocators associated with this membrane (see p. 15 and Table 2). In contrast, the outer membrane seems to be freely permeable to a wide range of substances, both charged and uncharged, up to a molecular weight of about 10,000. An important implication of this concept is that all low-molecular charged components present in isolated mitochondria, including various nucleotides, are located inside the inner membrane.

The most pronounced chemical features found so far that distinguish the two membranes, relate to their lipid composition [5, 9, 46-51]. The outer membrane contains, on the protein basis, 2 to 3 times more total phospholipid than the inner membrane, amounting to a phospholipid content of about 45% of the dry weight [5]. Qualitative differences in phospholipid composition are most striking in the case of cardiolipin, which occurs predominantly–if not exclusively–in the inner membrane, and phosphatidyl inositol, which is found more abundantly in the outer membrane [5]. Among neutral lipids, cholesterol has been reported to be 6 times more concentrated, on the protein basis, in the outer than in the inner membrane [46]; this explains the preferential action of digitonin on the outer membrane. Ubiquinone seems to be present only in the inner membrane [52].

Mitochondrial compartments

Before discussing the intramitochondrial distribution of enzymes, it may be appropriate to define the various mitochondrial compartments. According to our present notions, the mitochondrion contains four compartments where enzymes may be localized (Fig. 1): the outer membrane; the intermembrane space, i.e. the space between the outer and inner membranes, including the space bordered by the outer surface of the cristae; the inner membrane, including the cristae and their projecting subunits; and the matrix, i.e. the space within the inner surface of the inner membrane.

Occasionally, the terms "cristae space" and "intracristae space" have been used in the literature to denote what is defined here as the inner membrane and the intermembrane space, respectively. These terms seem to be somewhat unfortunate, since they relate to the infolding portions of the inner membrane and not to the inner membrane as a whole; for example, a swollen mitochondrion may have as much of an inner membrane space as an intact one, although it may be largely devoid of cristae, the latter having been unfolded as a result of the swelling. Also, the term "intracristae space" (i.e. the space *within*

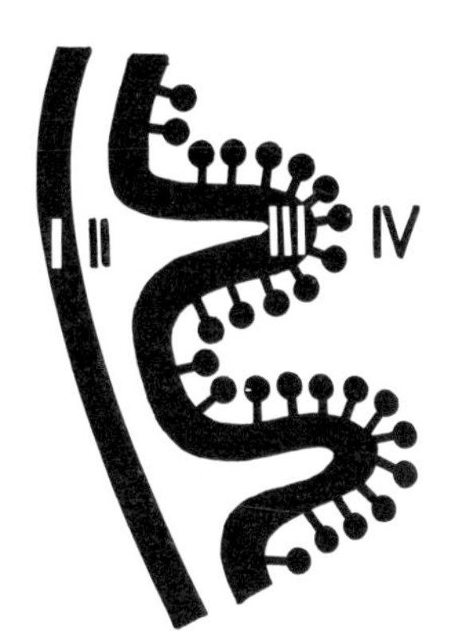

Figure 1. Definition of mitochondrial compartments.

the cristae) does not give a clear indication as to whether it refers to the space within the *in*foldings or the *out*foldings of the cristae, or, which would seem to be the most literal interpretation, within the cristae membrane itself.

Intramitochondrial distribution of catalysts

Current concepts regarding the localization of enzymes in the outer and inner mitochondrial membranes are based on results showing a concentration of these enzymes in isolated outer and inner membranes. In the case of enzymes localized in the intermembrane space or in the matrix, the conclusions are based, in addition to fractionation data, on certain assumptions. These involve considerations regarding, on the one hand, the possible leakage of enzymes from and through the membranes during the fractionation procedure, and, on the other hand, the permeability properties of the two membranes as established in studies with intact mitochondria. Some pertinent problems are illustrated in Fig. 2, which indicates the distribution and recovery of various enzyme activities upon subfractionation of rat-liver mitochondria by the procedure of Sottocasa *et al.* [19]. This procedure yields three subfractions: a "heavy", a "light", and a "soluble" subfraction. From electron microscopic examination, the heavy subfraction consists of inner membranes including part of the matrix, and the light subfraction consists of vesicles derived from the outer membrane. The soluble subfraction ought to include the contents of the intermembrane space, together with part of the matrix and any material released from the two membranes during the fractionation procedure.

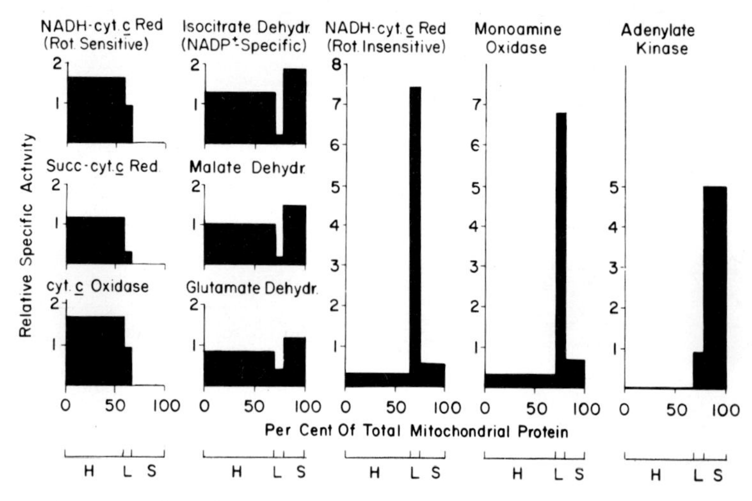

Figure 2. Distribution of some enzyme activities in the "heavy" (H), "light" (L), and "soluble" (S) fractions of rat liver mitochondria, prepared by the method of Sottocasa *et al.* [19].

As may be seen in Fig. 2, the enzymes investigated show four types of distribution pattern. One type, represented by cytochrome oxidase, succinate-cytochrome *c* reductase, and the rotenone-sensitive NADH-cytochrome *c* reductase, is concentrated in the heavy subfraction, and thus constitutes enzymes associated with the inner membrane. A second type, represented by the rotenone-insensitive NADH-cytochrome *c* reductase and monoamine oxidase, is concentrated in the light subfraction, and thus constitutes enzymes associated with the outer membrane. A third type of enzymes, exemplified by adenylate kinase, is concentrated in the soluble subfraction, with little activity found in the light subfraction, and practically none in the heavy one. By the above definition of the origin of the three subfractions, this type of enzyme may originate either from the intermembrane space or from one of the two membranes. It is unlikely to originate from the matrix, since, in that case, part of its activity ought to have been recovered in the heavy subfraction. Furthermore, the fact that adenylate kinase in the intact mitochondria is insensitive to atractylate as measured with external adenine nucleotides as substrates [53], eliminates its localization in the inner membrane, where the atractylate-sensitive ADP-ATP exchange system is located [42]. Thus adenylate kinase is most probably located in either the intermembrane space or as a loosely bound enzyme in the outer membrane.

A fourth type of distribution pattern is represented by the enzymes glutamate dehydrogenase, malate dehydrogenase, and the NADP-specific isocitrate dehydrogenase. These enzymes show a bimodal distribution, part of them being recovered in the heavy, and part in the soluble subfraction, with little activity appearing in the light subfraction. Furthermore, the specific activity of the soluble subfraction is higher than that of the heavy subfraction. This pattern of distribution is consistent with the conclusion that these enzymes are located in the matrix. The higher specific activity found in the soluble subfraction as compared with that in the heavy one is readily explained by the fact that the latter is made up to an appreciable portion by inner membrane material. If these dehydrogenases were located in the outer membrane or the intermembrane space, as postulated by Green and associates [21-33], one would not expect to find any significant activities in the heavy subfraction obtained by the present procedure.

Table 2 is a survey of current information concerning the intramitochondrial distribution of various catalysts including enzymes, coenzymes and translocators.

The enzymes so far found in the outer membrane represent a rather heterogeneous group from the functional point of view, and their localization in the outer membrane could not always be readily predicted from studies with intact mitochondria. For example, although the occurrence of monoamine oxidase in mitochondria has been known for a long time [54-57], it is only when the outer membrane has been isolated that its localization in this membrane has been recognized [9]. The function of this enzyme in mitochondria is not known. Kynurenine hydroxylase is another example of an enzyme of unexpected localization [58, 59], especially as most other NADPH-linked hydroxylases in the liver are associated with the endoplasmic reticulum (for review, see ref. 60). A rotenone-, amytal-, and antimycin A-insensitive NADH cytochrome *c* reductase [61-67], separate from the respiratory chain [61, 65, 67-70] and similar to the NADH-cytochrome b_5 reductase-cytochrome b_5 system of microsomes [71-74], has long been known to occur in mitochondria, and its recently-established localization in the outer membrane [17, 18] fits logically with its preferential reactivity with external NADH in the intact mitochondria. The function of this enzyme system or of its microsomal

Table 2. Localization of catalysts within rat-liver mitochondria

	Enzymes	Coenzymes*	Translocators
Outer membrane	NADH-cyt. b_5 reductase		
	Cyt. b_5		
	Kynurenine hydroxylase		
	Monoamine oxidase		
	ATP-dep. fatty-acyl-CoA synthetase		
	Fatty-acid-elongating system (C_{14}, C_{16})		
	Glycerol phosphate-acyl transferase		
	Lysophosphatidate-acyl transferase		
	Lysolecithin-acyl transferase		
	Cholinephosphotransferase		
	Phosphatidate phosphatase		
	Phospholipase A_{11}		
	Nucleoside diphosphokinase		
Intermembrane space	Adenylate kinase		
	Nucleoside diphosphokinase		
Inner membrane	Cyt. b, c_1, c, a, a_3	Ubiquinone	Anion- and cation-translocating systems
	NADH dehydrogenase		
	Succinate dehydrogenase		
	Resp. chain-linked phosph. system		
	β-Hydroxybutyrate dehydr.		

	Pyr. nucl. transhydrogenase	
	Choline dehydrogenase	
	Fatty-acyl CoA dehydr., ETF?	
	Fatty-acyl CoA-carnitine transferase	
	Fatty-acid-elongating system (C_{10})	
	Ferrochelatase	
	δ-Aminolevulinic acid synthetase?	
Matrix	Pyr. nucl.-linked dehydr.	NAD(H), NADP(H)
	(*except* β-OH-but. dehydr.)	CoA
	Citrate synthase	
	Aconitase	
	Fumarase	
	Transaminases	
	Succinyl thiokinase	
	Nucleoside monophosphokinase	
	GTP-dep. fatty-acyl-CoA synthetase	
	ATP-dep. fatty-acyl-CoA synthetase	
	Fatty-acyl-CoA dehydr., ETF?	
	δ-Aminolevulinic acid synthetase?	
	RNA-and protein-synthetizing systems	

* The list does not include coenzymes firmly associated with their apoproteins, such as FMN, FAD, α-lipoic acid, thiamine pyrophosphate, biotin, pyridoxal phosphate, which ought to be located along with the respective enzymes.

counterpart is not yet known, although, in the latter case, its possible involvement in the desaturation of fatty acids has recently been suggested [75].

The localization of certain ATP- and ADP-involving enzyme reactions in the outer membrane, including an ATP-dependent fatty-acyl-CoA synthetase [76, 77] (specific for long-chain fatty acids [78, 79]), adenylate kinase [10, 19, 20, 80, 81], and nucleoside diphosphokinase [80, 81] (the latter two possibly being located in the intermembrane space), is in accordance with the findings that these enzymes when assayed in the intact mitochondria with external ATP or ADP as substrate are insensitive to atractylate [53, 77, 82, 83]. Nucleoside monophosphokinase has been concluded to occur in the outer membrane or the intermembrane space, although the bulk of the enzyme is found in the inner membrane or the matrix [80]. A dual localization has also been reported in the case of the mitochondrial fatty acid elongating system, with a preferential activity for C_{14} and C_{16} fatty acids in the outer membrane, and for C_{10} fatty acid in the inner membrane [84]. The occurrence of a fatty acid synthetizing system in both the outer and inner membrane of rabbit heart mitochondria has been reported by Whereat *et al.* [85]. Pette [86] and Klingenberg and Pfaff [42] have concluded that creatine kinase in heart mitochondria has a localization similar to that of adenylate kinase in liver mitochondria. Rose and Warms [87] have found that the outer membrane of liver mitochondria contains the majority of the mitochondrial binding sites for hexokinase. In ascites tumour mitochondria, which contain bound hexokinase, the enzyme reaction is insensitive to atractylate with external ATP as substrate, indicating that it is located in the outer membrane.

The outer membrane also contains the enzymes glycerolphosphate- [51, 88], lysophosphatidate- [51], and lysolecithin- [51] acyl transferase, cholinephosphotransferase [51], as well as phosphatidate phosphatase [51] and phospholipase A_{II} [89] (the latter being distinct from the phospholipase A_I present in microsomes [90]), suggesting a role of the outer membrane in phospholipid metabolism. Such a function is further indicated by recent reports of Bygrave and associates who have found that the incorporation of [^{14}C]choline [50] and [^{14}C]serine [91] into the phospholipid of isolated rat-liver mitochondria takes place primarily in the outer membrane. This situation is in contrast to that found for the incorporation of amino acids into protein which takes place exclusively in the inner membrane [92-94].

In agreement with early morphological [2, 3] and histochemical [95] indications, the inner mitochondrial membrane is the site of the respiratory chain catalysts [4, 5, 10, 12-14, 17, 18, 20] as well as the associated phosphorylation process [10, 11, 14, 81]; the latter is probably located in the projecting subunits of Fernández-Morán [35], as revealed by studies of Racker and associates [96, 97]. A strict orientation of the respiratory chain within the inner membrane is indicated by the fact that NADH [61, 98] and succinate [99,

100] can reach the chain preferentially from the inside, and cytochrome *c* preferentially from the outside [101-103]. Whereas the succinate [17-20], NADH [17-20], choline [50] and glycerol phosphate [86] dehydrogenases are tightly bound to the membrane, other flavoproteins, such as lipoyl dehydrogenase [10, 104], sarcosine dehydrogenase [105], as well as the flavoenzymes involved in fatty acid oxidation [106] are either easily dissociated or located in the matrix. Firmly associated with the inner membrane are also the nicotinamide nucleotide transhydrogenase [107, 108] and β-hydroxybutyrate dehydrogenase [10, 76].

Of great interest are recent reports concerning the localization of ferrochelatase [109, 110] and δ-aminolevulinic acid synthetase [109] in the inner membrane. These findings suggest a role of the inner membrane in heme and porphyrin synthesis, with a possible regulatory function in the ribosomal synthesis of cytochromes.

As already mentioned, the inner membrane is believed to be the site of specific translocators for various charged molecules of physiological significance which cannot readily diffuse through this membrane. Specific anion translocators described up to now [111] include those for phosphate (and arsenate); malate and succinate (and certain non-physiological dicarboxylic acid, but not fumarate); α-ketoglutarate; citrate, isocitrate, and *cis*-aconitate; glutamate; aspartate; as well as the atractylate-sensitive ATP-ADP exchange carrier system. Also in this category of catalysts belongs, from the functional point of view, the enzyme fatty-acyl-CoA-carnitine transferase [76], which is associated with the inner membrane, and is responsible for the translocation of fatty acids. The inner membrane also contains the binding sites for various divalent cations involved in their energy-linked translocation across the inner mitochondrial membrane [112-114], as well as for different ionophoric antibiotics, such as valinomycin or gramicidin, that facilitate the penetration of univalent cations (for review, see ref. 115).

Enzymes believed to be localized in the matrix include those involved in the citric acid cycle (except succinate dehydrogenase) [5, 9, 10, 19, 20, 42] and related processes such as substrate-level phosphorylation, the pyruvate and phosphopyruvate carboxylase reactions [116], glutamate oxidation [10, 19, 20, 42, 76, 116], transamination [10], citrulline synthesis [10], a GTP- and an ATP-dependent fatty-acyl-CoA synthetase reaction (the latter specific for medium- and short-chain fatty acids [78, 79]), and fatty acid oxidation [20, 116a]. The evidence for the localization of these enzymes, which is in sharp contrast to the conclusions reached by Green and associates [21-33] (see below) rests, as already pointed out, on a combination of information from fractionation studies (to which reference is made above) and experience with intact mitochondria. As present in the intact mitochondria, these enzymes show so-called "latency", i.e. they reveal only limited activity towards externally added substrates or coenzymes which do not readily penetrate the inner mitochondrial membrane or the penetration of which, by way of a specific translocator, has been blocked. Upon removal of the outer membrane, part or the whole of these enzymes will remain in the inner membrane fraction (depending on the degree of damage to the inner membrane and consequent leakage of matrix occurring during the fractionation procedure) and may show latency just as in the intact mitochondria. Subsequent fragmentation of the inner membrane by suitable mechanical or chemical means will abolish the latency, and the bulk of the matrix enzymes will appear in the soluble fraction.

The matrix seems to be the site of mitochondrial DNA, as revealed by electron microscopic evidence [117]. Mitochondrial RNA has been found to be localized mainly in the "inner membrane fraction" (probably including part of the matrix), with a minor portion of the RNA being found in the outer membrane [92]. Although the localization of the mitochondrial DNA-dependent RNA polymerase and amino acid-activating enzyme system has not yet been established, the fact, already mentioned, that the amino acid incorporation into protein catalysed by isolated mitochondria is recovered exclusively in the inner membrane fraction [92-94], together with the localization of the mitochondrial DNA in the matrix [117], strongly suggests that the mitochondrial RNA- and protein-synthetizing system as a whole is located in the matrix space.

Are the citric acid cycle and related enzymes associated with the outer membrane?

As already pointed out, Green and associates [21-33] have based their methods for the separation and identification of the outer and inner mitochondrial membranes primarily on biochemical rather than morphological criteria. Their morphological definition of the two membranes has varied somewhat over the past few years, as illustrated in Fig. 3. In 1965, Green [21] visualized the mitochondrion as consisting of an outer double-membrane and a row of separate, closed inner membranes—a picture reminiscent of the 1953 model of Sjöstrand [1]. A year later Green and Perdue [22] adapted Palade's definition [3] of the mitochondrial structure, consisting of a single outer membrane surrounding a continuous, folded inner membrane; the latter is coated on its inner surface by the repeating subunits of Fernández-Morán [35], and the former on its outer surface by those described by Parsons [4]. The same year Green [118] proposed a further model, which may be denoted as a hybrid between the two earlier ones. According to this model "The mitochondrion appears to be built up of two membrane systems which closely interlock—an outer membrane system that encloses the mitochondrion and a system of inner membranes that radiate into the interior from the periphery. In the intact mitochondrion the two tubular systems are fused and the space within the tubules of one system is continuous with the space in the tubules of the other". In still a further modification [119] of this model, the area of fusion between the two tubular systems has been reduced to an orifice (similar to that earlier described by, e.g. Whittaker [120] and the subunits of the outer membrane have been moved from the outer surface to the space between the "outer" and "inner limiting membrane" components of the outer membrane system [30]. The two

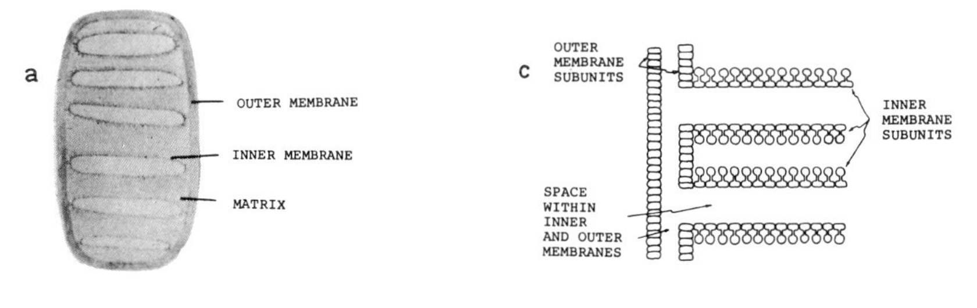

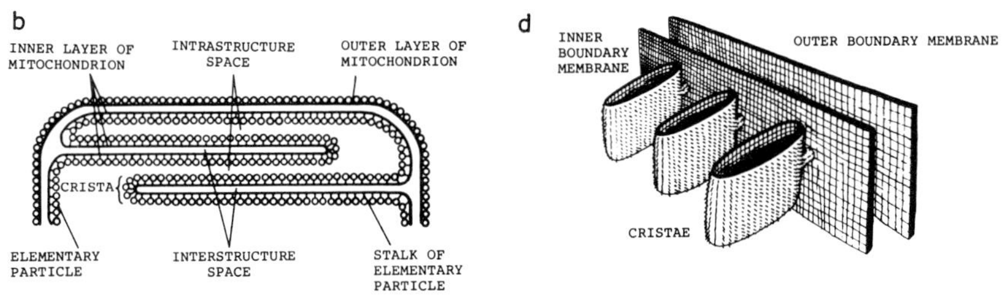

Figure 3. Mitochondrial membrane structures as envisaged by Green and associates 1965-68.
(a) Green [21].
(b) Green and Perdue [22].
(c) Green [118].
(d) Penniston et al. [119].

reasons given for regarding the "inner boundary membrane" as part of the outer membrane, in spite of its recognized continuity with the inner tubular system (i.e. the cristae), were: its postulated lack of projecting subunits, and its postulated identity in chemical composition with the outer limiting membrane. Both of these postulates, however, lack experimental support, and, indeed, it is difficult to see how a physical separation of the outer and inner mitochondrial membranes would ever be technically feasible if these latest models of Green and associates were correct. It is also noteworthy that these models, which have been derived from studies with beef heart mitochondria, are in striking disagreement with some excellent electron micrographs on the same material, and published from the same laboratory [121], which clearly show the detachment of a single outer membrane from the mitochondria upon swelling.

The biochemical criteria used by Green and associates [21-33] for the separation and identification of the two mitochondrial membranes were based essentially on three postulates: (1) that the inner membrane is the site of the respiratory chain; (2) that little if any mitochondrial protein is not membrane bound; and (3) that any enzyme found in the soluble form after subfractionation of beef heart or liver mitochondria by various procedures originates from the outer membrane. Based on these postulates, the conclusion was reached that all enzymes involved in the citric acid cycle (with the exception of succinate dehydrogenase), in fatty acid oxidation and prolongation, and in substrate-level phosphorylation, are in the intact mitochondrion associated with the outer membrane. It was further concluded that the outer membrane constitutes the permeability barrier for nicotinamide nucleotides as well as the site for the carnitine-mediated transport of fatty acids and the atractylate-sensitive translocation of ADP and ATP [122, 123].

While postulate (1) above is supported by ample evidence from other laboratories [2, 4, 5, 17, 18, 95], postulate (2) remains hypothetical and may, indeed, be difficult to prove experimentally. Regarding postulate (3), the evidence so far provided by Green and associates comes from attempts to recover citric acid cycle and related enzymes in a particulate submitochondrial fraction devoid of respiratory chain activity. Such a fraction, by definition, would represent the outer membrane. The seemingly strongest evidence of this kind published by Green and associates up to now comes from experiments [25] in which sonication, in the absence or presence of cholate, was used to disrupt beef heart mitochondria, and the resulting subfractions were separated by differential centrifugation. The postulated "outer membrane" subfractions, denoted as F_s as obtained in the absence of cholate and F_c in its presence, were reported to exhibit various citric acid cycle enzyme activities that were 2-3 times higher, on the protein basis, than that of the sonicated starting material. However, closer examination of the data reveals that the yield of the various enzyme activities recovered was relatively low, ranging between 8 and 23% in the case of F_s, and between 30 and 40% in the case of F_c. Moreover, cytochrome *a*, which was used as a marker for the respiratory chain, was recovered to an extent ranging between 4 and 20% in F_c, and between 23 and 38% in F_c. These data thus show no convincing concentration of citric acid cycle enzymes over those of the respiratory chain in the postulated "outer membrane" subfractions, and suggest that these subfractions rather represent inner membrane vesicles including some matrix.

In another instance, Green and associates, using phospholipase to disrupt mitochondria, described [24] a particulate subfraction devoid of cytochrome *a* but containing 50-60% of the mitochondrial α-ketoglutarate, pyruvate, and β-hydroxybutyrate dehydrogenase activities, while other citric acid cycle and related enzyme activities were found exclusively in the soluble subfraction.

These findings were interpreted by postulating that the two α-keto acid dehydrogenases, which are high molecular weight enzyme complexes, are firmly associated with the outer membrane, whereas other citric acid cycle enzymes, which represent single proteins of relatively low molecular weight, are easily detachable components of the outer membrane, residing in the postulated inner-surface subunits of the latter. While this remains a hypothetic possibility, the converse assumption, namely, that the two α-keto acid dehydrogenase complexes may become attached to, or cosediment with, the outer membrane during the fractionation procedure, can clearly not be excluded. An even more complicated situation seems to prevail in the case of the β-hydroxybutyrate dehydrogenase, where, as Green *et al.* [25] have pointed out, the enzyme is recovered in the outer membrane fraction only with the phospholipase procedure, whereas with the sonication method it is recovered in the inner membrane fraction. According to Green *et al.* [25] this inconsistency can be explained by an artefactual redistribution of the enzyme during the sonication procedure, due to a specific "relocation" of β-hydroxybutyrate dehydrogenase from the outer to the inner membrane. In view of the nature of the two procedures it would appear more logical to assume that the phospholipase treatment, which is a chemical intervention, is more likely to cause such a redistribution of an enzyme (in particular of a phospholipid-dependent enzyme such as β-hydroxydehydrogenase [124]) than is mechanical disruption by sonication.

Intramitochondrial localization of nicotinamide nucleotides

Perhaps the most compelling evidence in favour of the localization of the mitochondrial nicotinamide nucleotide-linked dehydrogenases inside the inner membrane comes from information relating to the localization of the mitochondrial nicotinamide nucleotides themselves. It is well established–and Green has been the pioneer in establishing this [125, 126]–that all mitochondrial nicotinamide nucleotide-linked dehydrogenase reactions, including the oxidations of the citric acid cycle, fatty acids, and glutamate, are catalysed by isolated, intact mitochondria at maximal rate via endogenous nicotinamide nucleotides, without a need for supplementation with externally added coenzymes. Therefore, if the mitochondrial nicotinamide nucleotides were located inside the inner membrane and incapable of rapidly penetrating the latter, then also the nicotinamide nucleotide-linked dehydrogenases ought to be located inside the inner membrane; if the dehydrogenases would reside outside the inner membrane, it would be impossible for them to interact with the mitochondrial nicotinamide nucleotides. The following lines of evidence strongly support the conclusion that, indeed, the mitochondrial nicotinamide nucleotides are located inside the inner membrane and do not readily penetrate the latter:

1. External nicotinamide nucleotides readily penetrate the "sucrose space" of mitochondria (i.e. the intermembrane space) [42], but exchange only very slowly with the intramitochondrial nicotinamide nucleotides [127].

2. Incubation of liver mitochondria at 30°C in a hypotonic medium containing EDTA results in an almost complete release of adenylate kinase–an enzyme located in the outer membrane or the intermembrane space–in less than 1 min; under the same conditions, practically no nicotinamide nucleotides are released from the mitochondria even after 3 h [128].

3. External nicotinamide nucleotides interact only slowly–if at all–with intramitochondrial dehydrogenases, and this interaction is greatly enhanced by agents that cause a swelling or disruption of the mitochondria [129-132]; contraction of the mitochondria restores the "latency" of these enzymes [130-132].

4. Contraction of swollen liver mitochondria–i.e. selective restoration of the inner membrane–results in a restricted accessibility of external nicotinamide nucleotides to the mitochondrial dehydrogenases [66]. When present during contraction, NAD is reincorporated into the mitochondria [133].

5. "Digitonin particles", which presumably represent derivatives of mitochondria devoid of outer membrane [9, 12, 16], contain endogenous NAD and catalyse the oxidation of β-hydroxybutyrate, but are inaccessible to external nicotinamide nucleotides [134].

6. Added cytochrome *c* enhances the aerobic oxidation of external NADH by intact liver mitochondria [66, 68-70], by a process that is insensitive to amytal [64-66], rotenone [67] and antimycin A [61-63], and that involves an NADH-cytochrome b_5 reductase–cytochrome b_5 system associated with the outer membrane [17, 18]. Addition of NAD and cytochrome *c* fails to restore amytal-inhibited respiration with NAD-linked substrates unless the respective dehydrogenase is also added from the outside [66].

The above lines of evidence, together with the conclusions already discussed regarding the localization of the various substrate-translocating systems in the inner membrane, constitute the strongest argument presently available at level of the intact mitochondrion against the localization of the citric acid cycle and related enzymes in the outer membrane.

Some problems related to the localization of the rotenone-insensitive NADH-cytochrome *c* reductase and monoamine oxidase

Recently Green and associates [29, 30] have seriously criticized the conclusions of other laboratories regarding the localization of the rotenone-insensitive NADH-cytochrome *c* reductase and of monoamine oxidase in the outer membrane of liver mitochondria. They claimed that both of these conclusions were based on experimental artefacts, arising, in one case, from the use of impure preparations of mitochondria, and in the other, from the application of an unsuitable enzyme assay.

The original conclusion [17, 18] that the rotenone-insensitive NADH-cytochrome *c* reductase is a true constituent of the outer membrane of liver mitochondria, rather than merely a microsomal contaminant–a conclusion later confirmed in several laboratories [5, 10, 13]–is supported by the following lines of evidence:

1. Twice-washed rat-liver mitochondria contains about 10 times higher activity, on the protein basis, of this enzyme than of other microsomal enzymes such as glucose-6-phosphatase or various NADPH-linked enzyme systems [17, 18]. Such a concentration of NADH-cytochrome *c* reductase in microsomes in relation to other enzymes has not been found by any physical procedure hitherto employed to subfractionate microsomes, including extensive sonication to produce vesicles of a diameter of one-tenth of the original microsomes [135].

2. The isolated outer membrane fraction exhibits little or no glucose-6-phosphatase or NADPH-linked enzyme activities [17, 18]. Its rotenone-insensitive NADH-cytochrome *c* reductase activity is higher, on the protein basis, than that of both microsomes and microsomal subfractions obtained after sonication.

3. The properties of the components of the outer-membrane NADH-cytochrome *c* reductase differ in several respects from those of the corresponding microsomal enzyme system, including sensitivity of the NADH-cytochrome b_5 reductase to dicoumarol [18], the firmness of the association of cytochrome b_5 with the membrane [18], the reducibility of the cytochrome by cysteine [18], and the low-temperature spectrum of the cytochrome [5].

The objection raised by Green *et. al.* [29, 30] against the above conclusion was based on the finding that further washing of twice-washed liver mitochondria, prepared and washed at relatively high centrifugal forces (10-15,000 x g), removed the rotenone-insensitive NADH-cytochrome *c* reductase activity along with that of glucose-6-phosphatase, i.e. along with the removal of residual microsomal contamination. We have recently carried out similar experiments, using both the centrifugal force employed in our earlier work (6,000 x g) [17, 18] and the high centrifugal force employed by Green *et al.* [29, 30]. The results are shown in Figs. 4 and 5. It may be seen that, with the 6,000 x g preparations, the rotenone-insensitive NADH-cytochrome *c* reductase activity remained virtually constant between 0 and 4 washes, while the glucose-6-phosphatase activity decreased exponentially; similar results were recently reported independently by Beattie [136]. After 2-4 washes, the rotenone-insensitive NADH-cytochrome *c* reductase activity of the mitochondria was 40-50%, and the glucose-6-phosphatase activity 4-5% of that of the microsomes (Fig. 5), which is in good agreement with the values earlier reported [17, 18] for the twice-washed mitochondria (37 and 4%, respectively). Furthermore, the NADPH-cytochrome *c* reductase activity of the 6,000 x g preparations decreased to practically nil after one wash. This finding suggests that contaminating microsomes are readily removed from the mitochondria by a single wash, and that the residual glucose-6-phosphatase activity may originate from non-microsomal—possibly lysosomal—material exhibiting presumably non-specific phosphatase activity. The 15,000 x g preparations exhibited considerably higher rotenone-insensitive NADH-cytochrome *c* reductase, glucose-6-phosphatase and NADPH-cytochrome *c* reductase activities than did those obtained at 6,000 x g, and these were not efficiently removed even after 4 washes. These preparations, which evidently were heavily contaminated with both microsomes and non-microsomal material, are apparently not suitable for studies requiring pure cell fractions, and any conclusions or criticisms based on the use of such preparations may be of doubtful validity.

The conclusion that monoamine oxidase is located in the outer membrane originates from studies of Schnaitman *et al.* [9], who have used the method of Tabor *et al.* [137] to assay the enzyme in submitochondrial fractions obtained by the digitonin procedure. This method is based on the spectrophotometric measurement of benzaldehyde formed from benzylamine in the course of the monoamine oxidase reaction. The conclusion of Schnaitman *et al.* [9] has been confirmed by Sottocasa *et al.* [19], using the same enzyme assay and submitochondrial fractions obtained by the swelling-contraction-sonication procedure. Green *et al.* [29, 30] have pointed out that the enzyme assay based on the determination of an aldehyde may give misleading results in the present connection, owing to the presence of aldehyde dehydrogenase in inner-membrane subfraction which would interfere with the demonstration of any monoamine oxidase activity associated with this fraction. By employing another enzyme assay, devised by McCaman *et al.* [138] and based on the use of [^{14}C]tyramine as the substrate for monoamine oxidase, they arrived at the conclusion that the enzyme is concentrated in the inner-membrane subfraction of liver mitochondria obtained by either phospholipase treatment of oleate-induced swelling and subsequent contraction in the presence of ATP.

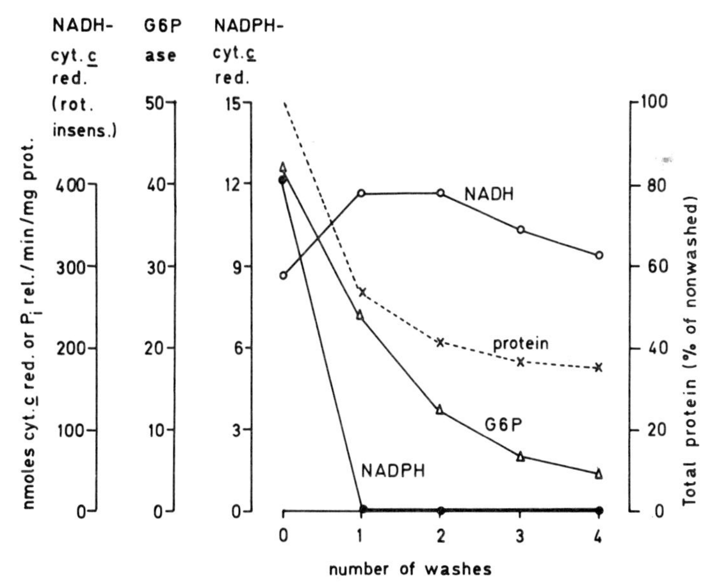

Figure 4. Effect of repeated washings on enzymatic activities of rat liver mitochondria. The microsomes and first mitochondrial pellet were prepared as described by Sottocasa *et al.* [18]. The resuspended mitochondria were sedimented by centrifugation at 6,000 x g for 15 min during the four washings. The enzymatic activities were assayed as described by Sottocasa *et al.* [18].

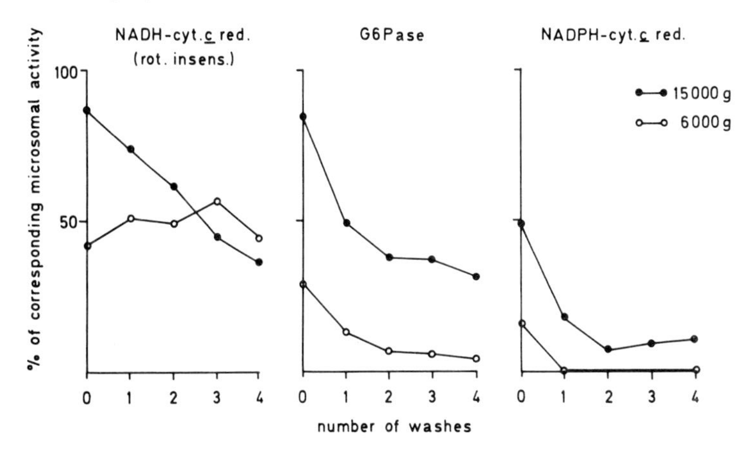

Figure 5. Effect of washing procedure upon the removal of microsomal enzymes from rat liver mitochondria. The 15,000 x g preparation of mitochondria was obtained from a 10% (w/v) liver homogenate in 0·25M sucrose. After sedimentation of the nuclear fraction at 600 x g for 10 min, mitochondria were sedimented from the supernatant by centrifugation at 15,000 x g for 10 min. The fluffy layer was carefully discarded. The pellet was washed four times by resuspension in 0·25M sucrose and centrifugation at 15,000 x g for 10 min. Preparation of microsomes, 6,000 x g preparation of mitochondria and enzyme assays as in Fig. 4.

Clearly, the criticism of Green *et al.* [29, 30] regarding the use of the enzyme assay based on aldehyde determination is perfectly valid and this assay *would* lead to false conclusions *if* the monoamine oxidase were located in the inner membrane. However, the data reported by Green *et al.* [29, 30] from which they conclude that the latter is indeed the case are, in turn, of doubtful validity, because of the lack of convincing documentation of their procedures employed for the subfractionation of mitochondria and of the identity and purity of the subfractions. Indeed, as shown in Table 3, using the fractionation procedure of

Table 3. Distribution of monoamine oxidase activity in subfractions of rat liver mitochondria

Fraction	Protein		nmoles tryptamine oxidized			
	mg	%	/min	%	/min /mg prot.	rel.
Mit. after swelling-contraction-sonication						
Before centrifugation	106·3	(100)	213	(100)	2·0	(1·00)
After centrifugation						
"Heavy" (inner membrane)	55·4	52·1	40	18·7	0·7	0·35
"Light" (outer membrane)	12·2	11·5	140	65·6	11·5	5·75
Soluble	37·1	34·9	31	14·6	0·8	0·40
Yield		98·5		99·0		

Activity was measured according to Wurtman and Axelrod [139], except that the substrate tryptamine-2-^{14}C (spec. act. = 10·3 mc/nmole) was added as bisuccinate. Mitochondrial subfractions were prepared according to Sottocasa *et al.* and separated on a 3-layer gradient [19].

Sottocasa *et al.* [19] and an enzyme assay similar to that employed by Green *et al.* [29, 30], involving the use of [^{14}C]tryptamine [139], we have obtained clear evidence for the concentration of monoamine oxidase in the outer membrane of rat-liver mitochondria, in agreement with the earlier conclusion of Schnaitman *et al.* [9]. Independently, Beattie [136] and Schnaitman and Greenawalt [10] have reported similar evidence, based on the [^{14}C]tyramine method. In addition, the latter authors have shown that chloral hydrate, in a concentration that inhibits aldehyde dehydrogenase to an extent of 80-90%, did not alter the distribution pattern of monoamine oxidase as determined by the benzaldehyde assay.

The criticism raised by Green *et al.* [29, 30] against the conclusions of other laboratories regarding the localization of rotenone-insensitive NADH-cytochrome *c* reductase and of monoamine oxidase thus seems to lack valid experimental ground.

Concluding remarks

For a long time, our knowledge of the chemical and enzymatic organization of mitochondria has been based almost exclusively on information from studies with intact mitochondria. Although sub-mitochondrial particle preparations of various kinds, exhibiting respiratory enzyme activity and identified with fragments of the inner mitochondrial membrane, have been studied extensively over the past years, relatively little attention has been paid to the question as to the chemical and enzymatic topography of the rest of the mitochondrion, namely the outer membrane, the intermembrane space, and the matrix.

Methods developed in recent years for the separation of the mitochondrial inner and outer membranes have made it possible to approach this problem. In fact, progress has been amazingly rapid in this field, and as a result, there is now information available on the intramitochondrial distribution of almost all enzymes and other chemical constituents so far known to be present in mitochondria. In so far as the methods employed have been adequately documented morphologically as well as biochemically, they have yielded results of remarkable reproducibility and agreement among various laboratories.

There is still some uncertainty regarding the localization of those enzymes which are recovered in the soluble form after the separation of the two membranes, and it is difficult in some cases to decide whether a given enzyme is present in the intact mitochondrion in the truly soluble form or as loosely attached to one of the membranes. On the other hand, the available information allows in most cases a clear distinction between enzymes located in the outer membrane or the intermembrane space, and those located in the inner membrane or the matrix. Moreover, the results of the majority of fractionation studies are in agreement with experience from work with intact mitochondria.

The inner membrane and the matrix constitute the mitochondrion in the classical functional sense, in that this "inner compartment" embodies the enzymic complements of substrate oxidation, respiration, and energy conservation. It also constitutes the mitochondrion as a functional entity, with its permeability barriers and specific ion-translocating systems, as well as its own machinery for DNA, RNA and protein synthesis, responsible for its genetic semi-autonomy. It is now quite obvious that ideas put forward in recent years concerning the possible bacterial origin of mitochondria (cf. refs. 140, 141) are pertinent to this inner mitochondrial compartment only.

The available information concerning the enzymic composition of the "outer compartment" of mitochondria suggests a rather diversified function, which in most cases lacks obvious metabolic relation to the function of the inner compartment. In fact, the ready permeability of the outer membrane to substances of low molecular weight suggests that the outer compartment of the mitochondria in the intact cell is in equilibrium with the cytoplasm with respect to free metabolities.

Several authors have pointed out similarities in enzyme composition between the outer mitochondrial membrane and the endoplasmic reticulum [5, 19, 51] and discussed these in relation to a possible phylogenetic and/or ontogenetic relationship between the two membranes [5, 19]. It would be interesting in the future to extend these studies to a comparison of the chemical and enzymic properties of the outer mitochondrial membrane with other intracellular membranes as well, such as the nuclear, lysosomal, and the Golgi membranes, and with the plasma membrane.

REFERENCES

1. Sjöstrand, F. S., *Nature, Lond.* **171** (1953) 30.
2. Palade, G. E., *Anat. Rec.* **114** (1952) 427.
3. Palade, G. E., *J. Histochem. Cytochem.* **1** (1953) 188.
4. Parsons, D. F., Williams, G. R. and Chance, B., *Ann. N.Y. Acad. Sci.* **137** (1966) 643.
5. Parsons, D. F., Williams, G. R., Thompson, W., Wilson, D. and Chance, B., *in* "Round Table Discussion on Mitochondrial Structure and Compartmentation" (edited by E. Quagliariello, S. Papa, E. C. Slater and J. M. Tager), Adriatica Editrice, Bari, 1967, p. 29.
6. Parsons, D. F. and Williams, G. R., *Meth. Enzym.* **10** (1967) 443.
7. Parsons, D. F., *Int. Rev. exp. Pathol.* **4** (1965) 1.
8. Wtodawer, P., Parsons, D. F., Williams, G. R. and Wojtczak, L., *Biochim. biophys. Acta* **128** (1966) 34.
9. Schnaitman, C., Erwin, V. G. and Greenawalt, J. W., *J. Cell Biol.* **32** (1967) 719.
10. Schnaitman, C. and Greenawalt, J. W., *J. Cell Biol.* **38** (1968) 158.
11. Caplan, A. I. and Greenawalt, J. W., *J. Cell Biol.* **31** (1966) 455; **36** (1968) 15.
12. Lévy, M., Toury, R. and André, J., *C.r. hebd. Séanc. Acad. Sci., Paris* Sér. D **262** (1966) 1593.
13. Lévy, M., Toury, R. and André, J., *C.r. hebd. Séanc. Acad. Sci., Paris* Sér. D **263** (1966) 1766.
14. Lévy, M., Toury, R. and André, J., *Biochim. biophys. Acta* **135** (1967) 599.
15. Hoppel, C. and Cooper, C., *Biochem. J.* **107** (1968) 367.
16. Morton, D. J., Hoppel, C. and Cooper, C., *Biochem. J.* **107** (1968) 377.
17. Sottocasa, G. L., Ernster, L., Kuylenstierna, B. and Bergstrand, A., *in* "Round Table Discussion on Mitochondrial Structure and Compartmentation" (edited by E. Quagliariello, S. Papa, E. C. Slater and J. M. Tager), Adriatica Editrice, Bari, 1967, p. 74.
18. Sottocasa, G. L., Kuylenstierna, B., Ernster, L. and Bergstrand, A., *J. Cell Biol.* **32** (1967) 415.

19. Sottocasa, G. L., Kuylenstierna, B., Ernster, L. and Bergstrand, A., *Meth. Enzym.* **10** (1967) 448.
20. Brdiczka, D., Pette, D., Brunner, G. and Miller, F., *Eur. J. Biochem.* **5** (1968) 294.
21. Green, D. E., *in* "Oxidases and Related Redox Systems" (edited by T. E. King, H. S. Mason and M. Morrison), Wiley, New York, 1965, Vol. II, p. 1061.
22. Green, D. E. and Perdue, J. F., *Ann. N.Y. Acad. Sci.* **137** (1966) 667.
23. Bachmann, E., Allmann, D. W. and Green, D. E., *Archs Biochem. Biophys.* **115** (1966) 153.
24. Allmann, D. W., Bachmann, E. and Green, D. E., *Archs Biochem. Biophys.* **115** (1966) 165.
25. Green, D. E., Bachmann, E., Allmann, D. W. and Perdue, J. F., *Archs Biochem. Biophys.* **115** (1966) 172.
26. Allmann, D. W. and Bachmann, E., *Meth. Enzym.* **10** (1967) 438.
27. Allmann, D. W., Galzinga, L., McCaman, R. E. and Green, D. E., *Archs Biochem. Biophys.* **117** (1966) 413.
28. Bachmann, E., Lenaz, G., Perdue, J. F., Orme-Johnson, N. and Green, D. E., *Archs Biochem. Biophys.* **121** (1967) 73.
29. Green, D. E., Allmann, D. W., Harris, R. A. and Tan, W. C., *Biochem. biophys. Res. Commun.* **31** (1968) 368.
30. Allmann, D. W., Bachmann, E., Orme-Johnson, N., Tan, W. C. and Green, D. E., *Archs Biochem. Biophys.* **125** (1968) 981.
31. Byington, K. H., Morey, A. V. and Smoly, J. M., *Fedn Proc. Fedn Am. Socs exp. Biol.* **27** (1968) 461.
32. Byington, K. H., Smoly, J. M., Morey, A. V. and Green, D. E., *Archs Biochem. Biophys.* **128** (1969) 762.
33. Smoly, J. M., Byington, K. H., Tan, W. C. and Green, D. E., *Archs Biochem. Biophys.* **128** (1969) 774.
34. Smith, D. S., *J. Cell Biol.* **19** (1963) 115.
35. Fernández-Morán, H., *Circulation* **26** (1962) 1039.
36. Fernández-Morán, H., Oda, T., Blair, P. V. and Green, D. E., *J. Cell Biol.* **22** (1964) 63.
37. Fleischer, S., Fleischer, B. and Stoeckenius, W., *J. Cell Biol.* **32** (1967) 193.
38. Thompson, J. E., Coleman, R. and Finean, J. B., *Biochim. biophys. Acta* **135** (1967) 1074.
39. Thompson, J. E., Coleman, R. and Finean, J. B., *Biochim. biophys. Acta* **150** (1968) 405.
40. Klingenberg, M., *in* "Energy-linked Functions of Mitochondria" (edited by B. Chance), Academic Press, New York, 1963, p. 381.
41. Pfaff, E., Klingenberg, M. and Heldt, H. W., *Biochim. biophys. Acta* **104** (1965) 312.
42. Klingenberg, M. and Pfaff, E., *in* "Regulation of Metabolic Processes in Mitochondria" (edited by J. M. Tager, S. Papa, E. Quagliariello and E. C. Slater), Elsevier, Amsterdam, 1966, B.B.A. Library, Vol. 7, p. 180.
43. Pfaff, E., "Unspezifische Permeabilität und spezifischer Austausch der Adeninnucleotide als Beispiel mitochondrialer Compartmentierung". Thesis, Phillips-Universität, Marburg, 1965.

44. Pfaff, E., *in* "Round Table Discussion on Mitochondrial Structure and Compartmentation" (edited by E. Quagliariello, S. Papa, E. C. Slater and J. M. Tager), Adriatica Editrice, Bari, 1967, p. 165.
45. Pfaff, E., Klingenberg, M., Ritt, E. and Vogell, W., *Eur. J. Biochem.* **5** (1968) 222.
46. Parsons, D. F. and Yano, Y., *Biochim. biophys. Acta* **135** (1967) 362.
47. Lévy, M. and Sauner, M.-T., *C. r. Séanc. Soc. Biol.* **161** (1967) 277.
48. Huet, C., Lévy, M. and Pascaud, M., *Biochim. biophys. Acta* **150** (1968) 521.
49. Newman, H. A. I., Gordesky, S. E., Hoppel, C. and Cooper, C., *Biochem. J.* **107** (1968) 381.
50. Kaiser, W. and Bygrave, F. L., *Eur. J. Biochem.* **4** (1968) 582.
51. Stoffel, W. and Schiefer, H.-G., *Hoppe Seyler's Z. physiol. Chem.* **349** (1968) 1017.
52. Sottocasa, G. L., *Abstr., 4th FEBS Meeting,* Oslo, 1967, p. 40; *Biochem. J.* **105** (1968) 1P.
53. Chappell, J. B. and Crofts, A. R., *Biochem. J.* **95** (1965) 707.
54. Cotzias, G. and Dole, V., *Proc. exp. biol. Med.* **78** (1951) 157.
55. Rodriguez de Lores Arnaiz, G. and de Robertis, E., *J. Neurochem.* **9** (1962) 503.
56. Baudhuin, P., Beaufay, H., Rahman-Li, Y., Sellinger, O., Wattiaux, R., Jacques, P. and de Duve, C., *Biochem. J.* **92** (1963) 179.
57. Gorkin, V., *Pharmac. Rev.* **18** (1966) 115.
58. Okamoto, H., Yamamoto, S., Nozaki, M. and Hayaishi, O., *Biochem. biophys. Res. Commun.* **26** (1967) 309.
59. Mayer, G., Ullrich, V. and Staudinger, H., *Hoppe Seyler's Z. physiol. Chem.* **349** (1968) 459.
60. Mason, H. S., *A. Rev. Biochem.* **34** (1965) 595.
61. Lehninger, A. L., *Harvey Lect.* **49** (1955) 176.
62. Pressman, B. C. and de Duve, C., *Archs int. Physiol.* **62** (1954) 306.
63. de Duve, C., Pressman, B. C., Gianetto, R., Wattiaux, R. and Appelmans, F., *Biochem. J.* **60** (1955) 604.
64. Ernster, L., Löw, H. and Lindberg, O., *Acta chem. scand.* **9** (1955) 200.
65. Ernster, L., Jalling, O., Löw, H. and Lindberg, O., *Expl. Cell Res.* Suppl. **3** (1955) 124.
66. Ernster, L., *Expl. Cell Res.* **10** (1956) 721.
67. Ernster, L., Dallner, G. and Azzone, G. F., *J. biol. Chem.* **238** (1963) 1124.
68. Lehninger, A. L., *J. biol. Chem.* **190** (1951) 345.
69. Lehninger, A. L., *Phosphorus Metabolism* **1** (1951) 344.
70. Maley, G. F., *J. biol. Chem.* **224** (1957) 1029.
71. Raw, I., Molinari, R., Ferreira do Amaral, D. and Mahler, H. R., *J. biol. Chem.* **233** (1958) 225.
72. Mahler, H. R., Raw, I., Molinari, R. and Ferreira do Amaral, D., *J. biol. Chem.* **233** (1958) 230.
73. Raw, I. and Mahler, H. R., *J. biol. Chem.* **234** (1959) 1867.
74. Raw, I., Petragnani, N. and Camargo-Nogueira, O., *J. biol. Chem.* **235** (1960) 1517.
75. Oshino, N., Imai, Y. and Sato, R., *Abstr. 7th Int. Congr. Biochem.*, Tokyo, 1967, p. 725.

76. Norum, K., Farstad, M. and Bremer, J., *Biochem. biophys. Res. Commun.* **24** (1966) 797.
77. Yates, D. W. and Garland, P. B., *Biochem. J.* **102** (1967) 40P.
78. Aas, M., *Abstr., 5th FEBS Meeting,* Prague, 1968, p. 151.
79. Aas, M. and Bremer, J., *Biochim. biophys. Acta* **164** (1968) 157.
80. Lima, M. S., Nachbaur, G. and Vignais, P. V., *C. r. hebd. Séanc. Acad. Sci., Paris* Sér. D **266** (1968) 739.
81. Schnaitman, C. and Pedersen, P. L., *Biochem. biophys. Res. Commun.* **30** (1968) 428.
82. Garland, P. B. and Yates, D. W., *in* "Round Table Discussion on Mitochondrial Structure and Compartmentation" (edited by E. Quagliariello, S. Papa, E. C. Slater and J. M. Tager), Adriatica Editrice, Bari, 1967, p. 385.
83. Van den Bergh, S. G., *in* "Round Table Discussion on Mitochondrial Structure and Compartmentation" (edited by E. Quagliariello, S. Papa, E. C. Slater and J. M. Tager), Adriatica Editrice, Bari, 1967, p. 400.
84. Colli, W., Hinkle, P. and Pullman, M. E., *Fedn Proc. Fedn Am. Socs. exp. Biol.* **27** (1968) 648.
85. Whereat, A. F., Orishimo, M. W. and Nelson, J. B., *Fedn Proc. Fedn Am. Socs exp. Biol.* **27** (1968) 362.
86. Pette, D., *in* "Regulation of Metabolic Processes in Mitochondria" (edited by J. M. Tager, S. Papa, E. Quagliariello and E. C. Slater), Elsevier, Amsterdam, 1966, B.B.A. Library, Vol. 7, p. 28.
87. Rose, I. A. and Warms, J. V. B., *J. biol. Chem.* **242** (1967) 1635.
88. Wojtczak, L. and Zborowski, J., *Abstr., 4th FEBS Meeting,* Oslo, 1967, p. 91.
89. Vignais, P. M., Nachbaur, J., André, J. and Vignais, P. V., this volume, p. 43.
90. van Deenen, L. L. M., van den Bosch, H., van Golde, L. M. G., Scherphof, G. L. and Waite, B. M., *in* "Symposium on Cellular Compartmentalization and Control of Fatty Acid Metabolism", Proc. 4th FEBS Meeting, Oslo, 1967 (edited by F. C. Gran), Universitetsforlaget, Oslo, and Academic Press, London and New York, 1968, p. 89.
91. Bygrave, F. L. and Bücher, Th., *Abstr., 5th FEBS Meeting,* Prague, 1968, p. 243.
92. Beattie, D. S., Basford, R. E. and Koritz, S. B., *Biochemistry, N.Y.* **6** (1967) 3099.
93. Neupert, W., Brdiczka, D. and Bücher, Th., *Biochem. biophys. Res. Commun.* **27** (1967) 488.
94. Work, T. S., *in* "Biochemical Aspects of the Biogenesis of Mitochondria" (edited by E. C. Slater, J. M. Tager, S. Papa and E. Quagliariello), Adriatica Editrice, Bari, 1968, p. 367.
95. Barrnett, R. J., *in* "Enzyme Histochemistry" (edited by M. S. Burstone), Academic Press, New York, 1962, p. 537.
96. Racker, E., Tyler, D. D., Estabrook, R. W., Conover, T. E., Parsons, D. F. and Chance, B., *in* "Oxidases and Related Redox Systems" (edited by T. E. King, H. S. Mason and M. Morrison), Wiley, New York, 1965, Vol. II, p. 1077.
97. Racker, E. and Horstman, L. L., *J. biol. Chem.* **242** (1967) 2547.
98. Lee, C. P., *Fedn Proc. Fedn Am. Socs exp. Biol.* 22 (1963) 527.

99. Harris, E. J., van Dam, K. and Pressman, B. C., *Nature, Lond.* **213** (1967) 1126.
100. Quagliariello, E. and Palmieri, F., *Eur. J. Biochem.* **4** (1968) 20.
101. Lenaz, G. and MacLennan, D. H., *J. biol. Chem.* **241** (1966) 5260.
102. Lee, C. P. and Carlson, K., *Fedn Proc. Fedn Am. Socs exp. Biol.* **27** (1968) 828.
103. Carafoli, E. and Muscatello, U., *Abstr., 5th FEBS Meeting,* Prague, 1968, p. 65.
104. Hassinen, I. and Chance, B., *Biochem. biophys. Res. Commun.* **31** (1968) 895.
105. Frisell, W. R., Patwardhan, M. V. and Mackenzie, C. M., *J. biol. Chem.* **240** (1965) 1829.
106. Garland, P. B., Chance, B., Ernster, L., Lee, C. P. and Wong, D., *Proc. natn. Acad. Sci. U.S.A.* **58** (1967) 1696.
107. Danielson, L. and Ernster, L., *Biochem. Z.* **338** (1963) 188.
108. Lee, C. P. and Ernster, L., *in* "Regulation of Metabolic Processes in Mitochondria" (edited by J. M. Tager, S. Papa, E. Quagliariello and E. C. Slater), Elsevier, Amsterdam, 1966, B.B.A. Library, Vol. 7, p. 218.
109. McKay, R., Druyan, R. and Rabinowitz, M., *Fedn Proc. Fedn Am. Socs exp. Biol.* **27** (1968) 774.
110. Jones, M. S. and Jones, O. T. G., *Abstr., 5th FEBS Meeting,* Prague, 1968, p. 92; *Biochem. biophys. Res. Commun.* **31** (1968) 977.
111. Chappell, J. B., *Br. med. Bull.* **24** (1968) 150.
112. Greenawalt, J. W., Rossi, C. S. and Lehninger, A. L., *J. Cell Biol.* **23** (1964) 21.
113. Brierley, G. P. and Slautterback, D. B., *Biochim. biophys. Acta* **82** (1964) 183.
114. Thomas, R. S. and Greenawalt, J. W., *J. Cell Biol.* **39** (1968) 55.
115. Lehninger, A. L., Carafoli, E. and Rossi, C. S., *Adv. Enzymol.* **29** (1967) 259.
116. Papa, S., Landriscina, C., Lofrumento, N. E. and Quagliariello, E., *Abstr., 5th FEBS Meeting,* Prague, 1968, p. 178.
116a. Beattie, D. S., *Biochem. biophys. Res. Commun.* **30** (1968) 57.
117. Nass, M. M. K. and Nass, S., *J. Cell Biol.* **19** (1963) 593.
118. Green, D. E., *in* "Comprehensive Biochemistry" (edited by M. Florkin and E. Stotz), Elsevier, Amsterdam, 1966, Vol. 14, p. 309.
119. Penniston, J. T., Harris, R. A., Asai, J. and Green, D. E., *Proc. natn. Acad. Sci. U.S.A.* **59** (1968) 624.
120. Whittaker, V. P., *in* "Regulation of Metabolic Processes in Mitochondria" (edited by J. M. Tager, S. Papa, E. Quagliariello and E. C. Slater), Elsevier, Amsterdam, 1966, B.B.A. Library, Vol. 7, p. 1.
121. Munn, E. A. and Blair, P. V., *Z. Zellforsch. mikrosk. Anat.* **80** (1967) 205.
122. Allmann, D. W., Harris, R. A. and Green, D. E., *Archs Biochem. Biophys.* **120** (1967) 693.
123. Allmann, D. W., Harris, R. A. and Green, D. E., *Archs Biochem. Biophys.* **122** (1967) 766.
124. Sekuzu, I., Jurtshuk, P. and Green, D. E., *J. biol. Chem.* **238** (1963) 975.
125. Cross, R. J., Taggart, J. V., Covo, A. G. and Green, D. E., *J. biol. Chem.* **177** (1950) 655.
126. Green, D. E., *Biol. Rev.* **26** (1951) 410.
127. Greenspan, M. D. and Purvis, J. L., *Biochim. biophys. Acta* **99** (1965) 191.

128. Pfaff, E. and Schwalbach, K., *in* "Round Table Discussion on Mitochondrial Structure and Compartmentation" (edited by E. Quagliariello, S. Papa, E. C. Slater and J. M. Tager), Adriatica Editrice, Bari, 1967, p. 346.
129. Hogeboom, G. H. and Schneider, W. C., *J. biol. Chem.* **204** (1953) 233.
130. Ernster, L. and Navazio, F., *Acta chem. scand.* **10** (1956) 1038.
131. Ernster, L. and Navazio, F., *Expl. Cell Res.* **11** (1956) 483.
132. Ernster, L., *Biochem. Soc. Symp.* **16** (1959) 54.
133. Hunter, F. E., Malison, R., Bridgers, W. F., Schultz, B. and Atchison, A., *J. biol. Chem.* **234** (1959) 693.
134. Cooper, C. and Lehninger, A. L., *J. biol. Chem.* **219** (1956) 489.
135. Dallman, P. R., Dallner, G., Bergstrand, A. and Ernster, L., *J. Cell. Biol.* **41** (1969) 357.
136. Beattie, D. S. *Biochem. biophys. Res. Commun.* **31** (1968) 901.
137. Tabor, C. W., Tabor, H. and Rosenthal, S. M., *J. biol. Chem.* **208** (1954) 645.
138. McCann, R. E., McCaman, M. W., Hunt, J. M. and Smith, M. S., *J. Neurochem.* **12** (1965) 15.
139. Wurtman, R. J. and Axelrod, J., *Biochem. Pharmac.* **12** (1963) 1439.
140. Lehninger, A. L., "The Mitochondrion", Benjamin, New York, 1965.
141. Roodyn, D. B. and Wilkie, D., "The Biogenesis of Mitochondria", Methuen, London, 1968.

FEBS Symposium, Volume 17, pp. 33-42

Recent Findings on the Biochemical and Enzymatic Composition of the Two Isolated Mitochondrial Membranes in Relation to their Structure

M. LÉVY, R. TOURY and M.-T. SAUNER
Laboratoire de Physiologie de la Nutrition, Faculté des Sciences, Paris, France, and

J. ANDRÉ
Laboratoire de Biologie Cellulaire 4, Faculté des Sciences, Orsay, France

It is chiefly due to their isolation and purification, that precise information has been gained on biochemical and enzymatical composition of mitochondrial membranes [1-5]. The present account summarizes the main results obtained with fragments of membranes which have been isolated and purified by a technique we have established [1]. This technique is founded on the successive action of digitonin and sonication on mitochondria. Under the conditions

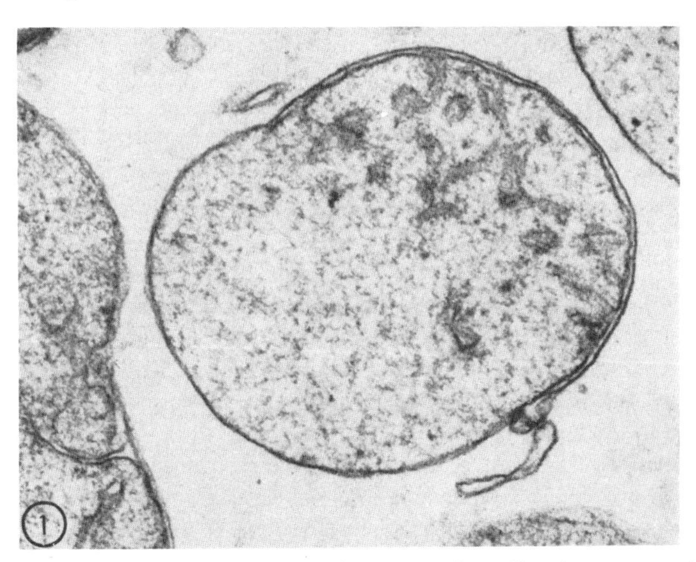

Figure 1. Isolated liver mitochondria in the process of expelling the outer membrane by digitonin action (x 45,000).

employed by us, digitonin selectively detaches the outer membrane (Fig. 1) [1]. This result has since been confirmed by Schnaitman *et al.* [5], and Hoppel and Cooper [6].

The specificity of digitonin action on the outer membrane is of great interest. We know, in fact, that it is able to form complexes with cholesterol. Therefore we can suppose that cholesterol is chiefly localized in the outer membrane or, at least, it is more accessible to digitonin action than the cholesterol of the inner membrane.

Table 1. Cholesterol/phospholipid molar ratio of mitochondrial membranes

Fraction	Molar ratio: Cholesterol / Phospholipid
Outer membrane:	
1. Real value*	1:9
2. Value measured from purified fragments	1:14
Inner membranes:	
Purified fragments	1:53

* The real value is obtained by the difference between values measured on intact mitochondria and on single-membrane mitochondria.

Cholesterol and phospholipid are determined by techniques previously described [7].

In a recent paper Lévy and Sauner [7] show that the outer membrane contains five times more cholesterol than the inner membrane. However, only one fraction of the cholesterol of the outer membrane is removed by digitonin action (Table 1). Pictures obtained by electron microscopy of the contamination separated from the crude outer membrane fraction by a subsequent purification show the cholesterol–digitonin complex so formed (Fig. 2). This looks like the one formed *in vitro* from pure commercial cholesterol and digitonin (Fig. 3). They both appear like short rods.

The cholesterol still remaining in the outer membrane is not accessible to a successive action of digitonin (Table 2). Thus, there could exist at least two types of cholesterol linkages in this membrane; one of the linkages is easily broken by digitonin action. Cholesterol of the inner membrane is entirely insensitive to this action (Table 2).

The separated fragments of outer membrane reorganize themselves into rigid and regularly rounded vesicles (Fig. 4). The inner membrane after sonication looks like flexible and undulating vesicles (Fig. 5) [1].

Measurements of density show that the outer membrane is lighter (1·10-1·12) than the inner membrane (1·15-1·16). These differences could be explained by a

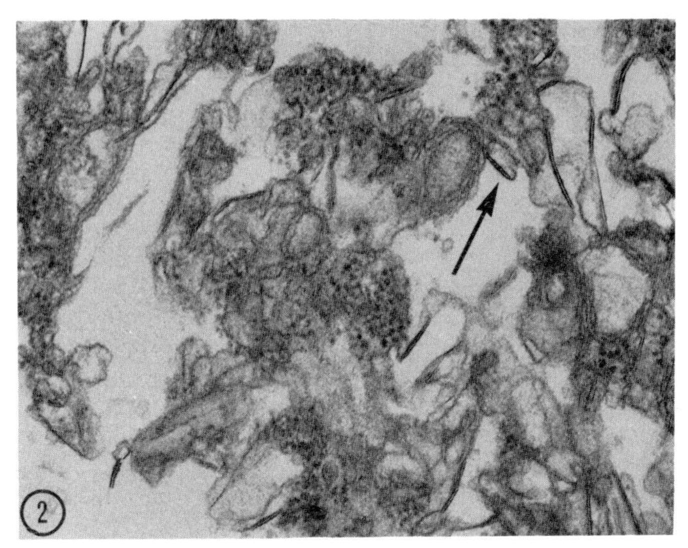

Figure 2. Contaminating material found after purification of the crude outer membrane fraction. The arrow points to a rod thought to be a digitonin–cholesterol complex (x 45,000).

higher content in phospholipids, compared to proteins, in the outer membrane (Table 3). These results are quite in agreement with those of Parsons *et al.* [2].

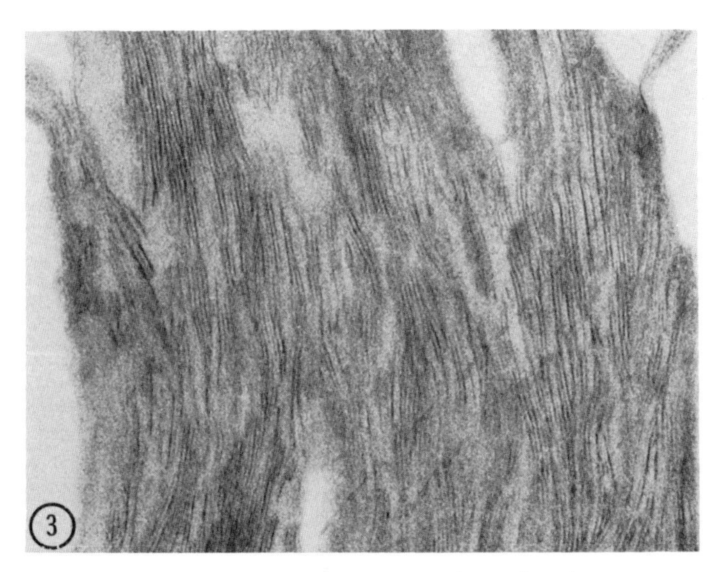

Figure 3. Digitonin–cholesterol complex obtained *in vitro* from pure commercial cholesterol (x 120,000).

Table 2. Digitonin action on purified mitochondrial membranes

Fraction	Digitonin	Molar ratio: Cholesterol / Phospholipid
Outer membrane	–	1:16
	+	1:18
Inner membrane	–	1:53
	+	1:50

Cholesterol and phospholipid are determined by techniques previously described [7]. Digitonin action is also described in [7].

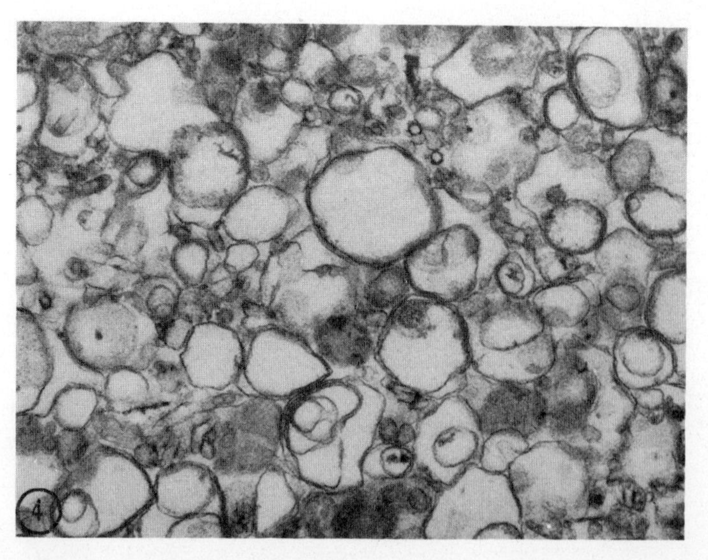

Figure 4. Purified outer membrane fragments. Most of the fragments are in the shape of rounded vesicles (x 30,000).

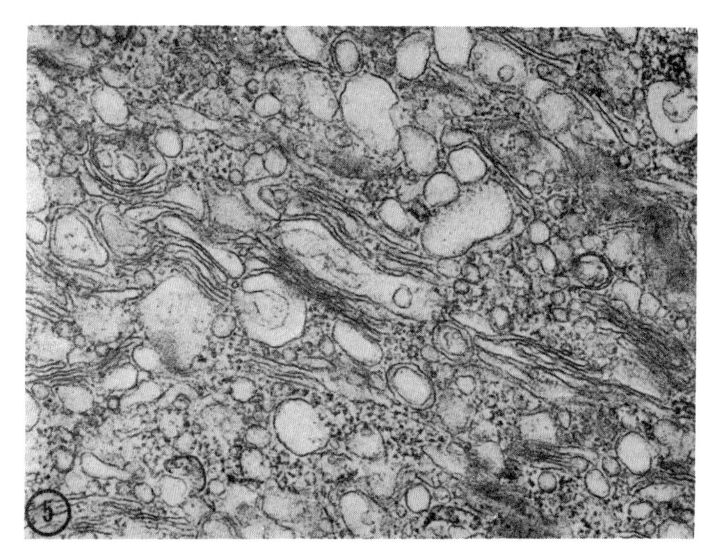

Figure 5. Inner membrane fragments. At variance with the outer membrane, a good number of the fragments here are in the shape of undulating vesicles (x 22,500).

Table 3. Relative proportions of phospholipid and protein in mitochondrial membranes

Fraction	Protein	Phospholipid
	(per cent of total protein + phospholipid)	
Outer membrane	62·6	37·4
Inner membrane	74·6	25·4

Protein and phospholipid are determined as previously described [7].

The relative proportions of the different phospholipids also vary from one to the other membrane (Table 4).

Finally, the value of the ratio of saturated fatty acids to unsaturated fatty acids of total phospholipids is characteristic of each membrane; it is more than one in the outer membrane and less than one in the inner membrane. This result is also true for each kind of isolated phospholipid, with only one exception: sphingomyelin. Cardiolipins show the most important difference (Table 5) [8].

Table 4. Phospholipid composition of mitochondrial membranes

Phospholipids	Outer membrane	Inner membrane
Lysophosphatidylcholine	3·4	3·1
Sphingomyelin + phosphatidylserine	7·5	3·4
Phosphatidylcholine	59·0	39·2
Phosphatidylinositol	4·8	6·6
Phosphatidylethanolamine	20·3	34·4
Phosphatidic Acid + cardiolipin	6·3	13·7

Phospholipids are determined as previously described [7].

Table 5. Saturated fatty acids/unsaturated fatty acids ratio of mitochondrial membranes

Phospholipids	Outer membrane	Inner membrane
Total phospholipids	1·20	0·72
Phosphatidic Acid + cardiolipin	1·28	0·36
Phosphatidylethanolamine	1·31	0·85
Phosphatidylcholine + phosphatidylinositol	1·28	0·87
Sphingomyelin + lysophosphatidylcholine + phosphatidylserine	1·84	1·36

Fatty acids are separated and determined by gas chromatography as previously described [8].

Thus, inasmuch as the lipid molecules posses a greater energy of interaction when they contain saturated chains [9], we are led to suppose that the outer membrane is more rigid than the inner one. The higher rigidity of this membrane could also be attributed [10] to its higher cholesterol content. This conception is in agreement with the morphological differences previously observed (Figs. 4 and 5).

Enzymatic studies made on purified fragments of mitochondrial membranes [1] have permitted us, on the one hand, to confirm that the inner membrane possesses the phosphorylating oxidative chains which are sensitive to the inhibitory effects of rotenone, antimycin A and KCN; and on the other hand, to show that the outer membrane is devoid of the phosphorylating oxidative chains but possesses an NADH-cytochrome *c* reductase. This enzymatic activity is non-phosphorylating and insensitive to the inhibitory effects of rotenone and antimycin A (Table 6). Our results are in agreement with those of Parsons *et al.* [2] and Sottocasa *et al.* [3].

Table 6. Oxidative activities of the mitochondrial outer membrane

Enzymes	Activity
Succinic-oxidase (oxygen uptake: mμ at./mg prot./min)	0
β-Hydroxybutyric-oxidase (oxygen uptake: mμ at./mg prot./min)	0·9
NADH-oxidase (oxygen uptake: mμ at./mg prot./min)	0·8
Cytochrome-oxidase (cytochrome *c* oxidized: mμ mol./mg prot./min)	41·3
NADH-cytochrome *c* reductase (cytochrome *c* reduced: mμ mol./mg prot./min)	1830·0

Enzymatic activities are determined as previously described [1].

Table 7. Activities of some enzymes of the tricarboxylic acid cycle in the mitochondrial outer membrane

Enzymes	Activity
Aconitase (mμ moles *cis*-aconitate appeared/mg prot./min)	0
Fumarase (mμ moles fumarate appeared/mg prot./min)	0
Isocitric-dehydrogenase (mμ moles NADPH appeared/mg prot./min)	0
Malate-dehydrogenase (mμ moles NADH disappeared/mg prot./min)	651·0

Aconitase and fumarase activities are determined by absorption increase at 240 mμ in a total volume of 3 ml that contains, in the case of aconitase: 0·03M sodium citrate, 0·05M phosphate buffer pH 7·4, 0·5-1·0 mg proteins. For determinations of fumarase activity, 0·05M sodium malate is substituted for sodium citrate. Isocitric-dehydrogenase activity is determined by absorption increase at 340 mμ in a total volume of 3 ml that contains: 0·006M D-L sodium isocitrate, 0·00135M NADP, 0·02M $MnSO_4$, 0·25M tris buffer pH 7·6, 0·8 mg proteins. Malate dehydrogenase activity is determined by absorption decrease at 340 mμ in a total volume of 3 ml that contains: 0·25 mM sodium oxalo-acetate, 0·07 mM NADH, 0·050M buffer phosphate pH 7·4, 0·07-0·1 mg proteins.

We have also shown recently that the outer membrane possesses a malate dehydrogenase activity. This enzyme is not associated with the other enzymes of the tricarboxylic acid cycle (Table 7). A malate dehydrogenase localized in the outer membrane seems at first surprising. It is less surprising if we consider that cytoplasm contains an isomalate dehydrogenase not associated with the other tricarboxylic acid cycle enzymes [11, 12]. The presence in the cytoplasm of this enzyme has led Krebs to postulate a mechanism to explain the oxidation of NADH formed during glycolysis [13]. We can suppose that a similar system could work at the mitochondrial level. In that case, the malate dehydrogenase of the outer membrane could play the same role as the cytoplasmic enzyme.

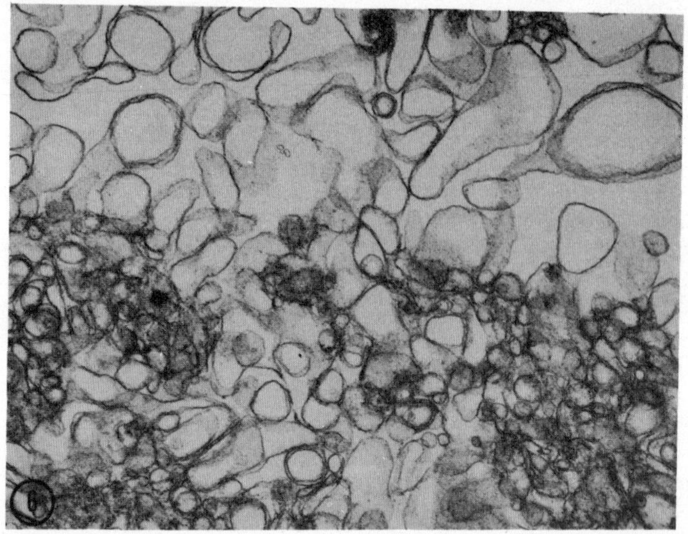

Figure 6. Outer membrane fragments after phosphate buffer treatment (× 30,000).

However, the malate dehydrogenase of outer membrane is easily detached by phosphate buffer (Table 8). There is a loss of about 40% of proteins and only about 10% of phospholipids accompanying the malate dehydrogenase removal. In spite of the important loss of proteins and non-negligible loss of phospholipids, the fragments of the outer membrane show after this treatment no appreciable difference when they are examined by electron microscopy (Figs. 4 and 6). After use of phosphate buffer, NADH-cytochrome *c* reductase remains firmly bound to the outer membrane (Table 8).

Other studies on either the mitochondrial inner membrane [14] or erythrocyte ghosts [15-17], have shown that use of phosphate, tris buffer or hypertonic saline solutions have the property of removing a variety of enzymes, while others remain firmly bound to these membranes.

Table 8. Effect of treatment with pH 7·4 phosphate buffer on malate-dehydrogenase and NADH-cytochrome *c* reductase activities of the mitochondrial other membrane

Fraction	Malate dehydrogenase		NADH-cytochrome *c* reductase	
	Activity (mμ mol. NADH oxid./min/ mg prot.)	Percentage of activity decrease	Activity (mμ mol. NADH oxid./min/ mg prot.)	Percentage of activity decrease
Outer membrane	651·0	–	1830·0	–
Outer membrane washed with 0·2M phosphate buffer	58·0	91·0	2055·0	0

Malate dehydrogenase activity is determined as described in Table 7. NADH-cytochrome *c* reductase is determined spectrophotometrically as previously described [1].

All of these findings raise the problem of the existence of different types of linkages for enzymes and non-enzymatic proteins inside membranes. We can propose, in agreement with Green and Perdue [18], Omura *et al.* [19], and Dallner *et al.* [20], that membranes, including the mitochondrial outer membrane, could be built by macromolecular elements, some of them being easily removed, while others, more firmly linked, could constitute the skeleton of the membrane.

REFERENCES

1. Lévy, M., Toury, R. and André, J., *Biochim. biophys. Acta* **135** (1967) 599.
2. Parsons, D. F., Williams, G., Thompson, W., Wilson, D. and Chance, B., *in* "Round Table Discussion on Mitochondrial Structure and Compartmentation" (edited by E. Quagliariello, S. Papa, E. C. Slater, and J. M. Tager), Adriatica Editrice, Bari, 1967, p. 29.
3. Sottocasa, G. L., Kuylenstierna, B., Ernster, L. and Bergstrand, A., *J. Cell Biol.* **32** (1967) 415.
4. Allmann, D., Bachmann, E. and Green, D. E., *Archs Biochem. Biophys.* **115** (1966) 165.
5. Schnaitman, C., Erwin, V. G. and Greenawalt, J. W., *J. Cell Biol.* **32** (1967) 719.
6. Hoppel, C. and Cooper, C., *Biochem. J.* **107** (1968) 367.
7. Lévy, M. and Sauner, M.-T., "Chemistry and Physics of Lipids" (1968) in press.
8. Huet, C., Lévy, M. and Pascaud, M., *Biochim. biophys. Acta* **150** (1968) 521.
9. Salem, L., *Can. J. Biochem. Biophys.* **40** (1962) 1287.
10. van Deenen, L. L. M., *in* "Progress in the Chemistry of Fats and other Lipids (edited by R. T. Holman) Pergamon Press, Oxford, 1965, Vol. III, Part I, p. 16.
11. Thorne, C. J. R., *Biochim. biophys. Acta* **42** (1960) 175.
12. Siegel, L. and Englard, S., *Biochim. biophys. Acta* **64** (1962) 101.
13. Krebs, H. A., *in* "Biochemistry of Mitochondria" (edited by E. C. Slater, Z. Kaniuga and L. Wojtczak), Academic Press, New York and London, 1967, p. 105.
14. Beenakkers, A. M. T. and Klingenberg, M., *Biochim. biophys. Acta* **84** (1964) 205.
15. Mitchell, C. D., Mitchell, W. B. and Hanahan, D. J., *Biochim. biophys. Acta* **104** (1965) 348.
16. Green, D. E., Murer, E., Hultin, O., Richardson, S. H., Salmon, B., Brierley, G. P. and Baum, H., *Archs Biochem. Biophys.* **112** (1965) 635.
17. Mitchell, C. D. and Hanahan, D. J., *Biochemistry, N.Y.* **5** (1966) 51.
18. Green, D. E. and Perdue, J. F., *Proc. natn Acad. Sci. U.S.A.* **55** (1966) 1295.
19. Omura, T., Siekevitz, P. and Palade, G. E., *J. biol. Chem.* **242** (1967) 2389.
20. Dallner, G., Siekevitz, P. and Palade, G. E., *J. Cell Biol.* **30** (1966) 73, 97.

FEBS Symposium, Volume 17, 1969, pp. 43-58

A Critical Approach to the Study of the Localization of Phospholipase-A in Mitochondria

P. M. VIGNAIS, J. NACHBAUR and P. V. VIGNAIS

Biochimie, Centre d'Etudes Nucléaires et Faculté de Médecine de Grenoble, Grenoble, and,

J. ANDRÉ

Laboratoire de Biologie Cellulaire 4, Faculté des Sciences, Orsay, France

INTRODUCTION

A phospholipase, A_2, catalysing the hydrolysis of the 2-acyl group of phospholipids was recently identified in mitochondrial fractions obtained from rat liver homogenates by differential centrifugation [1-5]. Although this finding was of considerable interest with regard to the regulation of the dynamic properties of mitochondrial phospholipids, its bearing had to be interpreted with caution in relation to the purity of the mitochondrial fractions used as a source of phospholipase. It must be recalled that mitochondria obtained by the classical differential centrifugation method [6] are contaminated by microsomes and lysosomes and that both types of particles exhibit phospholipase activities. The microsomal phospholipase catalyses the hydrolysis of the 1-acyl group of phospholipids at neutral pH [3, 7] and the lysosomal phospholipase cleaves both fatty acid ester linkages of phospholipids with an optimum activity at pH 4·5 [5, 8].

By contrast with microsomal contamination, lysosomal contamination of mitochondrial preparations is not substantially lowered by repeated washings with isotonic sucrose. Lysosomal contamination may be very critical as far as the determination of phospholipase activity is concerned since the specific activity of phospholipase in lysosomes is markedly higher than in mitochondria (40 times according to Mellors and Tappell [8]).

The purpose of this study was to examine, by enzymatic and morphological criteria, the contribution of the lysosomal contamination to the phospholipase activity of mitochondrial fractions isolated from rat liver homogenates. With mitochondrial fractions substantially freed of lysosomes, it has become possible to ascribe to mitochondria the phospholipase-A_2 activity of liver cell. The properties of the mitochondrial phospholipase-A_2 as well as its distribution in the mitochondrial spaces are reported in this paper.

RESULTS AND DISCUSSION

In this study, the phospholipase assays were carried out either on endogenous or on added phospholipids and the phospholipase activity was usually evaluated in terms of fatty acids released. Although liver mitochondria exhibit a lipase activity [3], the above assay seems to be justified since we found a satisfactory correlation between the amount of fatty acids released and the amount of

Table 1. Formation of lysophosphatides by hydrolysis of endogenous phospholipids

	Percentages of total Phosphorus		Δ
	Zero time	1 h	
CL	10·6	9·8	– 0·8
PE	38·4	23·6	– 14·8
PC + LysoPE	42·1	51·3	(+ 9·2)
LysoPC	2·6	6·0	+ 3·4

Incubation conditions: 2·5 mM $CaCl_2$, 0·02M triethanolamine buffer, pH 8·0, mitochondria from E-fraction (5·2 mg protein)–final volume 2 ml. Reaction time 1 h at 37°C. The lipids were extracted twice by 4 ml of chloroform/methanol (2/1) and separated by thin layer chromatography, first using the chloroform/light petroleum (b.p. 60°-80°C)/acetic acid (65:35:2 v/v) system, and then the chloroform/methanol/water (65:35:4 v/v) system [3]. Phospholipids detected on thin layer chromatograms by staining with iodine vapour were extracted with methanol, digested in 10N H_2SO_4 and their phosphorus content determined by the method of Bartlett [9]. On a basis of 190 nmoles of phospholipids per mg of mitochondrial protein it can be calculated that the amount of phosphatidylethanolamine (PE) which has disappeared in mitochondria during their incubation is 0·145 μmole and that the amount of lysophosphatidylcholine (LPC) formed is 0·033 μmole. Assuming that phosphatidylethanolamine has disappeared to form lysophosphatidylethanolamine (LPE), the amount of LPE + LPC equals 0·178 μmole. The amount of free fatty acids recovered and measured by gas liquid chromatography was 0·200 μmole.

lysophosphatides which are accumulated under the conditions used to measure phospholipase activity. In fact, as shown in Table 1, under optimal conditions (Ca^{2+} added, pH 8·1) 90% of the release of fatty acids from the mitochondrial lipids is accounted for by the hydrolysis of phospholipids into lysophosphatides. The hydrolysed phospholipids are essentially phosphatidylethanolamine and phosphatidylcholine. The endogenous cardiolipins remain unattacked.

Wattiaux *et al.* [10] have recently published a method of separation of rat liver lysosomes in a continuous sucrose gradient based on the finding that lysosomes filled with Triton WR-1339 are rendered lighter than mitochondria. This method, which was adapted for our experiments aiming at the isolation of liver mitochondria free of lysosomes, is now briefly described. Four to six days

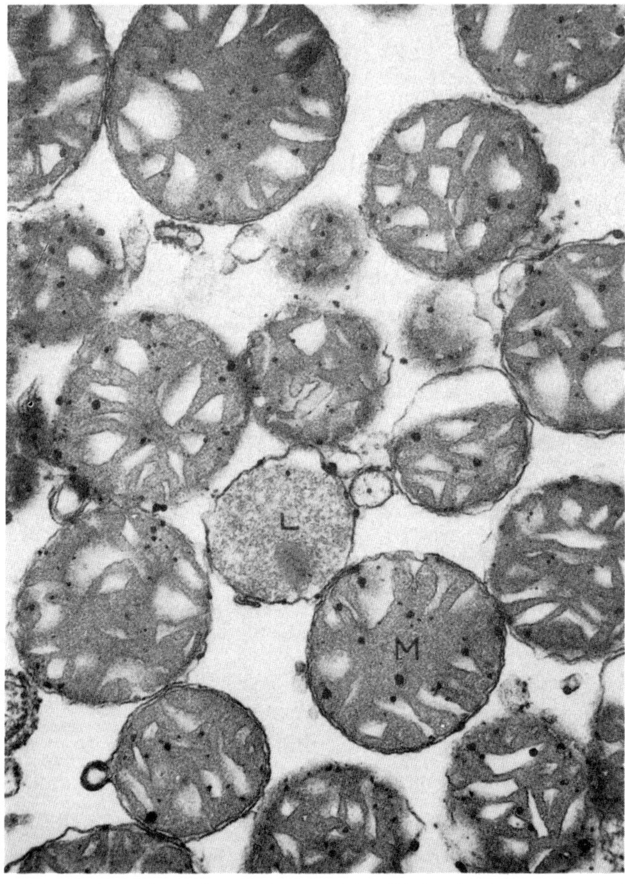

Figure 1. Thin section of rat liver mitochondria (M) fixed with osmium tetroxide. Rats were previously injected with Triton WR-1339 (x 30,000).

before sacrifice, rats were injected intraperitoneously with 170 mg of Triton WR-1339. In a first step a mitochondrial fraction was obtained by differential centrifugation between 10,000 **g**/min and 53,000 **g**/min from a 10% homogenate of liver in 0·27M sucrose. This fraction was washed three times with half the initial volume of sucrose solution. In a second step, mitochondria (Fig. 1) were separated from Triton-filled lysosomes (H-fraction) by centrifugation in a two-layer sucrose gradient as shown in Fig. 2A. The distribution pattern of

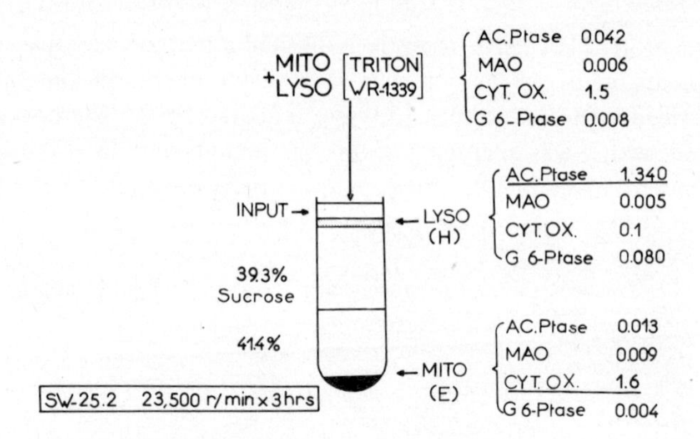

Figure 2A. Separation on sucrose gradient of mitochondria and Triton-filled lysosomes. The crude mitochondria preparation formed the input to the sucrose gradient and was spun at 23,000 rev/min x 3 h. Particles floating on top of the gradient (H-fraction) are Triton-filled lysosomes. Mitochondria are found at the bottom of the tube (E-fraction). Acid phosphatase was estimated with *p*-nitrophenylphosphate as substrate [11] and the activity was measured at 37°C after unmasking with Triton X-100. Glucose 6-phosphatase was determined by the amount of inorganic phosphate formed at 37°C and at pH 8·5 [12]. Monoamine oxidase (MAO) was measured at pH 7·5 and at 25°C by following the formation of benzaldehyde from benzylamine [13]; identical distribution of MAO was found by following the disappearance of kynuramine [14]. Cytochrome oxidase was assayed spectrophotometrically at 25°C [15]. Since the oxidation of reduced cytochrome *c* is of first order with respect to its concentration, the rate constant *k* was deduced from a logarithmic plot of the kinetics of the reaction. The rate of oxidation of reduced cytochrome *c* was calculated from the rate constant. All enzymatic activities are expressed as μmoles of substrate used per min per mg protein.

Table 2. Effect of the purification of a mitochondrial fraction on the phospholipase activity at pH 4·5 and at pH 8·1

Mitochondrial preparation	pH	Fatty acids released* Total nmoles	Unsaturated per cent	Acid† phosphatase
Classical	4·5	53·3	48·2	83
	8·1	13·4	64·1	
Purified	4·5	22·6	45·1	8
	8·1	21·0	72·7	

* nmole/h/mg prot. (37°C).

† nmole/min/mg prot. (pH 4·8, 37°C).

The incubation medium contained: ultrasonicated egg phosphatidylethanolamine (0·4 mg), 0·02M triethanolamine buffer, pH 8·1, or 0·02M acetate buffer, pH 4·5, and 2 mM Ca^{2+} in a final volume of 4 ml. Incubation at 37°C for one hour with shaking.

typical enzymatic activities in E- and H-fractions (Fig. 2A) clearly indicates that the E-fraction consists of mitochondria substantially free of lysosomes.

Data in Table 2 illustrate typical differences in phospholipase activity at pH 4·5 and pH 8·1 with liver mitochondria. The lowering of the acid phosphatase activity in lysosome-free mitochondria (E-fraction) is accompanied by an

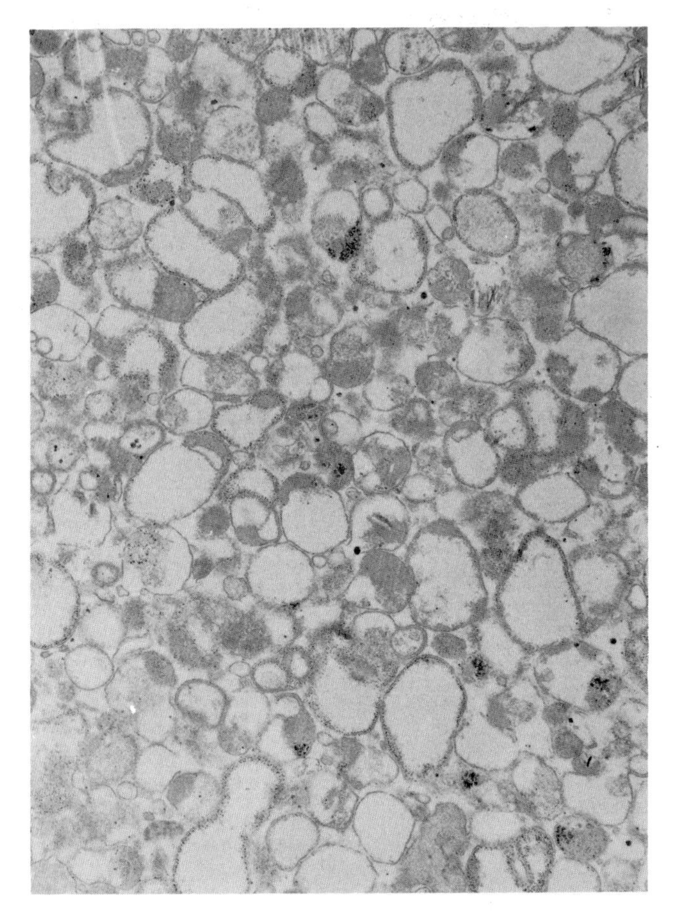

Figure 2B. Thin section of rat liver lysosomes (H-fraction) fixed with osmium tetroxide (× 15,000).

enhancement of the release of unsaturated fatty acids at pH 8·1, and on the contrary, by a decrease of both unsaturated and saturated fatty acids at pH 4·5. This emergence of a phospholipase-A_2 activity of pH 8·1 in mitochondrial fraction free of lysosomes is taken as a proof that phospholipase-A_2 is a mitochondrial enzyme. The phospholipase activity which is detected at pH 4·5 and which concerns the release of both saturated and unsaturated fatty acids may reflect a residual contamination of mitochondria by lysosomes. This result, which is perfectly compatible with the high specific activity of lysosomal phospholipase, justifies, *a posteriori,* the requirement of highly purified mitochondrial fractions for assessing the activity and the properties of mitochondrial phospholipase.

As shown in the next experiment, outer membranes of mitochondria cannot be separated from lysosomal membranes in the three-layer sucrose gradient described by Parsons *et al.* [16] if membranes of both types are present in the suspension which forms the input to the gradient. A fraction enriched in Triton-loaded lysosomes (H-fraction) was prepared, as described previously [5], by flotation on a 39·3% (w/v) sucrose solution (Fig. 2B). This fraction of

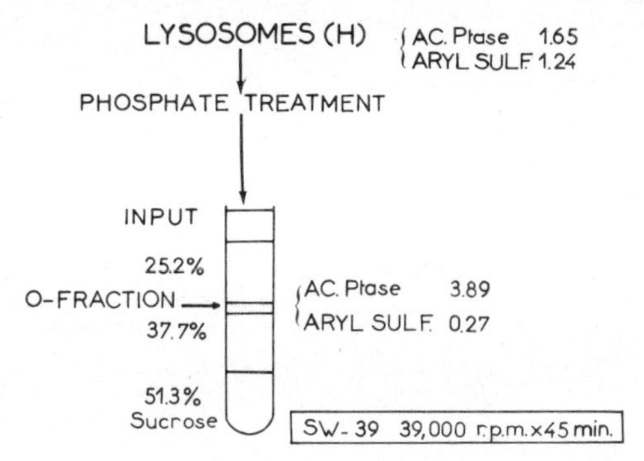

Figure 3. Isolation of lysosomal membranes on a discontinuous sucrose gradient. A suspension of Triton-filled lysosomes treated by phosphate formed the input to a discontinuous sucrose gradient and was spun at 39,000 rev/min × 45 min. A particulate material (O-fraction) was recovered at the first interface. Enzymic assay conditions as in Fig. 2. Arylsulphatase was estimated with nitrocatechol sulphate as substrate at pH 5·2 at 37°C [17] and expressed as μmoles of substrate used per h per mg protein.

lysosomes was treated exactly as were mitochondria for the isolation of outer mitochondrial membranes; it was homogenized with 20 mM phosphate, pH 7·3, left for 20 min at 0°C, placed on the three-layer sucrose gradient described by Parsons *et al.* [16] and spun at 38,000 rev/min for 40 min in a Spinco SW-39 rotor. The fraction which was recovered at the first interface of the gradient (O-fraction), i.e. at the same place as outer membranes prepared from mitochondria, was characterized by a higher acid phosphatase activity and a lower aryl sulphatase activity than the H-fraction (Fig. 3). A thin section micrograph of the O-fraction (Fig. 4) shows membranes with open structures which contrast with the vesicular structure of outer mitochondrial membranes (Fig. 5). These membranes, identified as lysosomal membranes by their bound acid phosphatase, display a phospholipase activity whose characteristics are typical of lysosomal phospholipase: 4·5 as optimal pH, release of saturated and unsaturated fatty acids in equal proportions and insensitivity to calcium ions.

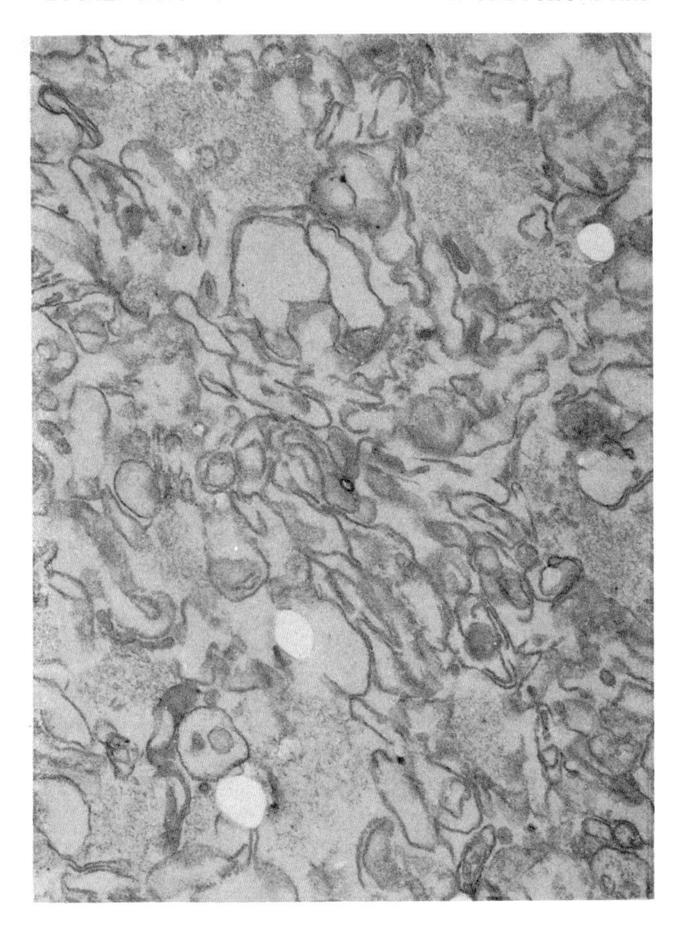

Figure 4. Thin section of lysosomal membranes (O-fraction) fixed with osmium tetroxide (× 30,000).

As far as the preparation of mitochondrial membranes itself is concerned, we will show now that inner mitochondrial membranes devoid of matrix can partly sediment at the same place as outer mitochondrial membranes in the three-layer sucrose gradient described by Parsons *et al.* [16]. A fraction of "inner membrane + matrix" (thin section micrograph in Fig. 6) obtained from purified mitochondria (E-fraction) by treatment with phosphate followed by a low speed centrifugation [16] was resuspended in 0·27M sucrose and exposed to sonic oscillation at 0°C (twice 1 min at 10 A in a Branson sonifier). The sonicated suspension was then centrifuged through the three-layer sucrose gradient already mentioned (Fig. 7). Some of the particulate material was gathered at the first interface (thin section micrograph in Fig. 8), i.e. at the same level as outer mitochondrial membranes (cf. above). Because of its high cytochrome oxidase activity and its origin, this material was identified as the inner membrane of mitochondria. These data suggest that outer mitochondrial membrane and inner membrane devoid of matrix by sonication have about the same density. Actually, the amount of phospholipids which is 0·19 μmole of lipid phosphorus/mg protein in "inner membrane + matrix" raises to 0·42 in the inner membrane devoid of matrix, a value which is close to that found for outer

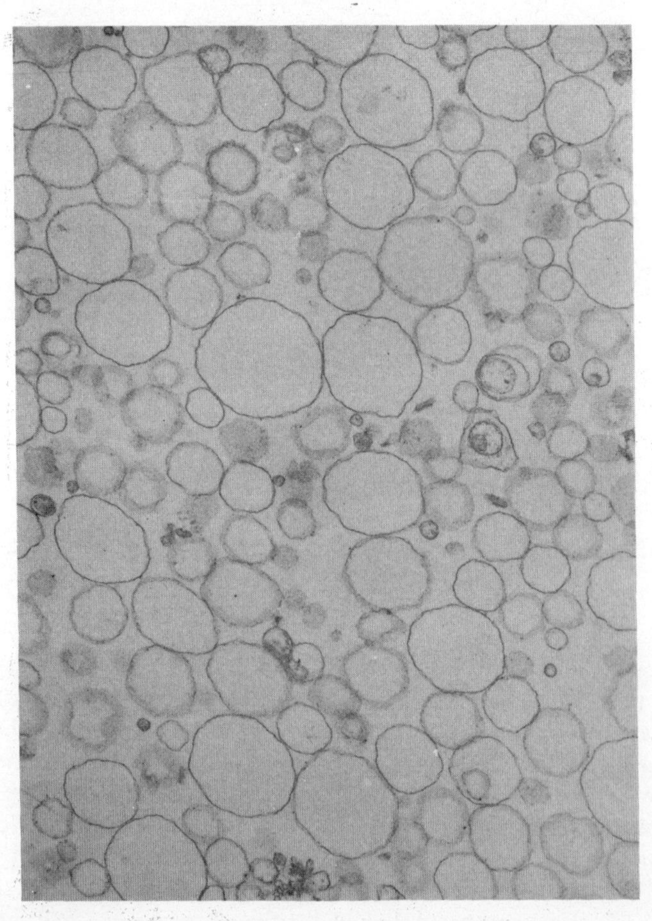

Figure 5. Thin section of outer mitochondrial membranes fixed with osmium tetroxide (× 30,000). (Courtesy of M. Satre.)

membrane (0·51 μmole P/mg protein). From this series of experiments it is clear that no clean separation of inner and outer membranes of mitochondria can be achieved by centrifugation in a sucrose gradient when both membranes are ruptured at the very first step by inadequate treatment of mitochondria.

Investigations concerning the localization of phospholipase-A_2 in rat liver mitochondria have shown that less than 10% of phospholipase is solubilized during membrane fractionation. The specific activity of phospholipase-A_2 (Table 3) is four times higher in the outer membrane than in the inner membrane + matrix fraction.

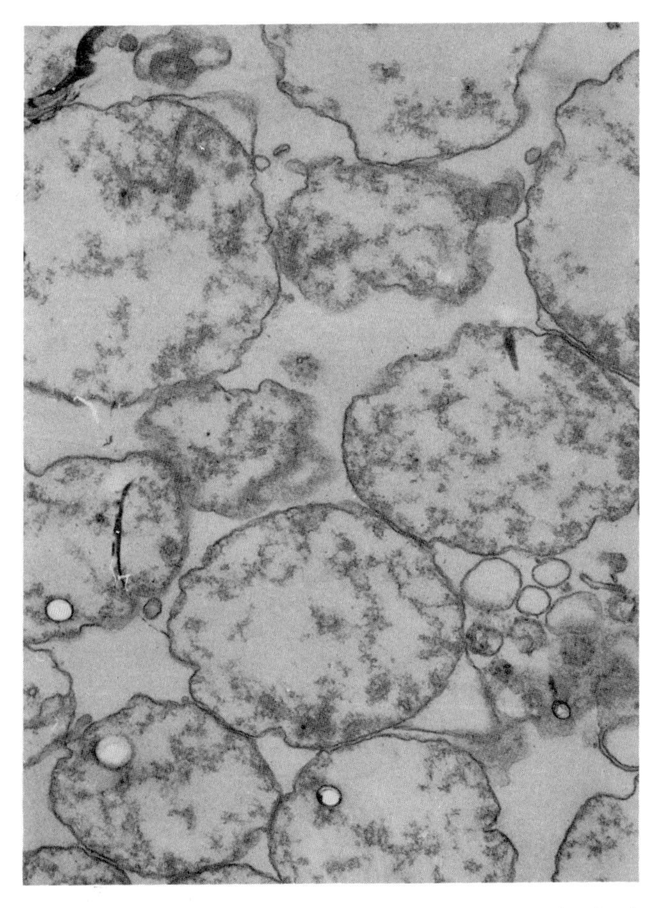

Figure 6. Thin section of inner mitochondrial membranes + matrix. Osmium tetroxide fixed (x 30,000).

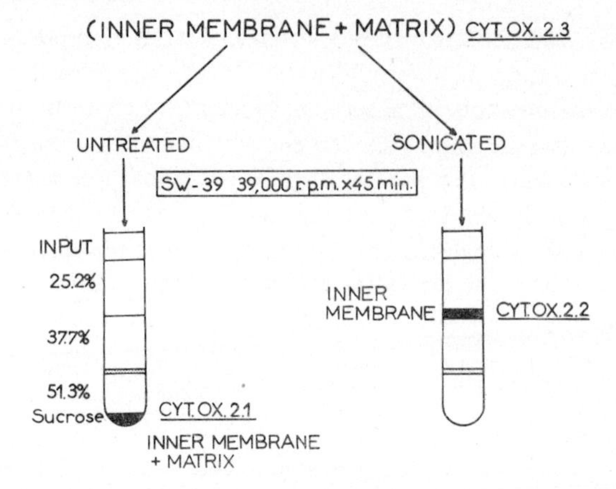

Figure 7. Centrifugation in a discontinuous sucrose gradient of the fraction "inner membrane + matrix" before and after sonication. It is to be noted that at the centrifuge force used most of the particulate material, after sonication, gathered at the first interface. However, after longer periods of centrifugation, the inner membrane gathered at the second interface. For determination of cytochrome oxidase activity, cf. Fig. 2.

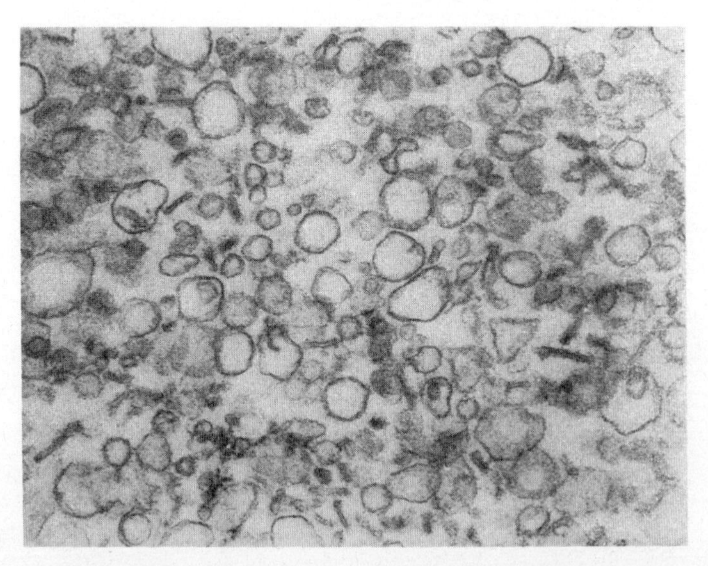

Figure 8. Thin section of inner mitochondrial membrane devoid of matrix. Osmium tetroxide fixed (x 42,000). (Courtesy of Dr Alix.)
(Note the small size of the vesicles.)

Table 3. Effect of pH on the phospholipase activity of mitochondrial and lysosomal membranes

Fraction	Fatty acid released*	
	Spec. Act. at pH 8·1	Spec. Act. at pH 8·1 / Spec. Act. at pH 4·5
Inner membrane + matrix	19	8·0
Outer membrane	72	4·2
Lysosomal membrane	118	0·4

* nmole/mg protein.

The incubation medium contained: ultrasonicated egg phosphatidylethanolamine (0·4 mg), 0·02M triethanolamine buffer, pH 8·1, 2 mM $CaCl_2$ and 0·3 to 4 mg of protein in a final volume of 2 ml. One hour incubation at 37°C in a shaking waterbath. The fatty acids extracted by 4 ml of chloroform/methanol (2/1) were isolated by chromatography on silicic acid column, methylated and analysed by gas liquid chromatography.

Since the effect of phospholipase-A_2 on mitochondrial phospholipids results in the formation of lysophosphatides, one may wonder whether the reverse process, i.e. the reacylation of lysophosphatides, does occur in mitochondria. Two different enzymatic mechanisms account for the conversion of lysophosphatides into phosphatides. The first requires the participation of fatty acyl-CoA derivatives [18-20], and the second consists of a transacylation reaction between two molecules of lysophosphatides [21].

An acyl-CoA:phospholipid acyl transferase activity has been detected in mitochondrial fractions prepared by differential centrifugation from various tissue homogenates [22, 23]. As shown in Table 4, rat liver mitochondria practically devoid of microsomes and lysosomes and supplemented with ATP, CoA and Mg^{2+} incorporate [^{14}C] oleic acid in their endogenous phosphatidylethanolamine and phosphatidylcholine, a result which shows unambiguously

Table 4. Incorporation of [^{14}C] oleic acid into mitochondrial phospholipids

Additions	Values are expressed as percentages of total radioactivity			
	LPC	PC	PE	CL
None	0·9	17·2	9·8	–
NaF (200 μmoles)	0·4	15·4	19·0	0·4
NaF (400 μmoles)	2·0	25·8	37·4	0·3

The incubation medium contained: 0·2 μmole of [^{14}C] oleic acid (1200 c/μmole/min), 35 μmoles of ATP, 1 μmole of CoASH, 5 μmoles of $CaCl_2$, 50 μmoles of $MgCl_2$, 100 μmoles of triethanolamine buffer, pH 8, rat liver mitochondria (E-fraction), 11 mg protein. Final volume 2 ml–temp. 37°C–incubation for 30 min.

that reacylation of lysophosphatides is achieved in mitochondria. In agreement with Webster and Alpern [22], fluoride markedly enhances the incorporation of [^{14}C] oleic acid, possibly by reducing the adenosinetriphosphatase activity of mitochondria. It is worth noting that the incorporation of radioactivity occurs essentially in the phospholipids which are preferentially attacked by mitochondrial phospholipase-A_2 (namely phosphatidylethanolamine and phosphatidylcholine).

Although the experiments which have been reported above allow us to establish unambiguously the presence in rat liver mitochondria of a phospholipase-A_2 together with an acylating enzyme, many problems concerning

Table 5. Hydrolysis of cardiolipin by lysosomal phospholipase

Fatty acid designation	Fatty acids released by lysosomes (nmoles/h/mg prot.)		Fatty acid composition of beef heart CL (moles %)
	without CL	with CL	
14:0	11·9	2·0	1·5
16:0	17·7	13·8	–
16:1	0·7	20·1	2·3
18:0	34·3	42·9	–
18:1	15·2	47·3	7·0
18:2	20·8	339·1	77·3
18:3	–	31·8	11·8

Lysosomes from H-fraction were frozen and thawed 10 times before being shaken in 0·04M acetate buffer, pH 4·5, at 37°C for one hour. 0·45 mg protein was used in a final volume of 2 ml. Beef heart cardiolipin (0·4 mg) was first emulsified by ultrasonication in a Branson Sonifier.

the specificity of action and the function of mitochondrial phospholipase remain to be solved. For instance, cardiolipin, a phosphatide specifically located in the inner membrane of mitochondria, is not hydrolysed by the mitochondrial phospholipase (Table 1). This finding may be contrasted with the fact that lysosomal phospholipase hydrolyses cardiolipin as shown by the selective release of linoleic acid, a fatty acid which represents 77% of the fatty acid content of cardiolipin (Table 5).

In the remaining part of this paper, we shall briefly describe identification of a phospholipase-A activity in yeast mitochondria. Although the presence of various phospholipases in yeast cells has been reported by other authors [24-26] no reference has been given as to its subcellular distribution. In experiments described here, it was found that a marked phospholipase activity could be developed in mitochondria from *Torulopsis utilis* [27]. These mitochondria isolated according to the procedure described for *S. Cerevisiae* by Mattoon and Sherman [28] (thin section micrograph in Fig. 9) were able to efficiently couple

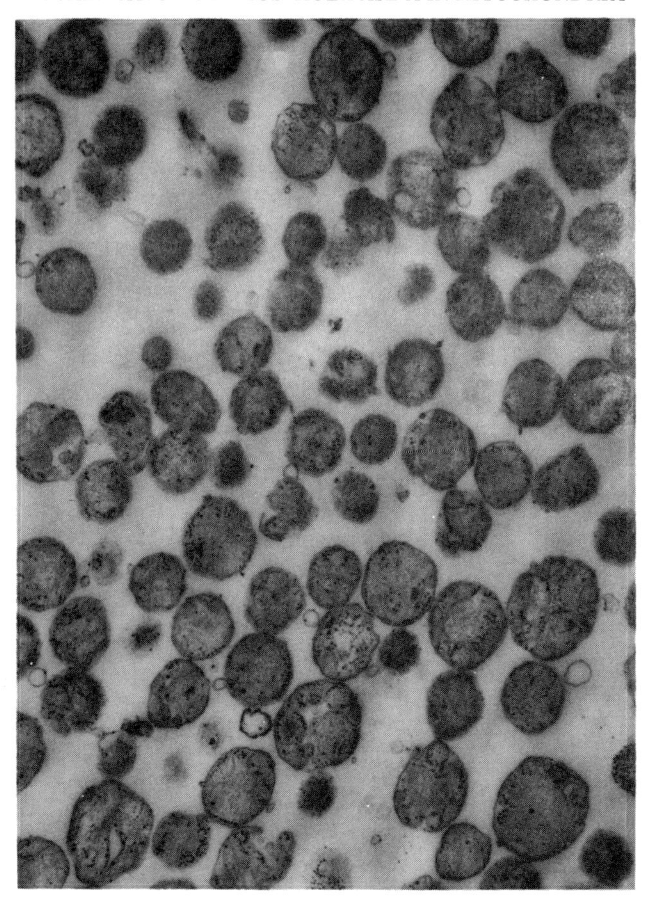

Figure 9. Thin section of a mitochondrial pellet of *Torulopsis utilis* fixed with osmium tetroxide (x 16,000).

Table 6. Effect of Ca^{2+} on phospholipase-A of yeast mitochondria

Expt.	Prot. (mg)	Additions	PE	Released total (nmoles)	Fatty acids unsaturated %
1	0·84	EGTA	+	529	61
		Ca^{++}	+	877	57
2	1·26	EGTA	–	140	92
		Ca^{++}	–	275	92

Yeast mitochondria were shaken for 30 min at 37°C in 0·04M acetate buffer in a final volume of 2 ml. When used, egg phosphatidylethanolamine (0·4 mg) had been emulsified by ultrasonic irradiation. EGTA: 0·1 mM, $CaCl_2$: 4 mM.

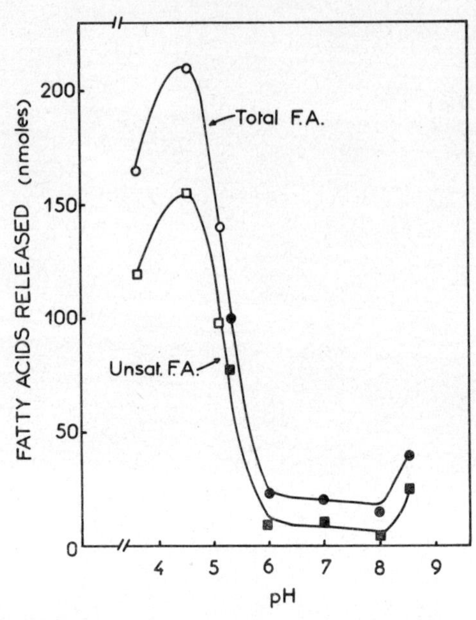

Figure 10. Effect of pH on the hydrolysis of phosphatidylethanolamine by yeast mitochondria. Mitochondria from *T. utilis* (0·4 mg) were incubated with ultrasonicated egg phosphatidylethanolamine (0·4 mg) in 0·04M Tris acetate buffer (○, □) or 0·04M Tris maleate buffer (●, ■) at the appropriate pH. Final volume 2 ml. Incubation at 37°C for 30 min in a shaking waterbath.

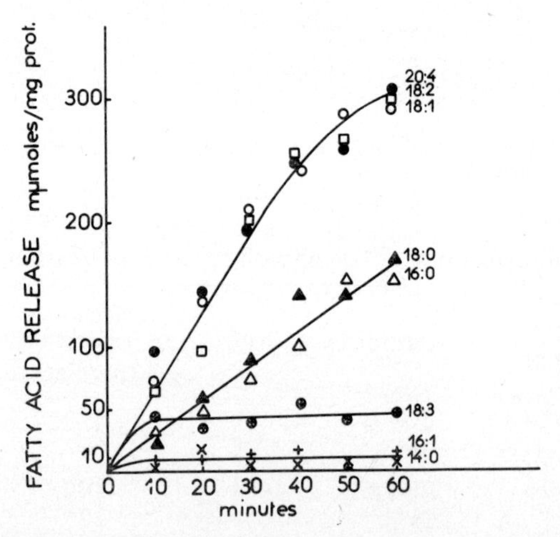

Figure 11. Effect of time on the release of saturated and unsaturated fatty acids from phosphatidylethanolamine by yeast mitochondria. 0·04M acetate buffer, pH 4·5, 2 mM $CaCl_2$, ultrasonicated egg phosphatidylethanolamine (0·4 mg). 0·5 mg of protein of mitochondria from *T. utilis.* Temperature 37°C.

phosphorylation and respiration; for instance, pyruvate + malate are oxidized with a P/O ratio higher than 2 and an index of respiratory control of 3 to 4. Since the presence, in yeasts, of particles rich in hydrolytic enzymes, such as ribonuclease, has been recently reported [29], the contamination of mitochondrial fractions by such particles was investigated in order to assess the significance of their phospholipase activity. On the basis of ribonuclease and acid phosphatase assays, no detectable contamination of the mitochondrial preparation by extramitochondrial particles with hydrolytic activities was found.

Phospholipase from *T. utilis* mitochondria is characterized by a pH optimum of 4·5 (Fig. 10) and by a Ca^{2+} requirement for maximal activity (Table 6). A small but reproducible activity is detected around pH 8·5. The kinetic data summarized in Fig. 11 indicate that phospholipase of yeast mitochondria cleaves both fatty acid ester linkages of egg phosphatidylethanolamine but that unsaturated fatty acids are released twice as fast as the saturated ones. This type of hydrolysis may be due to distinct phospholipase-A_1 and -A_2 acting separately, or to phospholipase-A_2 and lysophospholipase [25] acting sequentially. The demonstration of the presence of phospholipase in yeast mitochondria not only extends the concept of mitochondrial phospholipase but also raises the problem of how phospholipase activity in mitochondria is regulated inasmuch as lysophosphatides and unsaturated fatty acids are critical factors involved in membrane permeability and efficiency of oxidative phosphorylation, respectively.

ACKNOWLEDGEMENTS

Yeast mitochondria were prepared by Mrs J. Huet. The authors wish to thank her for her friendly cooperation. They thank Dr M. Faure for gifts of beef heart cardiolipin. The skilful technical assistance of Miss J. Baranne, Mr R. Cesarini and Mrs Balmefrezol is also gratefully acknowledged. This research has been supported by grants from the "Centre National de la Recherche Scientifique" (RCP 21, ERA No. 36), the "Fondation de la Recherche Médicale" and the "Délégation Générale à la Recherche Scientifique".

REFERENCES

1. Rossi, C. R., Sartorelli, L., Tato, L., Baretta, L. and Siliprandi, N., *Biochim. biophys. Acta* **98** (1965) 207.
2. Scherphof, G. L. and van Deenen, L. L. M., *Biochim. biophys. Acta* **98** (1965) 204.
3. Waite, M. and van Deenen, L. L. M., *Biochim. biophys. Acta* **137** (1967) 498.

4. Bjørnstad, P., *J. Lipid Res.* **7** (1966) 612.
5. Vignais, P. M. and Nachbaur, J., *Bull. Soc. Chim. biol.* **50** (1968) 1473.
6. Hogeboom, G. H., *in* "Methods in Enzymology" (edited by S. P. Colowick and N. O. Kaplan), Academic Press, London and New York, 1955, Vol. 1, p. 6.
7. Bjørnstad, P., *Biochim. biophys. Acta* **116** (1966) 500.
8. Mellors, A. and Tappell, A. L., *J. Lipid Res.* **8** (1967) 479.
9. Bartlett, G. R., *J. biol. Chem.* **234** (1959) 466.
10. Wattiaux, R., Wibo, M. and Baudhuin, P., *in* "Lysosomes" (edited by A. V. S. de Reuck and M. P. Cameron), Ciba Foundation Symposium, 1963, p. 176.
11. Linhardt, K. and Walter, K., *in* "Methods of Enzymatic Analysis" (edited by H. U. Bergmeyer), Verlag Chemie, 1965, 2nd Edn., p. 783.
12. de Duve, C., Pressman, B. C., Gianetto, R., Wattiaux, R. and Appelmans, F., *Biochem. J.* **60** (1955) 604.
13. Tabor, C. W., Tabor, H. and Rosenthal, S. M., *in* "Methods in Enzymology" (edited by S. P. Colowick and N. O. Kaplan), Academic Press, London and New York, 1955, Vol. 2, p. 390.
14. Weissbach, H., Smith, T. E., Daly, J. W., Witkop, B. and Udenfriend, S., *J. biol. Chem.* **235** (1960) 1160.
15. Appelmans, F., Wattiaux, R. and de Duve, C., *Biochem. J.* **59** (1955) 438.
16. Parsons, D. F., Williams, G. R., Thompson, W., Wilson, D. and Chance, B., *in* "Round Table Discussion on Mitochondrial Structure and Compartmentation" (edited by E. Quagliariello, S. Papa, E. C. Slater and J. M. Tager), Adriatica Editrice, Bari, 1967, p. 29.
17. Dodgson, K. S., Spencer, B. and Thomas, J., *Biochem. J.* **59** (1955) 29.
18. Lands, W. E. M., *J. biol. Chem.* **235** (1960) 2223.
19. Lands, W. E. M. and Merkl, I., *J. biol. Chem.* **238** (1963) 898.
20. Merkl, I. and Lands, W. E. M., *J. biol. Chem.* **238** (1963) 905.
21. Erbland, J. F. and Marinetti, G. V., *Biochim. biophys. Acta* **106** (1965) 128.
22. Webster, G. R. and Alpern, R. J., *Biochem. J.* **90** (1964) 35.
23. Turkki, P. R. and Glenn, J. L., *Biochim. biophys. Acta* **152** (1968) 104.
24. Kokke, R., Hooghwinkel, G. J. M., Booij, H. L., van den Bosch, H., Zelles, L., Muller, E. and van Deenen, L. L. M., *Biochim. biophys. Acta* **70** (1963) 351.
25. van den Bosch, H., van den Elzen, H. H. and van Deenen, L. L. M., *Lipids* **2** (1967) 279.
26. Letters, R., *in* "Aspects of Yeast Metabolism" (edited by A. K. Mills and H. A. Krebs), Blackwell Scientific Publications, Oxford, 1967, p. 303.
27. Vignais, P. M., Huet, J., Nachbaur, J. and Vignais, P. V., in preparation.
28. Mattoon, J. R. and Sherman, F., *J. biol. Chem.* **241** (1966) 4330.
29. Matile, P. and Niemken, A., *Arch. Mikrobiol.* **56** (1967) 148.

Note added in proof

Complementary data have been published in the following notes:
Vignais, P. M. and Nachbaur, J., *Biochem. biophys. Res. Commun.* **33** (1968) 307.
Nachbaur, J. and Vignais, P. M., *Biochem. biophys. Res. Commun.* **33** (1968) 315.
Nachbaur, J., Colbeau, A. and Vignais, P. M., *FEBS letters* **3** (1969) 121.

FEBS Symposium, Volume 17, 1969, pp. 59-77

Control of Adenine Nucleotide Translocation

M. KLINGENBERG, R. WULF and H. W. HELDT
Lehrstuhl für Physikalische Biochemie der Universität München, West Germany, and

E. PFAFF
Institut für Physikalische Biologie und Elektronenmikroskopie der Universität Marburg, West Germany

The adenine nucleotide translocation across the inner membrane of mitochondria is limited by its specificity to a mole-by-mole counter exchange of ADP and ATP only. Apparently all combinations of the counter exchange between ADP and ATP are possible (Fig. 1). However, depending on the state of

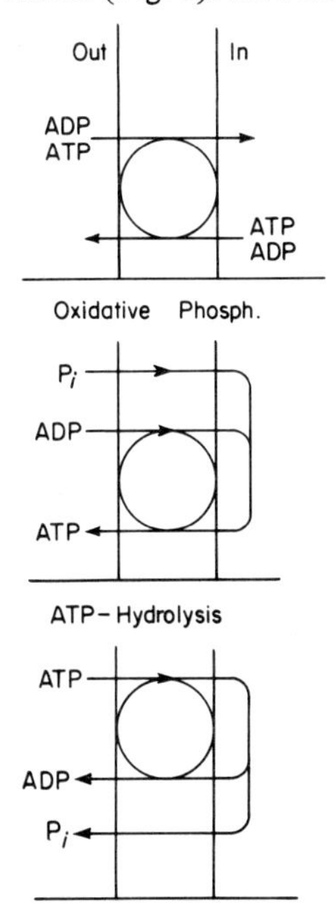

Figure 1. The modes of adenine nucleotide exchange. ADP-ATP exchange in oxidative phosphorylation, ATP-ADP exchange in ATP hydrolysis.

the energy transfer in mitochondria, the *a priori* equal specificity of the translocation for both ADP and ATP may be modified [1]. This control of the specificity may influence the two major functions of the AdN* exchange: in the synthesis of ATP, the uptake of ADP in exchange for ATP, and in the reverse reaction (ATP hydrolysis), the uptake of ATP in exchange for ADP. The characteristics and the mechanism of control of the translocation are the subject of this paper.

Principal results on the specificity of the AdN translocation

The essential characteristics of the AdN translocation are summarized in Table 1, comprising the specificity, kinetic properties, temperature dependence, etc. [2]. It is seen that the kinetic properties as well as the specificity vary according to the functional state of the mitochondria. This extensively documented observation (cf. [3]) forms the basis for a proposed control of the AdN translocation. The mechanism and the consequences of this control will be discussed in this paper. First, a short evaluation of the basic properties will be given.

In coupled mitochondria the exchange by exogenous ATP has been found to be slower than the exchange by ADP. On the addition of uncoupler, the ATP exchange can be increased to the level of the ADP exchange. Thus the ATP exchange can be considered to be inhibited in the controlled state as compared to the ADP exchange. The stimulation is already effective at a very low concentration of uncoupler (CCP), which also abolishes oxidative phosphorylation but does not yet stimulate ATP hydrolysis [1, 3]. This, and the ineffectiveness of oligomycin, indicate that the control of the ATP exchange is linked to the primary site of uncoupler action in the energy transfer and is not connected to the hydrolysis of the added ATP.

Various features of the control of the exchange are briefly comprised in a competition experiment where ADP and ATP are added simultaneously for exchange (Fig. 2). Under these conditions the exchange by ATP is successfully suppressed by that of ADP. On the addition of CCP, however, the situation is reversed. On the addition of valinomycin and phosphate, the ATP exchange is also stimulated, though to a smaller degree. This suggests that a depletion of intermediate energy by uncoupling or induced ion transport removes the inhibition of the ATP exchange.

Inhibition of respiration, combined with energy-depleting treatment of mitochondria, stimulates the ATP exchange, however, less than in the uncoupled state (Table 2). In combination with oligomycin a stronger stimulation is observed. Oligomycin obviously prevents the added ATP from causing an inhibition by generating the inhibitory energized state at the membrane in a

* Non-standard abbreviations: AdN, adenine nucleotide; CCP, carbonylcyanidphenylhydrazone; OM, oligomycin.

Table 1. Characteristics of adenine nucleotide translocation (liver mitochondria)

Specificity	
Controlled state:	*Relative rate (ADP = 1)*
exo → endo reaction	ATP: 0·2-0·5; AMP: $\approx 10^{-2}$
endo → exo reaction	ATP: ≈ 1; AMP: $\leqslant 10^{-1}$
Competitive:	
exo → endo reaction	ATP: $\approx 0{\cdot}1$
exo → endo reaction	dADP: $\approx 0{\cdot}3$; dATP: 0·1 dAMP: $< 10^{-2}$
Uncoupled state:	ATP: 1 dATP: = dADP: $\approx 0{\cdot}3$
Anaerobic state:	ATP: 0·3-0·5 dATP: = dADP: = 0·3
Kinetic characteristics	
zero order translocation (k) = first order exchange of endogenous (ADP + ATP) pool k_{ADP} (μmoles/min/g prot.) = 20 (at 5°C), = 100 (at 15°C)	
Concentration dependence	
$v = f$ (ADP, ATP) only approximately hyperbolic K_m = 1·5 to 4 μM ADP 3 μM ATP, 10 μM ATP-Mg	
Temperature dependence	
log v linear versus 1/T (0-20°C) E $\approx$ 24 kcal	
Specific inhibitor	Atractyloside
pH Dependence	Essentially no pH dependence (pH 5-8)

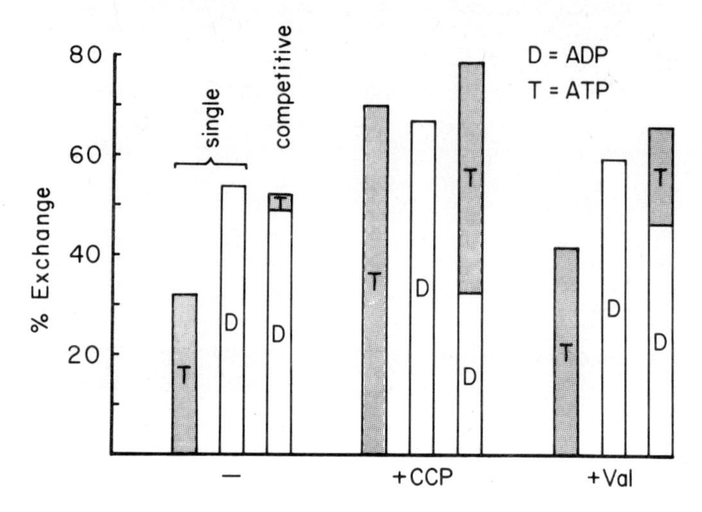

Figure 2. Competition of ADP and of ATP exchange. Separate exchange by either ATP or ADP is compared with the combined addition of ADP + ATP in the controlled state, uncoupled state and K^+-permeable state of the mitochondria [3]. The exchange was measured by first labelling all endogenous adenine nucleotides with [^{3}H] ADP and then measuring the back exchange by the addition of ATP and [^{14}C] ADP. The loss of [^{3}H] AdN from the mitochondria gives the total exchange by ADP plus ATP, and the incorporation of ^{14}C-label gives the portion of ADP exchange. Incubation of mitochondria in 0·13M KCl at 9°C, exchange time 9 s. Additions, 5 μM CCP, 1 mg/ml valinomycin (Val.), 4 mM P.

Table 2. Inhibition of the ATP exchange by coupled respiration or by reverse energy transfer

Addition	% Exchange	
	ATP	ADP
—	24	36
Oligomycin	29	58
CCP	46	38
Succinate	12	44
Succinate + CCP	50	58

Liver mitochondria prelabelled with [^{14}C] ADP were incubated with 0·2 μM CCP and 5μ M rotenone for 15 min (2 mg prot/ml). Then CCP was removed by addition of 1·5 mg serum albumin. The exchange was measured by the atractyloside stopping method. The additions were 5 μM CCP, 2 μg oligomycin/ml, 4 mM succinate and the AdN concentration was 500 μM.

reversed reaction. The subsequent supply of respiratory energy by succinate to the energy-depleted mitochondria exerts a strong inhibition on the ATP exchange, whereas the ADP exchange remains largely active. The fact that uncoupler also stimulates the ATP exchange in an energy-depleted state with oligomycin, indicates that the inhibition is built up to some extent by the ATP exchange itself rather than by a reverse energy supply from ATP. The stimulation of the ATP uptake in energy-depleted mitochondria has been proposed also by Tager *et al.* [4] on the basis of the ATP-dependent citrulline synthesis.

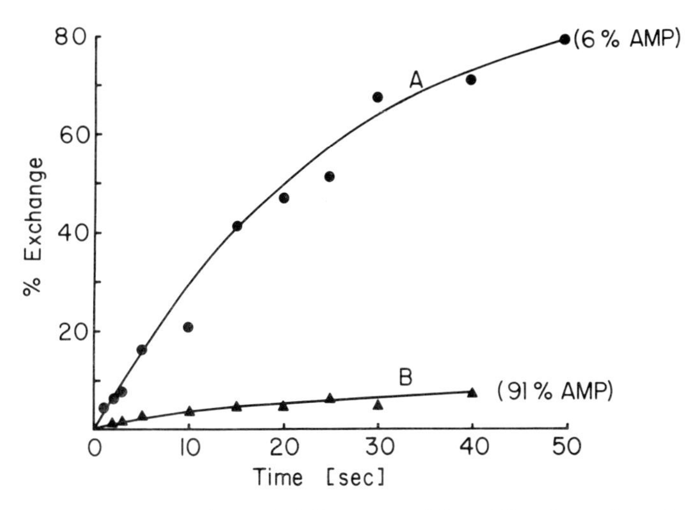

Figure 3. Dependence on the internal AMP content of the AdN exchange [5]. Measuring of the back exchange by the atractyloside stopping method in a rapid sampling apparatus. The mitochondria were preincubated in case (A) with 3 mM ketoglutarate and 3 mM P_i, (B) with 3 mM arsenate. Exchange performed in sucrose-EDTA medium at 10°C by the addition of 0·3 mM ADP.

This regulation of the AdN exchange concerns only the specificity from the outside. The specificity of the AdN exchange from the inside is more difficult to determine, since the endogenous enzyme activities readily interconvert the three forms of endogenous AdN into each other. The intramitochondrial AMP, however, is relatively inert and the non-activity of endogenous AMP in the exchange can clearly be demonstrated [2]. Thus, the endogenous AMP content may be varied over a wide range by influencing the substrate level phosphorylation which is most closely associated with the AMP level. As shown in a kinetic experiment (Fig. 3), the short time exchange up to 60 s approaches only the level of ADP + ATP present in the mitochondria.

The kinetics of the AdN exchange, referred to the amount of endogenous

ADP + ATP, were found to follow first order kinetics [5]. Thus, the homogeneity of the endogenous ADP + ATP pool could be recognized.

The main interest for the control concerns the specificity of the back exchange for ADP and ATP. The fact that the AdN exchange, as referred to ADP + ATP, follows homogeneous kinetics, indicates that both endogenous ADP and ATP exchange at equal rates. Furthermore, it can be shown that ADP and ATP are released to the outside during the exchange in about the same proportion as the endogenous AdN pattern. This is demonstrated in Fig. 4 for

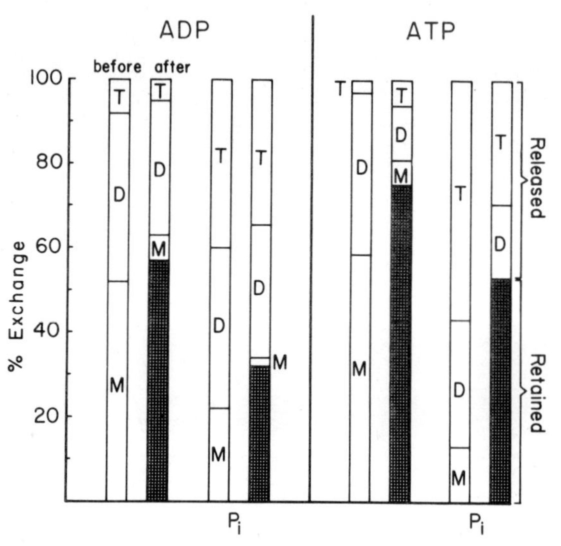

Figure 4. Influence of the endogenous adenine nucleotide pattern on the exchange by added ADP or ATP. Different phosphorylation patterns of the endogenous AdN were obtained by incubation for 5 min at 15°C with and without 2 mM glutamate, 2 mM succinate and 4 mM phosphate. The exchange was measured for 15 s as back exchange by pressure filtration. The phosphorylation patterns were determined by column chromatography immediately before the start of the exchange by addition of ADP or ATP and after the short time exchange (15 s). (For details cf. [4].)

varying endogenous contents of ADP and ATP. No endogenous AMP appears outside besides that formed by exogenous adenylate kinase. Finally, the computation of the phosphorylation rate of exogenous ADP is in good agreement with the experimental data when equal specificity for the exchange to ADP and ATP is assumed [6]. It may be concluded that the specificity difference for the back reaction between ADP and ATP is not greater than 20%. This holds good both for the coupled and uncoupled states.

Thus, the inhibition of the ATP exchange in the coupled state superimposes on the translocation an asymmetry which appears to be a consequence of an energy-dependent state of the membrane and which does not reflect an asymmetric specificity of the exchange carrier *per se.* This is in agreement with basic thermodynamic laws.

The AdN translocation as an anion exchange

An explanation for the specificity differences between ATP and ADP has been based on the fact that the AdN exchange is considered to be an exchange of anions across the membrane [2]. Therefore, problems of electrical charge differences or deficiencies and of membrane potential should influence this

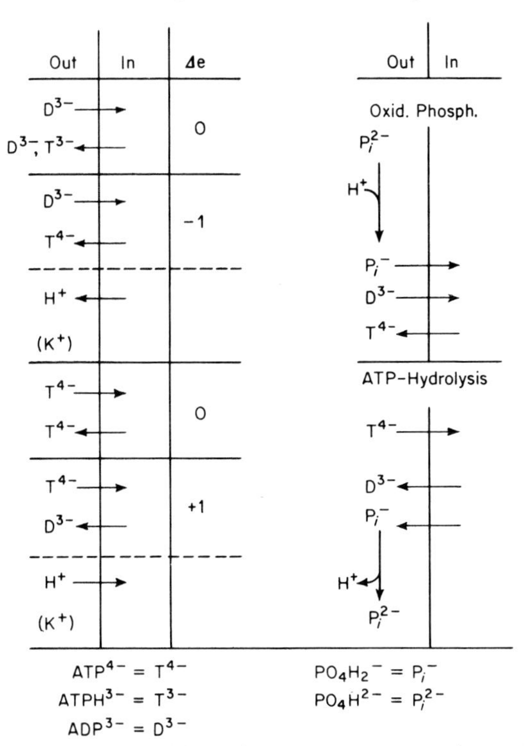

Figure 5. The adenine nucleotide exchange as an anion exchange. The various ionic modes of the exchange, alone, in oxidative phosphorylation, and in ATP hydrolysis.

exchange. A variety of conditions for this exchange can be considered employing various charged forms of the anions, some cases of which are depicted in Fig. 5. In the ADP exchange, charge neutrality would result if ADP^{3-} or $ATPH^{3-}$ were released. A negative charge excess would result in the ADP exchange if ATP were released in the preponderant form at pH 7 as ATP^{4-}. In the opposite case—the ATP^{4-} against ADP^{3-} exchange—a positive charge is formed outside. If we assume that primarily ADP^{3-} is released during ADP or ATP exchange, a charge discrepancy would only result in the latter case, which should lead to an inhibition of the ATP exchange unless a charge-compensating movement of other ions takes place. This function can be fulfilled by the uncouplers which may facilitate the uptake of H^+ through the membrane in the ATP-ADP exchange. Thus, the activation of the ATP exchange by uncoupling would remove the charge discrepancy created in the ATP-ADP exchange. Other auxiliary H^+ shuttles can also be proposed, such as a facilitated movement of P_i^- in the presence of valinomycin and K^+. However, these shuttles are relatively inert.

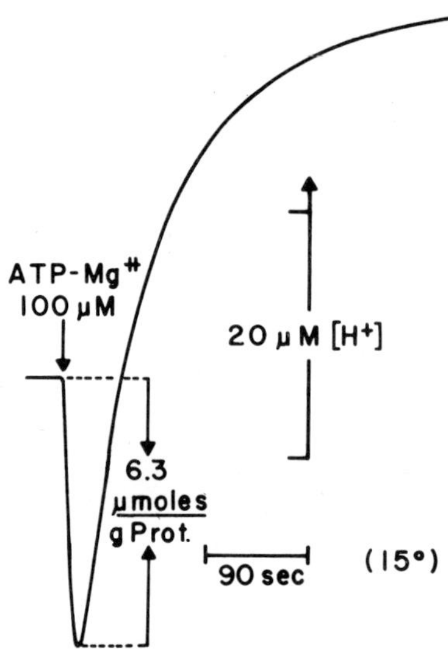

Figure 6. The H^+ uptake on addition of ATP to mitochondria in the uncoupled state. Liver mitochondria prepared in unbuffered sucrose medium. To the stock suspension (about 50 mg prot/ml) 2 mg serum albumin were added. For the experiments the mitochondria were incubated in 0·15M KCl, 2·2 mM glycylglycine, pH 7·13, 15°C under addition of 20 μM rotenone, 10 μM FCCP, 1 μM oligomycin.

The existence of the AdN translocation raises the question as to which ionic species of the AdN and of Pi permeate through the inner membrane during the synthesis or hydrolysis of ATP. In other words, on which side of the membrane is H^+ used or liberated in these reactions? In the chemiosmotic hypothesis, which implies the impermeability of the inner membrane for H^+, the movement of charges due to the translocation of AdN or P_i through the membrane had previously not been considered [7]. In a recent study on ATP hydrolysis by Mitchell and Moyle [8], however, this problem was recognized and put to experimental test. It was concluded that no charge difference is seen in the ATP hydrolysis and that the H^+ is delivered on the outside of the inner membrane. This is in agreement with our earlier proposed model that ATP^{4-} enters the mitochondria in exchange for ADP^{3-} and P_i^- [2]. The P_i^- then dissociates on the outside into P^{2-} and H^+. Other models could also suit the main requirements that the H^+ is taken up or released on the outside and therefore does not have to penetrate the intramitochondrial membrane.

The mechanism proposed earlier for the control of ATP translocation can be tested experimentally by following the movements of H^+ during ATP translocation. It has been noticed by Mitchell and Moyle [8] that on addition of ATP to mitochondria in the presence of uncoupler, a small alkalization in the mitochondrial suspension precedes the acidification due to ATP hydrolysis. This could correspond to the H^+uptake postulated by us for the ATP-ADP exchange

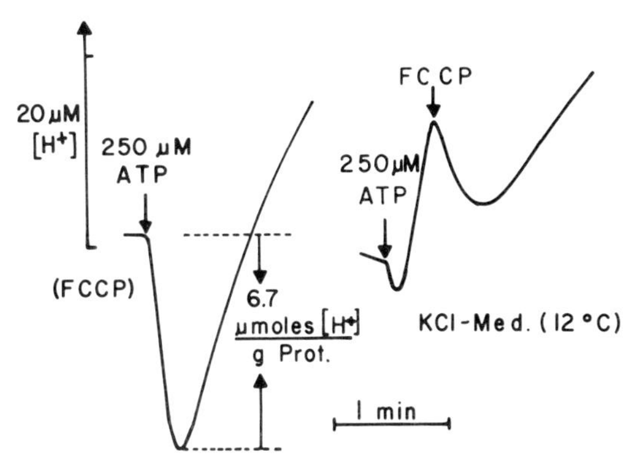

Figure 7. Influence of uncoupler on the ATP-induced H^+uptake. Conditions as in Fig. 6 (addition of 10 μM FCCP as indicated), 12°C.

[2] and was also discussed by Mitchell and Moyle in this context [8]. However, the observed effect is so small that its relation to the AdN exchange appears to be questionable, all the more so since unrelated small changes of pH on addition of ATP to the mitochondria occur in control experiments [9].

In the present studies, an ATP-induced H^+ uptake was observed that was of a magnitude in accordance with an ATP-ADP exchange and 5- to 10-fold larger than previously reported by Mitchell and Moyle [8]. The effect can easily be recorded and subjected to a detailed analysis of its kinetics, stoichiometry, etc. As shown in Fig. 6, the initial H^+ uptake is rapidly reversed by a subsequent ATP hydrolysis which leads to a net H^+ release. The problem is now to establish that this H^+ uptake is a result of the ATP-ADP exchange. The amount of H^+ taken up by the mitochondria corresponds to roughly half the amount of endogenous AdN, as might be expected if more than 50% of the endogenous AdN were present in the form of ADP^{3-} in the presence of uncoupler. The true maximum may be diminished by the H^+ back leakage and the ATP hydrolysis.

The H^+uptake is largely dependent on the presence of uncoupler, as shown in an experiment where the uncoupler is added after the prior addition of ATP

(Figs. 7 and 8). The experiment also shows that in the absence of uncoupler the generation of H^+ is much faster than after the addition of uncoupler. This can be interpreted to be caused by the rapid uptake of K^+ and concomitant H^+ release. In the KCl medium a large part of the ATP hydrolysis is insensitive to oligomycin. Less interference by ATP hydrolysis is observed in a sucrose-EDTA medium. There is appreciable H^+ uptake on ATP addition even in the absence of uncoupler which is then increased by FCCP. This H^+ uptake upon uncoupling is only partially due to increased ATP uptake and is more likely the known effect of alkalization by the influx of H^+ [9a].

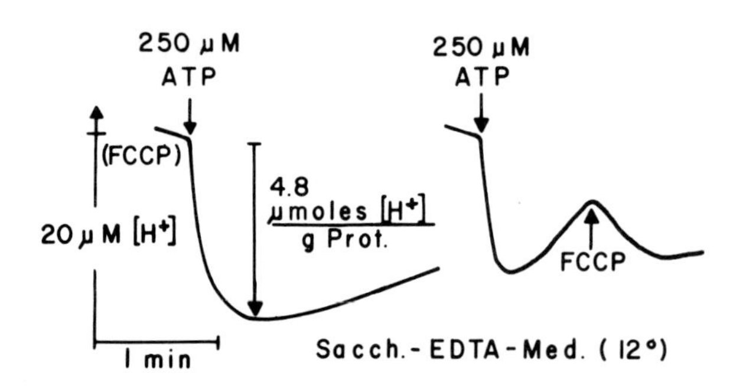

Figure 8. The H^+ uptake on addition of 250 μM ATP in sucrose-EDTA medium.

At a lower temperature the H^+ uptake is slower than the response time of the glass electrode ($t_{1/2}$ = 2 s). Therefore in a simultaneous experiment the kinetics of the H^+ uptake can be followed and compared with the kinetics of the AdN exchange, measured by rapid sampling (Fig. 9). The kinetics of the H^+ uptake closely follow those of the exchange. The uptake of H^+ corresponds approximately to the ATP exchanged. In agreement, the amount of ATP exchanged corresponds to the amount of endogenous ADP initially present, as there is almost no endogenous ATP available for the exchange. The exchange reaches about 50% of the total endogenous AdN pool, in accordance with the content of about 44% AMP. Thus, all the exchange can be considered to be an ATP against ADP exchange and the result corresponds to the postulate that one H^+ per exchange should be taken up.

The H^+ uptake after ATP addition is sensitive to atractyloside (Fig. 10). At any time during the H^+ uptake, the addition of atractylate will stop the reaction completely.

The termination of the exchange by addition of atractyloside at different times gives the possibility of varying over a wide range the amount of ATP taken up. Thus the H^+ uptake can be correlated with the amount of H^+ finally released

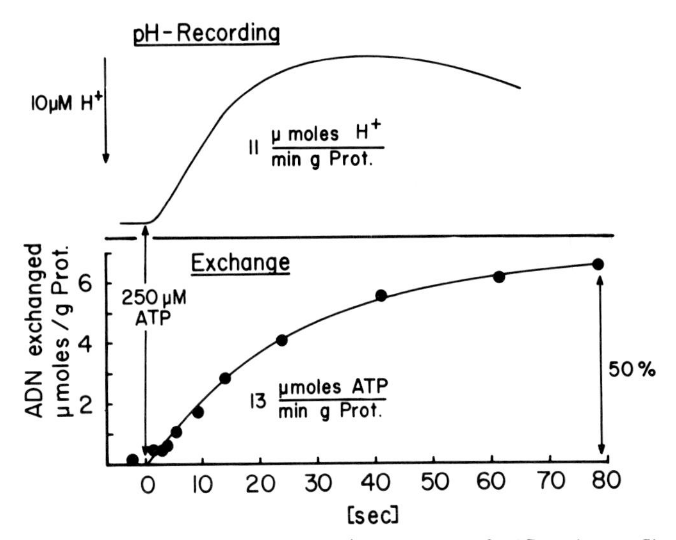

Figure 9. Comparison of the kinetics of H^+ uptake and of ATP exchange. Simultaneous measurement of both functions in a rapid sampling apparatus. 5·3 mg/ml liver mitochondria incubated in 0·15M KCl medium at pH 7·13, 5 °C. The AdN composition was measured in the sample drawn shortly before the start of the experiments: AMP 44%, ADP 55%, ATP 1%. Total content of AdN: 13 μmoles/g prot.

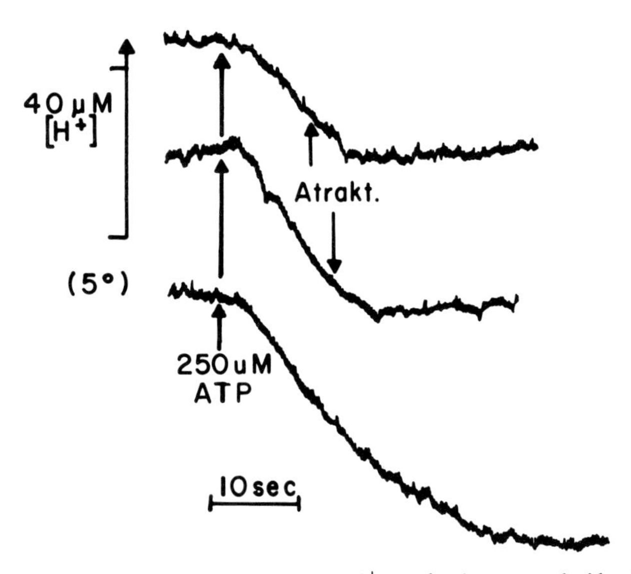

Figure 10. Inhibition of the ATP-induced H^+ uptake by atractyloside. Liver mitochondria (4 mg prot/ml) incubated in 0·15M KCl medium at 5 °C. Addition of 5 μM atractyloside. Conditions, cf. Fig. 6.

after hydrolysis (Fig. 11), and affords a test as to whether the H^+ uptake is stoichiometrically related to the amount of ATP exchanged, as only this ATP can undergo subsequent hydrolysis. Figure 11 shows that an approximate 1:1 correlation between the initial H^+ uptake and the subsequent H^+ release is observed. When atractyloside is added too late or omitted, more H^+ is released than taken up, since then the endogenous AdN pool is turned over more than once.

The dependence of the H^+ uptake on the concentration of ATP also gives a further indication as to its relation to the AdN exchange (Fig. 12). Saturation is reached as soon as the amount of ATP approximates the endogenous ADP

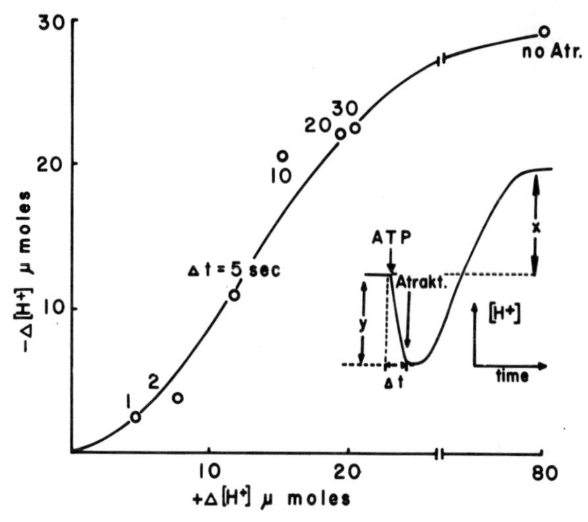

Figure 11. Comparison of the H^+uptake due to ATP translocation and of the H^+release due to ATP hydrolysis. The extent of the ATP exchange is varied by the addition of 5 μM atractyloside at different times. For incubation conditions see Fig. 6.

content of mitochondria. This saturation is found at 50 μM ATP, reaching 5 μmoles H^+/g prot, corresponding to 12 μmoles ATP/mg prot and to 0·4 H^+ per ATP.

In the range of saturating concentration of ATP, the H^+ uptake is proportional to the amount of mitochondrial protein (Fig. 13). The slope of this relation corresponds to 6 μmoles H^+/g prot. This value is again in approximate agreement with the other results.

The previous findings indicate stoichiometric agreement between the H^+ uptake and the size of the endogenous AdN pool. This, however, may be fortuitous unless the relation still holds when the AdN pool is varied. By a controlled leakage of the endogenous AdN after pre-incubation with P_i and Mg^{2+} for a varying time period, the endogenous AdN content is diminished, as

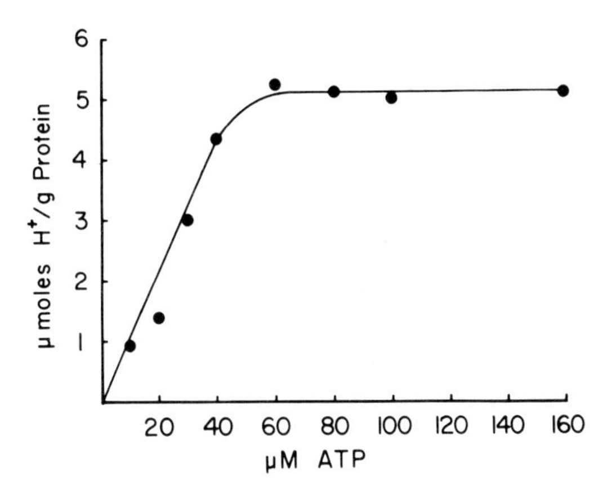

Figure 12. The ATP-induced H^+ uptake dependence on the concentration of ATP. 3·7 mg mitochondria under incubation conditions as described for Fig. 6.

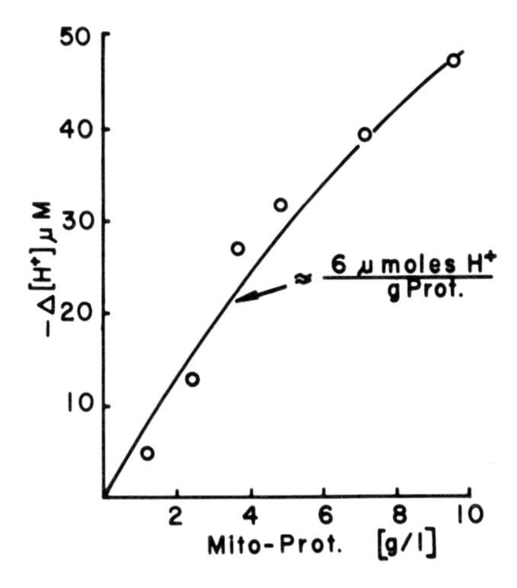

Figure 13. The ATP-induced H^+ uptake as a function of the concentration of mitochondria. Addition of 250 μmoles ATP. For incubation conditions cf. Fig. 6.

demonstrated by Meisner and Klingenberg [10]. As shown in Fig. 14, the H^+ uptake decreases strongly with the endogenous AdN content and reaches nearly zero at 2 μmoles AdN/mg prot. The residual content may consist mainly of AMP. The slope of the curve is about 0·6 H /AdN which may indicate the proportion of ADP present in these mitochondria.

It may be emphasized that the correlation of the H^+ uptake to the size of the endogenous AdN pool is a more convincing demonstration of the correlation to the AdN exchange than, for example, the inhibition by atractyloside or the correlation of the kinetics. Other intramitochondrial hypothetical reactions with ATP which take up H^+ would also depend on the translocation, and would thus

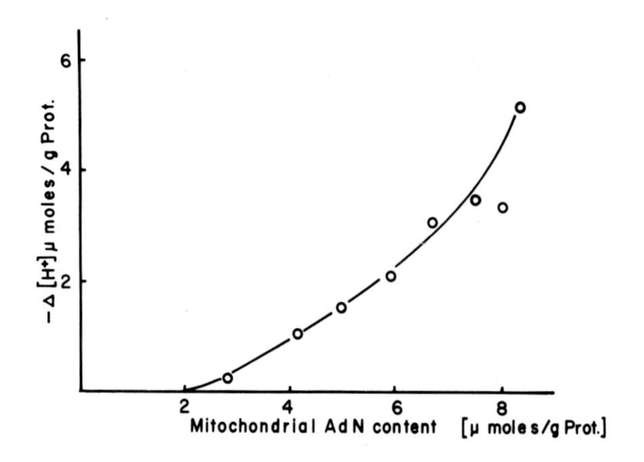

Figure 14. ATP-induced H^+ uptake as a function of the adenine nucleotide content of mitochondria (250 μM ATP). Mitochondria are partially depleted of endogenous AdN by incubation for various periods of time between 1 to 20 min at 15°C in sucrose medium under addition of 5 mM P_i and 4 mM $MgCl_2$. Each sample was centrifuged and washed before measuring the H^+ uptake.

be sensitive to atractyloside or correlate to the kinetics of the exchange. However, it is difficult to explain why the H^+ uptake follows closely the size of the endogenous AdN pool, other than by an exchange as a direct cause of the H^+ uptake.

In conclusion it may be pointed out that these studies substantiate the hypothesis [2] that the exchange of ATP^{4-} against ADP^{3-} is linked to the uptake of H^+ in the uncoupled state. The inhibition of H^+ uptake in the controlled state is one factor controlling the ATP specificity. Once ATP hydrolysis is activated, charge neutrality during the translocation can be expected when the release of P_i^- accompanies that of ADP in the exchange.

The formation of a phosphorylation potential difference by the adenine nucleotide translocation

During the process of oxidative phosphorylation of exogenous ATP, ADP^{3-} exchanges against endogenous ATP^{4-}, creating a negative charge excess at the outside. The reaction can be charge neutral if coupled to the uptake of P_i^-. The negative charge excess generated in the ADP^{3-} against ATP^{4-} exchange should also cause an inhibition of this exchange unless the charge-compensating movements are possible. This, however, is in contrast to observations that ADP exchange with endogenous ATP is not inhibited in the controlled or uncoupled state. It may indicate that another force is superimposed on the translocation which permits release of negative charge excess out of the mitochondria but inhibits the uptake. The force might be identified with the membrane potential which is poised positive outside [7, 11]. It would drive the translocation in favour of an ADP^{3-} against ATP^{4-} exchange and prevent the ATP^{4-} against ADP^{3-} exchange.

The asymmetry for ATP in the controlled state of the forward and of the backward translocation and the symmetry for ADP should lead to a larger ratio of ATP/ADP outside than inside. In equilibrium the following relation should hold:

$$\left(\frac{\mathrm{ATP}}{\mathrm{ADP}}\right)_{\mathrm{out}} \Big/ \left(\frac{\mathrm{ATP}}{\mathrm{ADP}}\right)_{\mathrm{in}} = \frac{k_{\mathrm{D}}^{\rightarrow}}{k_{\mathrm{T}}^{\rightarrow}} \Big/ \frac{k_{\mathrm{D}}^{\leftarrow}}{k_{\mathrm{T}}^{\leftarrow}} \tag{1}$$

$k^{\rightarrow}, k^{\leftarrow}$ = rate constant out → in, and vice versa.
With $k_{\mathrm{T}}^{\leftarrow} = k_{\mathrm{D}}^{\leftarrow}$:

$$\left(\frac{\mathrm{ATP}}{\mathrm{ADP}}\right)_{\mathrm{out}} = \frac{k_{\mathrm{D}}^{\rightarrow}}{k_{\mathrm{T}}^{\rightarrow}} \left(\frac{\mathrm{ATP}}{\mathrm{ADP}}\right)_{\mathrm{in}} \tag{2}$$

From the rates measured, the ratio ATP/ADP should be 5 to 10 times larger outside than inside. In the uncoupled state ($k_{\mathrm{T}}^{\rightarrow} = k_{\mathrm{D}}^{\rightarrow}$) the inner and outer ratios should be equal.

By filtration centrifugation experiments, the difference in the internal and external ratios could be demonstrated to be in accordance with these predictions (Table 3) [5]. The dependence of the equilibration between the inner and outer ATP/ADP on the uncoupler concentration is the same as the stimulation of the ATP exchange by uncoupler [2].

So far, the AdN exchange and its control have been considered from the kinetic point of view. The difference between the ratios ATP/ADP in the inner and outer space indicates that thermodynamic considerations have to be taken into account. Thus, through the preferred uptake of ADP and the release of ATP, energy is taken up into the system which may originate from the net transport of one negative charge to the outside through a membrane potential gradient. In equilibrium this maximum energy could be $F.\Delta E$ and the resulting factor for the difference between the inner and outer ratio ATP/ADP, considering eqn. 2, is:

$$e^{F.\Delta E/RT} = k_{\vec{D}}/k_{\vec{T}} \tag{3}$$

As discussed below, only a portion (β) of the ATP-ADP exchange may be electrogenic:

$$\frac{k_{\vec{D}}}{k_{\vec{T}}} = e^{\beta.F.\Delta E/RT},\ \beta \langle\ 1$$

Table 3. Equilibration by uncoupling of the internal and external ATP/ ADP

CCP (μM)	ATP/ADP	
	external	internal
0	6·3	0·7
0·03	4·8	2·0
0·09	3·8	2·5
0·15	3·1	3·1

Rat liver mitochondria (5 mg prot/ml), pretreated for 1 min at 0°C with oligomycin (final concentration 20 μM) and rotenone (final concentration 10 μM) were incubated for 60 s at 12°C in the presence of ADP + ATP = 1 mM without substrate or phosphate added. CCP was added to the incubation as indicated. Incubation was stopped by filtering centrifugation and adenine nucleotides assayed enzymatically in the extracts corresponding to the incubation medium and the sedimented mitochondria.

In the overall process of oxidative phosphorylation the ADP-ATP exchange should be coupled to the uptake of one P_i. For the sake of simplicity it is assumed that the monovalent anion P_i^- is taken up (cf. Fig. 15). In this case the overall translocation is electroneutral and the membrane potential is without influence on the ATP/ADP ratio. However, the ADP-ATP exchange is only "statistically" coupled to the P_i-uptake, as the ADP-ATP exchange is not dependent on the presence of P_i and vice versa. P_i is probably taken up by a separate carrier which may be assumed to link the uptake of P_i^- with H^+. The resulting electroneutral uptake of P_i^- leads to an electrogenic ADP-ATP exchange. In principle other cations such as K^+ may accompany the uptake of P_i^-.

This model requires that positive charges such as H^+ or K^+ are pumped outside the mitochondria in equal amounts, at least for the H^+ and K^+ which are taken up with P_i^-. These H^+ may originate from the respiratory chain either directly, according to the model proposed by Mitchell [7], or from an energy-rich intermediate [12]. Here the chemiosmotic model shall be preferred for the sake of simplicity. In this case the uptake of P_i^- competes with the ATP synthesis for H^+.

The problem may be described in a different way. If the overall reaction of the translocation for ATP synthesis is electroneutral, and if P_i is taken up only in a neutralized form (e.g. H^+ plus P_i^-), then the ADP-ATP must also be electroneutral (e.g. ADP^{3-} against $ATPH^{3-}$). No net H^+ movements would result from this overall reaction. However, as discussed above, the ADP-ATP exchange does not appear to be electroneutral. Consequently, half of H^+ generated by the respiratory chain should be utilized for the uptake of P_i^- and only half of the H^+ are left for ATP synthesis by the ATPase system of the chemiosmotic

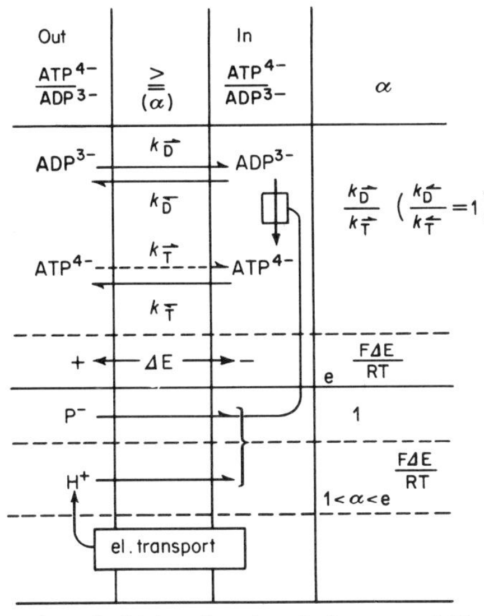

Figure 15. The generation of a higher phosphorylation potential inside than outside by the adenine nucleotide exchange. The kinetic and thermodynamic derivations for the relation between the ratios ATP/ADP inside and outside are given.

hypothesis. This difficulty is overcome by the model discussed before, i.e. that there is an alternative "statistical" coupling of P_i^- uptake either to the ADP^{3-}-ATP^{4-} exchange or to an H^+uptake. Thus, considerably less than 50% of the H^+ may be utilized in the uptake of P_i^- and effectively a small fraction of the membrane potential is utilized in the generation of the ATP/ADP difference. The fraction β, which is used in translocation, can be expected to increase with the electrochemical H^+ gradient across the membrane. It also follows from the model that the energy liberated by the back leakage of H^+ is recovered by the ATP-ADP exchange in building up a higher ATP/ADP ratio outside.

The model is to be considered mainly to afford a mechanism for the generation of higher phosphorylation potential $[\Delta G\pi = G\pi - RT \ln(ATP/ADP.P_i)]$ outside of the mitochondria as compared to that found inside. This holds under the provision that $(P_i)_{in} > (P_i)_{out}$.

$(P_i)_{in} \approx 5$ mM can be estimated from the content of $P_i \approx 10\,\mu$ moles/g prot [13]; however, the true concentration may be somewhat lower, but still higher than $(P_i)_{out}$. The AdN exchange would thus generate a potential difference:

$$G_T \geqq RT\ln\left(\frac{ATP}{ADP}\right)_{out} \Big/ \left(\frac{ATP}{ADP}\right)_{in}$$

The model gives a simple relation between the translocation energy (G_T) and the measured P/O ratio. The consumption of energy equivalents represented here as the rate of back flux of H^+, $[H^+]^{\bullet}$, can be divided into the part utilized for the formation of ATP and the part utilized by the ADP-ATP exchange, $[H^+]^{\bullet}_T$, and an undefined back leakage, $[H^+]^{\bullet}_L$.

$$[H^+]^{\bullet} = [H^+]^{\bullet}_P + [H^+]^{\bullet}_T + [H^+]^{\bullet}_L$$

The total potential of the energy-rich intermediate (e.g. electrochemical potential of H^+) may be divided as follows:

$$G_{\pi out} = G_{\pi in} + G_T + G_L$$

$G_{\pi in}$ = phosphorylation potential of endogenous ATP

G_T = free energy used by AdN translocation

G_L = free energy leakage

$$\frac{G_{\pi out}}{G_{\pi in}} = \frac{[H^+]^{\bullet}_t}{[H^+]^{\bullet}_P} = \frac{(P/O)_{th}}{(P/O)_m} \quad \text{with} \quad \frac{[H^+]^{\bullet}}{2[O]^{\bullet}} = \frac{P}{O}$$

th = theoretical

m = measured

From this follows (with $G_L > 0$):

$$G_T \leqq \frac{(P/O)_{th} - (P/O)_m}{(P/O)_m} \cdot G_{\pi in}$$

The term G_L contains the free energy of H^+ which, in the back leakage, may be coupled to other anion translocations or irreversibly lost to heat.

The model affords a tentative explanation for the puzzling fact that the endogenous phosphorylation potential in mitochondria is lower than the exogenous phosphorylation potential. The adenine translocation contributes a significant part of the phosphorylation energy of ATP delivered from the mitochondria which is generated during the transport through the membrane in addition to the energy generated in the mitochondria by the formation of the anhydride bond. Thus, the "control" of the adenine translocation reflects its active participation in the generation of phosphorylation energy for extra-mitochondrial consumption. According to the model, a fraction of intermediate phosphorylation energy equivalents are consumed for the generation of ATP at a higher phosphorylation potential under sacrifice of the theoretically maximum ATP yield at a lower potential.

ACKNOWLEDGEMENTS

This research was supported by a grant from the Deutsche Forschungsgemeinschaft. We thank Professor T. Ajello, University of Palermo, for supplying us with some atractyloside.

REFERENCES

1. Klingenberg, M. and Pfaff, E., *in* "Regulation of Metabolic Processes in Mitochondria" (edited by J. M. Tager, S. Papa, E. Quagliariello and E. C. Slater), Elsevier, Amsterdam, 1966, B.B.A. Library, Vol. 7, p. 180.
2. Klingenberg, M. and Pfaff, E., *in* "The Metabolic Roles of Citrate" (edited by T. Goodwin), Academic Press, London, 1968, p. 105.
3. Pfaff, E. and Klingenberg, M., *Eur. J. Biochem.* **6** (1968) 66.
4. Graafmans, W. D. J., Charles, R. and Tager, J. M., *Biochim. biophys. Acta* **153** (1968) 916.
5. Klingenberg, M., Heldt, H. W. and Pfaff, E., *in* "The Energy Level and Metabolic Control in Mitochondria" (edited by J. M. Tager, S. Papa, E. Quagliariello and E. C. Slater), Adriatica Editrice, Bari, 1969, p. 237.
6. Heldt, H. W. and Pfaff, E., *Eur. J. Biochem.* in press.
7. Mitchell, P., *Biol. Rev.* **41** (1966) 445.
8. Mitchell, P. and Moyle, J., *Eur. J. Biochem.* **4** (1968) 530.
9. Wulf, R., Diploma thesis, München, 1969.

9a. Mitchell, P. and Moyle, J., *in* "Biochemistry of Mitochondria" (edited by E. C. Slater, Z. Kaniuga, L. Wojtczak), Academic Press, London, 1967, p. 53.

10. Meisner, H. and Klingenberg, M., *J. biol. Chem.* **243** (1968) 3631.
11. Mitchell, P., *Fedn Proc. Fedn Am. Socs exp. Biol.* **26** (1967) 1370.
12. Slater, E. C., *Eur. J. Biochem.* **1** (1967) 317.
13. Heldt, H. W. and Klingenberg, M., *Biochem. Z.* **343** (1965) 433.

FEBS Symposium, Volume 17, 1969, pp. 79-92

Effect of Fatty Acids on Energy Metabolism and the Transport of Adenine Nucleotides in Mitochondria and other Cellular Structures

L. WOJTCZAK, K. BOGUCKA, M. G. SARZAŁA and H. ZAŁUSKA

Department of Biochemistry, Nencki Institute of Experimental Biology, Warsaw, Poland

Long chain fatty acids occupy an important position in the energy metabolism of the cell. They are good respiratory substrates whose oxidation is coupled to the generation of ATP [1], but they can also be potent uncouplers of oxidative phosphorylation [2-7]. It has therefore been suggested that fatty acids may play a role in the regulation of the degree of coupling between the respiration and the synthesis of ATP. Such a regulatory role of fatty acids has been recently postulated for the brown adipose tissue [8, 9]. The aim of this paper is to describe the effects of fatty acids on the energy-producing and some energy-requiring processes in mitochondria and some other membranous structures, as well as on the permeability of mitochondria towards adenine nucleotides.

Figure 1 shows the effect of oleate on oxidative phosphorylation and the ATP-P_i exchange reaction, and Fig. 2 its effect on the oxidation of succinate in state 4 (in the absence of phosphate acceptor) and under uncoupled conditions. It can be seen that oleate in the amount of 50 nmoles/mg mitochondrial protein fully uncouples oxidative phosphorylation and blocks the ATP-P_i exchange as well as maximally stimulates succinate oxidation in the absence of phosphate acceptor but partly inhibits the oxidation uncoupled by 2,4-dinitrophenol. At very low levels, 2-5 nmoles/mg protein, oleate substantially stimulates the exchange reaction. This stimulation [10], as well as the inhibition by higher amounts of oleate [5, 6, 10], have already been observed by other workers. The inhibition by fatty acids of mitochondrial respiration in the active (+ ADP) and uncoupled states has also been described [5, 11-13].

With regard to the strong absorption of fatty acids by mitochondria [13, 14] and other cellular structures [14], the amount of fatty acids in experiments shown in Figs. 1 and 2, as well as in all subsequent experiments, is expressed in nmoles per mg mitochondrial (or other particle) protein rather than in terms of absolute concentration.

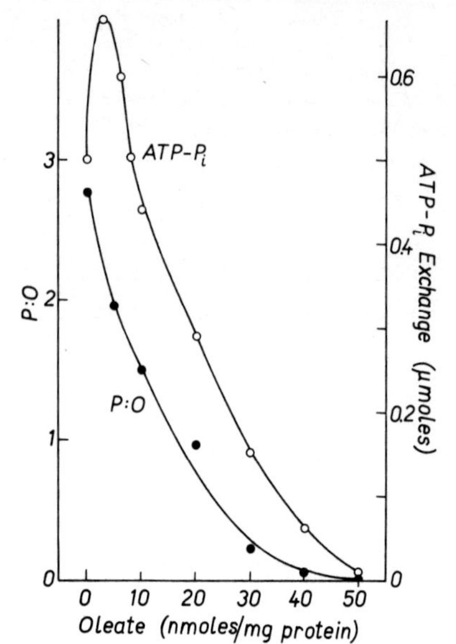

Figure 1. Effect of oleate on oxidative phosphorylation and the ATP-P_i exchange reaction in rat liver mitochondria.

Incubation medium for P:O ratio: 70 mM KCl, 32 mM Tris-Cl, pH 7·4, 6 mM $MgSO_4$, 2·5 mM EDTA, 1·8 mM P_i (containing ^{32}P), 0·24 mM ATP, 20 mM glucose, 200 units hexokinase, 7 mM β-hydroxybutyrate, and 10 mg mitochondrial protein. Total volume 3 ml. Temperature 30°C. Incubation time 6-9 min. Oxygen uptake was measured polarographically and phosphate esterification isotopically.

Incubation medium for ATP-P_i exchange: 74 mM KCl, 38 mM Tris-Cl, pH 7·4, 3 mM $MgCl_2$, 1 mM EDTA, 3 mM P_i (containing ^{32}P), 8 mM ATP, and 2·7 mg mitochondrial protein. Total volume 1 ml. Temperature 20°C. Incubation time 15 min.

●—● P:O ratio; ○—○ ATP-P_i exchange.

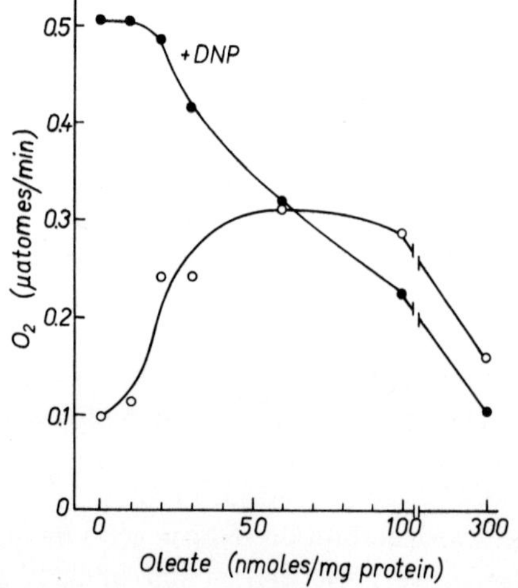

Figure 2. Effect of oleate on succinate oxidation by rat liver mitochondria.

Incubation medium: 120 mM KCl, 20 mM Tris-Cl, pH 7·4, 3 mM $MgCl_2$, 3·3 mM P_i, 3·3 mM succinate, 1 μM rotenone, and 3 mg mitochondrial protein. Total volume 3 ml. Temperature 26°C. O_2 was measured polarographically.

○—○ State 4; ●—● uncoupled state (+0·1 mM 2,4-dinitrophenol).

Fatty acids are also powerful swelling-producing agents for mitochondria [4, 15, 16]. It is noteworthy that the uncoupling action of fatty acids is correlated with their swelling action. Both effects strongly depend on the length of the carbon chain and the presence or absence of double bonds, most active being unsaturated fatty acids and, among the saturated acids, those containing 12 and 14 carbon atoms [2, 7, 10, 15, 16].

Peculiar, and not typical for other uncouplers of oxidative phosphorylation, is the effect of fatty acids on mitochondrial ATPase [7, 17-21]. This is shown in Fig. 3. In the presence of Mg^{2+}, increasing amounts of oleate gradually stimulate the latent ATPase. Occasionally, a small maximum can be observed at about 25 nmoles/mg protein, followed by a sharp decrease at 50 nmoles/mg protein, but this transient maximum is not always seen. Further increase in the amount of oleate produces an increase in the ATPase activity, attaining its maximum usually at 400-500 nmoles/mg protein (Figs. 3 and 4). In the absence of Mg^{2+}, or better, in the presence of a magnesium chelator, e.g. EDTA, oleate does not stimulate the ATPase at any concentration used and even inhibits the small ATPase activity present in freshly isolated mitochondria (not shown in Fig. 3). On the contrary, the stimulation of mitochondrial ATPase by 2,4-dinitrophenol and other typical uncouplers does not require the addition of magnesium and can be observed even in the presence of low concentrations of EDTA [22]. The dinitrophenol-stimulated ATPase is strongly inhibited by small concentrations of oleate (Figs. 3 and 4). The minimum is attained at about 50 nmoles oleate/mg mitochondrial protein and is followed by an increase of the activity usually exceeding, at 500 nmoles/mg protein, the value obtained in the absence of oleate. This increase in the activity is, however, observed only in the presence of Mg^{2+}. In its absence or in the presence of EDTA or other magnesium chelators, a further decrease occurs (Fig. 3). This has also been observed by other authors [7, 17, 20, 21].

The dependence of the oleate-stimulated and dinitrophenol-stimulated ATPases on the concentration of Mg^{2+} in the medium is shown in Fig. 5. It can be concluded that small amounts of oleate, between 25 and 150 nmoles/mg protein (occasionally between 25 and 75 nmoles, e.g. in Fig. 3) are inhibitory to the dinitrophenol-stimulated ATPase irrespective of whether Mg^{2+} is present or absent in the medium. On the other hand, the increase of the activity at higher amounts of oleate is absolutely dependent on the presence of Mg^{2+}. These effects of oleate strikingly resemble the action on mitochondrial ATPase of another surface-active agent, deoxycholate [22].

It is a characteristic feature of the ATPase stimulated by higher amounts of oleate, both in the presence and absence of 2,4-dinitrophenol, that it is no longer sensitive to atractyloside (Fig. 4).

The inhibition by low amounts of oleate of the dinitrophenol-stimulated ATPase (Figs. 3 and 4) is characteristic for intact mitochondria and cannot be

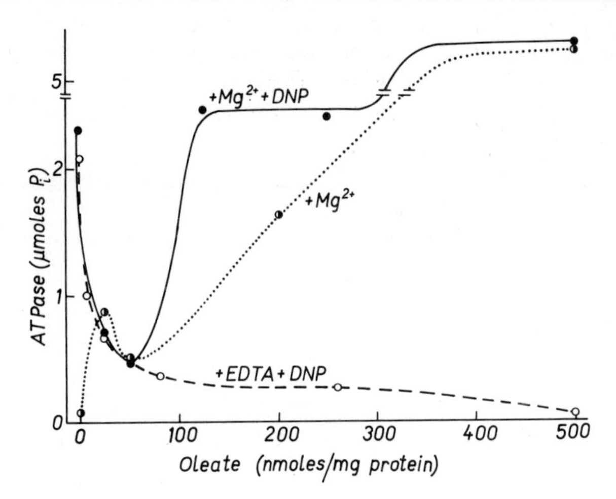

Figure 3. Effect of oleate on the ATPase activity of rat liver mitochondria.

Incubation medium: 75 mM KCl, 40 mM Tris-Cl, pH 7·4, 3·3 or 6·7 mM ATP, and 2 mg mitochondrial protein. Total volume 1·5 ml. Temperature 20°C. Incubation time 15 min. ◑······◑ 1 mM $MgCl_2$; ●——● 1 mM $MgCl_2$ + 0·1 mM 2,4-dinitrophenol; ○------○ 1 mM EDTA + 0·1 mM 2,4-dinitrophenol.

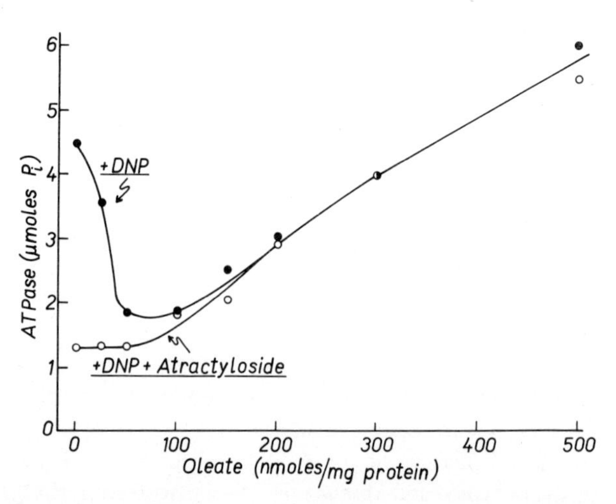

Figure 4. Effect of atractyloside on the oleate-stimulated ATPase of rat liver mitochondria.

Incubation medium as in Fig. 3 plus 1 mM $MgCl_2$ and 0·1 mM 2,4-dinitrophenol. ●——● Control; ○——○ 12·5 nmoles atractyloside.

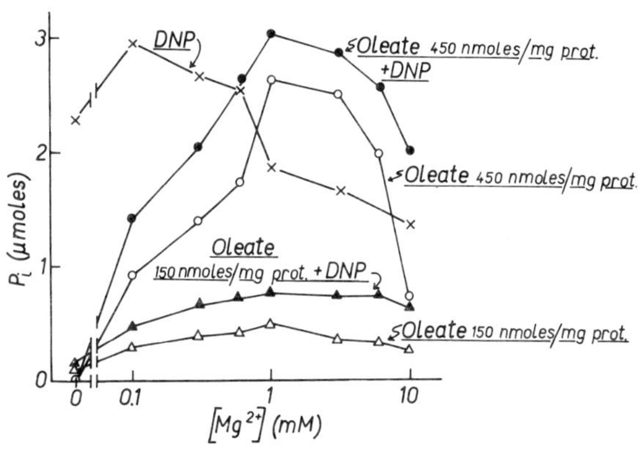

Figure 5. Effect of Mg^{2+} on mitochondrial ATPase. Conditions as in Fig. 3. ×——× 0·1 mM 2,4-dinitrophenol; △——△ oleate 150 nmoles/mg mitochondrial protein; ▲——▲ oleate 150 nmoles/mg protein + 0·1 mM 2,4-dinitrophenol; ○——○ oleate 450 nmoles/mg protein; ●——● oleate 450 nmoles/mg protein + 0·1 mM 2,4-dinitrophenol.

observed, or is largely abolished, when mitochondria are disrupted, e.g. by sonication. Small amounts of oleate also inhibit substrate-level phosphorylation (in the presence of 2,4-dinitrophenol to eliminate respiratory chain-linked phosphorylation, Table 1) and the hydrolysis of ATP stimulated by gramicidin + K^+, but are without effect on mitochondrial adenylate kinase. Oleate thus simulates at low concentrations the effect of atractyloside. It has therefore been proposed [23] that a permeability factor may be involved in this process. It has been shown [23] that small amounts of oleate strongly inhibit the translocation of adenine nucleotides (ATP and ADP) through mitochondrial membranes. A typical experiment is illustrated by Fig. 6, which also shows that the effect is not dependent on the presence or absence of Mg^{2+} and is also not influenced by oligomycin [23]. In contrast to oleate, typical uncouplers, like 2,4-dinitrophenol, do not influence the translocation [24]. The inhibition by oleate of ATP translocation can be reversed by serum albumin (Fig. 7). A

Table 1. Effect of oleate on substrate-level phosphorylation in rat liver mitochondria

Oleate (nmoles/mg protein)	Esterification of P_i (μmoles)
0	1·50
25	0·89
150	0·07

Incubation medium: 125 mM KCl, 10 mM Tris-Cl, pH 7·4, 6 mM $MgCl_2$, 2·5 mM EDTA, 1·7 mM P_i (containing ^{32}P), 0·25 mM ATP, 10 mM glucose, 150 units hexokinase, 5 mM α-ketoglutarate, 2·5 mM malonate, 0·05 mM 2,4-dinitrophenol, 0·6 mM sodium azide (to inhibit ATPase), and 10 mg mitochondrial protein. Total volume 2 ml. Temperature 22°C. Incubation time 30 min.

similar inhibition is also produced by palmitate, although at a much higher amount, and by deoxycholate (Fig. 7). The latter observation suggests that it may be connected with a physical mechanism of action rather than with a chemical effect.

In experiments shown in Figs. 6 and 7 and in the previous paper [23], mitochondria were incubated with [^{14}C] ATP and the translocation reaction was stopped by the addition of atractyloside followed by the separation of mitochondria by centrifugation and a careful washing with 0·25M sucrose. Control tests have now shown that the amounts of oleate used in these

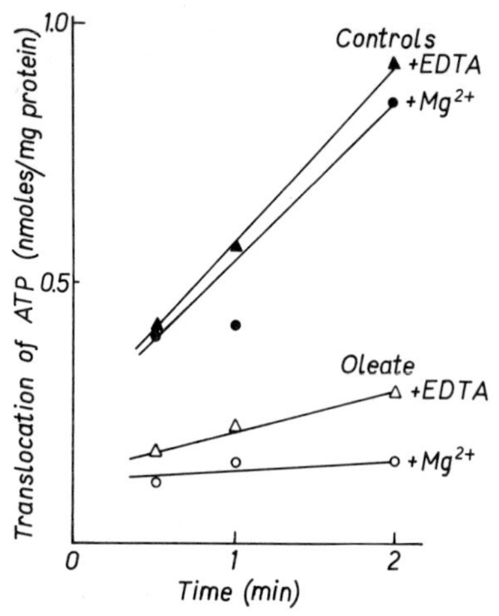

Figure 6. Effect of oleate on ATP translocation in rat liver mitochondria.
Incubation medium: 120 mM KCl, 20 mM Tris-Cl, pH 7·4, 25 mM sucrose, 60 μM ATP (containing [^{14}C] ATP), and 8 mg mitochondrial protein. Total volume 2·2 ml. Temperature 0°C. The incubation was stopped by the addition of 60 nmoles atractyloside.
●——●Control + 1·1 mM $MgCl_2$; ▲——▲ control + 1 mM EDTA; ○——○ oleate 50 nmoles/mg protein + 1·1 mM $MgCl_2$; △—△ oleate 50 nmoles/mg protein + 1 mM EDTA.

experiments, when added after atractyloside, do not change the amount of labelled ATP accumulated in mitochondria; in other words, a decrease in the radioactivity of mitochondria incubated in the presence of oleate is due to a true inhibition of the translocation and not to the elution of [^{14}C] ATP after the reaction is stopped by atractyloside. This was confirmed using a different procedure of studying the translocation. It was based on the centrifugation-filtration procedure introduced by Klingenberg *et al.* [25] with the modification that no silicone oil was used and the layers were separated only due to the difference in their density (Fig. 8). By this procedure a strong inhibition of ATP translocation by oleate could be observed as well (Table 2). Similar results were also obtained using filtration of mitochondrial suspension through membrane filters.

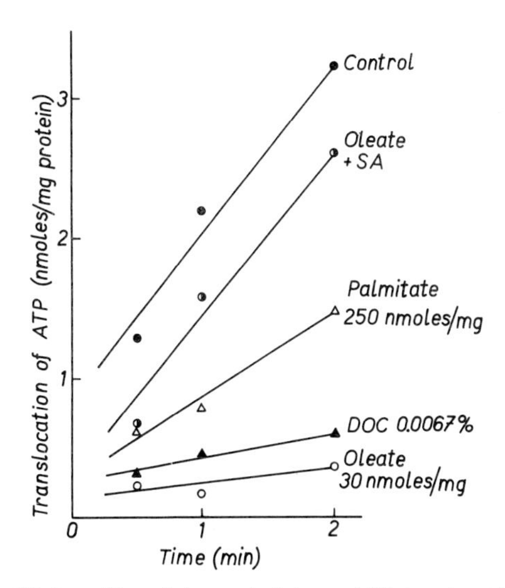

Figure 7. Effect of fatty acids and deoxycholate on ATP translocation.
Incubation conditions as in Fig. 6; the medium contained 1·1 mM $MgCl_2$ and 9·4 mg mitochondrial protein.
●——● Control; ○——○ oleate 30 nmoles/mg protein; ◑——◑ oleate 30 nmoles/mg protein followed by serum albumin 1 mg/mg mitochondrial protein; △——△ palmitate 250 nmoles/mg protein; ▲——▲ deoxycholate 0·0067%.

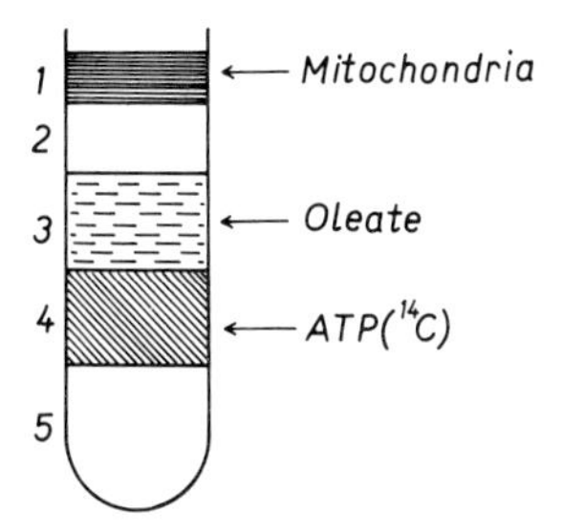

Figure 8. Schematic representation of the centrifugation-filtration procedure.
The medium in all layers contained 120 mM KCl and 20 mM Tris-Cl, pH 7·4. Density gradient was obtained by dissolving dextran in amounts of 30 mg/ml, 40 mg/ml, 50 mg/ml and 60 mg/ml in media forming layers 2 to 5 respectively. Tubes were centrifuged in SW39 rotor of a Spinco ultracentrifuge during 5 min at 8000 rev/min (5000 x g average) followed by 5 min at 16,000 rev/min (20,000 x g average). Mitochondrial pellet sedimented at the bottom of the tubes was analysed for radioactivity.

Figures 6 and 7 show (see also Fig. 2 in [23]) that the translocation is linear between 30 s and 2 min at 0°C. However, the extrapolation to zero time does not pass through the origin of the graph. This is most likely a reflection of the fact recently observed by Winkler *et al.* [26] that the rate of the translocation is much higher during the first few seconds of the incubation than during a later period. Although the initial rapid phase of translocation was not studied in detail in the present investigation, the data of Figs. 6 and 7 suggest that both the rapid and the slow phases of the translocation are inhibited by fatty acids and deoxycholate.

Table 3 shows the effect of oleate on the accumulation of succinate in mitochondria. It is evident that the inhibition of this accumulation parallels the

Table 2. Effect of oleate on ATP translocation in rat liver mitochondria studied by the centrifugation-filtration method (Fig. 8)

Oleate (nmoles/mg protein)	Translocation (relative values)		
	Expt. 1 (0°C)	Expt. 2 (0°C)	Expt. 3 (20°C)
0	100	100	100
25	14	28	35
160	3		

uncoupling of oxidative phosphorylation (cf. Fig. 1) and can also be obtained with other uncouplers [27, 28]. This effect is therefore due to simple uncoupling and thus differs from the inhibition of translocation, which is an energy-independent process.

The effect of oleate on the release of intramitochondrial constituents is shown in Fig. 9. It can be seen that magnesium, protein and malate dehydrogenase are released by increasing amounts of oleate. No such release could be obtained with typical uncouplers, like 2,4-dinitrophenol or carbonyl-cyanide *m*-chlorophenyl-hydrazone. A release of magnesium on ageing of mitochondria has already been observed by Chefurka [18] and a release of protein under the effect of oleate by Drahota and Honová [30].

These observations enable us to explain the peculiar effect of oleate (as well as other fatty acids and some detergents) on mitochondrial ATPase. The inhibition of the dinitrophenol-stimulated ATPase by small amounts of oleate is apparently due to the inhibition of the transport (translocation) of ATP into the mitochondrion. On the other hand, higher amounts of oleate (and other detergents) increase in an unspecific way the permeability of mitochondrial membranes. This is manifested by the leakage of intramitochondrial magnesium,

Table 3. Effect of oleate on the accumulation of succinate in rat liver mitochondria

Additions	Concentration of succinate in mitochondria* (mM)
None	1·35
Oleate 5 nmoles/mg protein	0·91
Oleate 7·5 nmoles/mg protein	0·71
Oleate 10 nmoles/mg protein	0·58
Oleate 20 nmoles/mg protein	0·37
Oleate 50 nmoles/mg protein	0·34
2,4-Dinitrophenol 0·1 mM	0·27

* Based on the assumption that intramitochondrial water equals the amount of protein.

Incubation medium: 230 mM sucrose, 10 mM KCl, 19 mM Tris-Cl, pH 7·4, 1 mM EDTA, 6 mM ascorbate, 0·3 mM tetramethyl-*p*-phenylenediamine, 0·2 mM succinate (containing [^{14}C] succinate), 1 μM rotenone, 15 μg antimycin A, and 14 mg mitochondrial protein. Temperature 20°C. Succinate accumulation in mitochondria was studied by the silicone oil-filtration procedure [25].

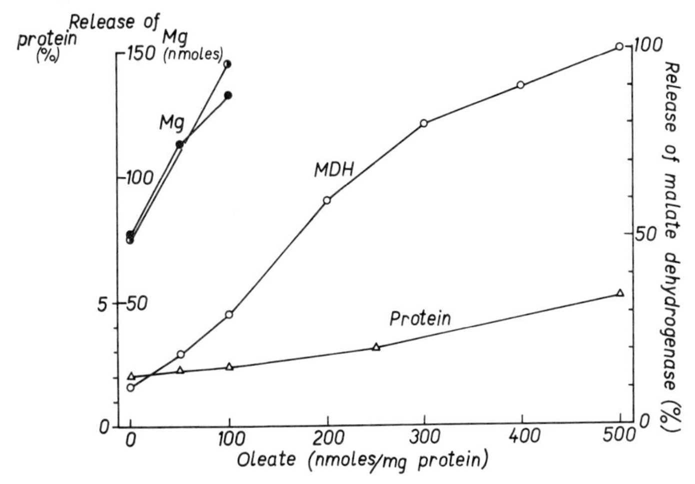

Figure 9. Release of magnesium, protein and malate dehydrogenase from rat liver mitochondria treated with various amounts of oleate.

Mitochondria (10 mg protein) were suspended in the medium containing 75 mM KCl, 40 mM Tris-Cl, pH 7·4, 1 mM $MgCl_2$, and oleate as indicated, and were incubated for 5 min at 0°C. The mixture was then centrifuged and the clear supernatant was analysed for protein (biuret method), malate dehydrogenase (spectrophotometrically with NADH and oxaloacetate) and magnesium (complexometric titration with Eriochrome black T as indicator [29]). Protein release is expressed in percentage of total mitochondrial protein and malate dehydrogenase as per cent of total activity released on sonication. Total content of Mg in mitochondria was 220 nmoles/10 mg protein. Typical uncouplers, e.g. carbonyl-cyanide *m*-chlorophenylhydrazone (CCCP), had no effect on the release; protein: control 2·1%, with CCCP 2·1%; malate dehydrogenase: control 7·1%, with CCCP 7·4%; magnesium: control 77 nmoles, with CCCP 66 nmoles.

△—△ Protein; ○—○ malate dehydrogenase, ●—● ◑—◑ magnesium (2 experiments).

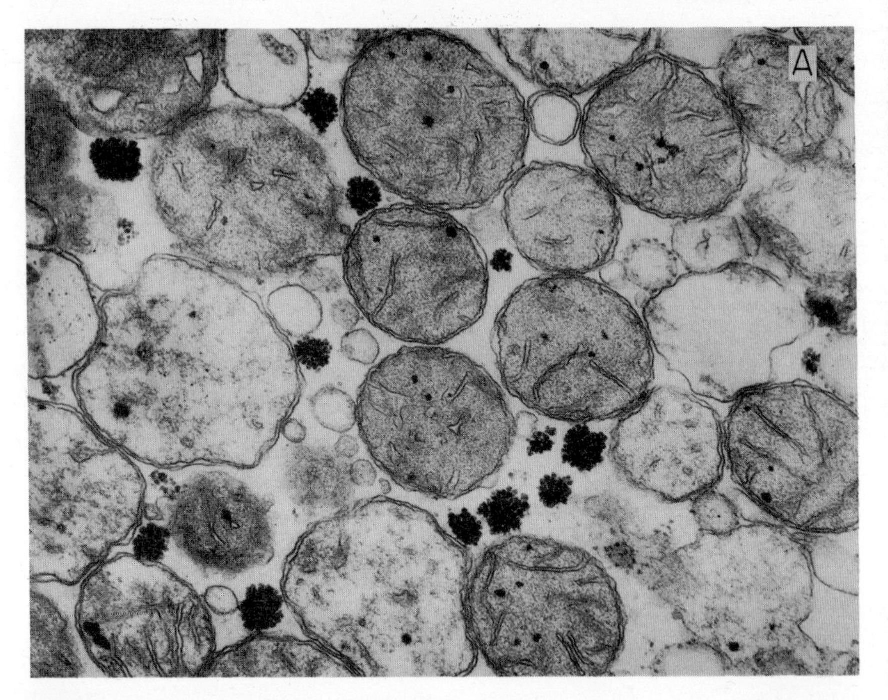
A

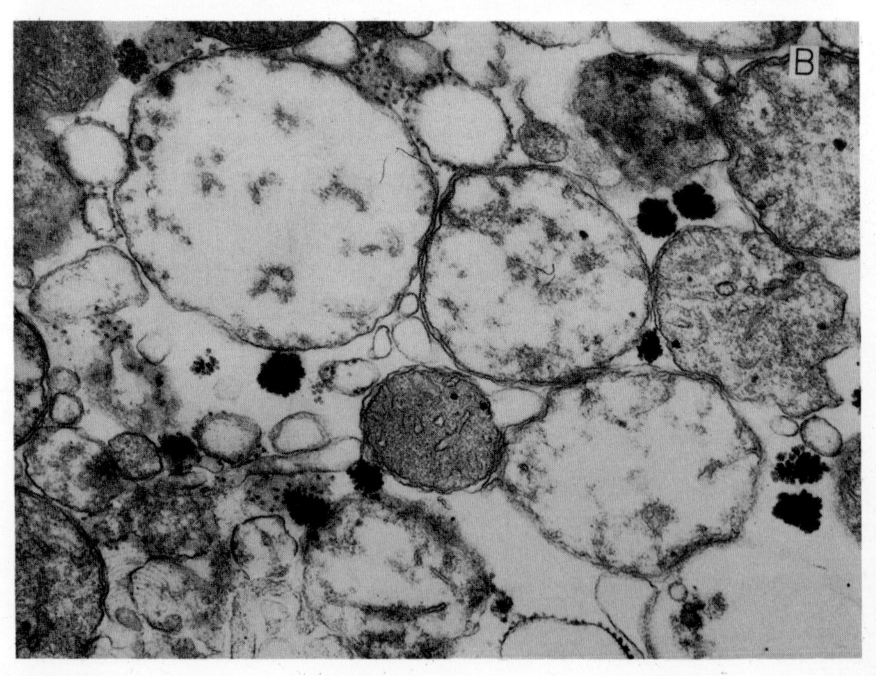
B

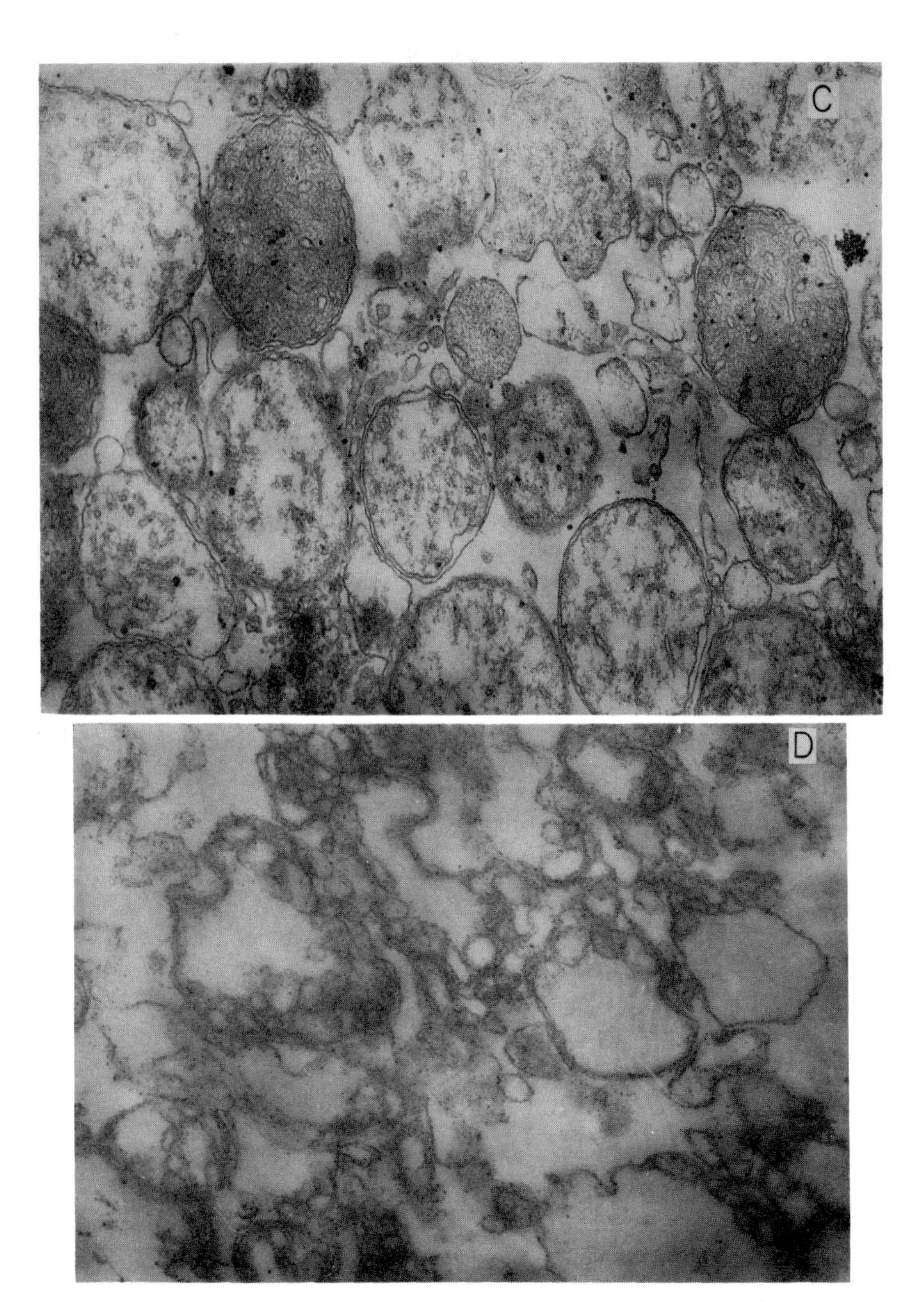

Figure 10. Electron micrographs of rat liver mitochondria treated with various amounts of oleate.

A, no oleate; B, oleate 40 nmoles/mg mitochondrial protein; C, oleate 200 nmoles/mg protein; D, oleate 500 nmoles/mg protein. (× 26,000)

Black granules outside mitochondria are glycogen; smooth and rough endoplasmic reticulum is also seen occasionally. The photographs were made by Dr Wiesława Biczysko and Mr M. Walski, Department of Pathological Anatomy, Medical School of Warsaw.

protein and malate dehydrogenase (supposed to be located in mitochondrial matrix). Thus, higher amounts of oleate probably "open" mitochondria to ATP and the penetration of this nucleotide is no longer dependent on the translocation. This is supported by the fact that the ATPase is then no longer sensitive to atractyloside. Under these conditions ATP has probably a free access to the inner side of the cristae membrane. The release of magnesium is presumably responsible for the dependence of the oleate-stimulated ATPase on the addition of Mg^{2+}.

The increased permeability of mitochondria brought about by higher amounts of oleate can be easily correlated with morphological changes of these organelles. As shown by electron micrographs (Fig. 10D), oleate in the amount of 500 nmoles/mg protein produces a potent swelling (cf. [31]) and a complete disorganization of mitochondrial structure. On density gradient centrifugation, such mitochondria form a band at the interface of the densities 1·108 to 1·126, while normal mitochondria and mitochondria treated with 150 nmoles oleate/mg protein sediment from 1·25M sucrose (density 1·167). On the other hand, the correlation between the morphology of mitochondria and the inhibition of the translocation at low oleate concentrations is less certain. Nevertheless, a comparison of electron micrographs of untreated mitochondria (Fig. 10A) and mitochondria treated with 40 nmoles oleate/mg protein (Fig. 10B), the amount about maximally inhibiting the translocation, shows an increased proportion of swollen (but not disrupted) mitochondria in the latter preparation. On further increasing the amount of oleate, the proportion of swollen mitochondria increases, but at the same time more and more particles become disrupted (Fig. 10C), the fact which is most probably responsible for the activation of the atractyloside-insensitive ATPase.

It is interesting to note that the interference of fatty acids with the energy metabolism is not limited to mitochondria. Figure 11 shows the effect of oleate on calcium binding by muscle microsomes. The maximum inhibition of the ATP-dependent binding is produced by about 100 nmoles oleate/mg microsomal protein, while the ATP-independent binding is slightly increased. Concomitantly, there is an inhibition of the Ca-stimulated microsomal ATPase and a stimulation of the "latent" ATPase, giving a cross-over point at about 200 nmoles oleate/mg protein (Fig. 12). The inhibitory effect of oleate on Ca binding by muscle microsomes has already been observed by Hasselbach [32, 33].

In conclusion, oleate, and to a smaller or greater degree other long chain fatty acids, and at least some detergents, strongly interfere with energy-producing and some energy-requiring processes, as well as with transport phenomena in which adenine nucleotides participate. This effect is most likely due to some physical rather than chemical interaction of these substances with the molecular arrangement of biological membranes. It may have a regulatory importance in the energy metabolism of the cell.

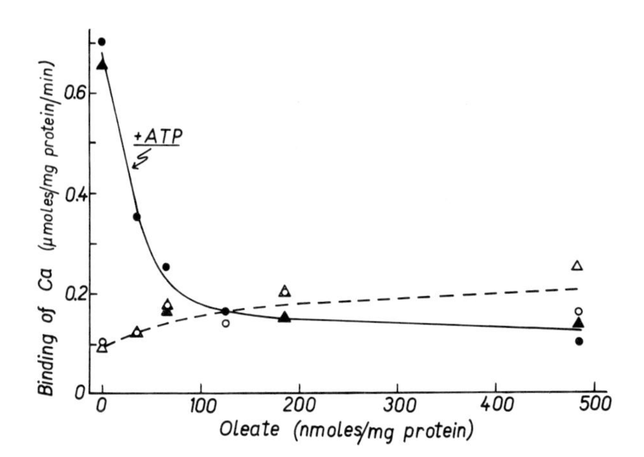

Figure 11. Effect of oleate on calcium binding by rabbit skeletal muscle microsomes.
Incubation medium: 10 mM KCl, 10 mM histidine-HCl buffer, pH 7·2, 5 mM $MgCl_2$, 5 mM oxalate, 0·1 mM $CaCl_2$ (containing ^{45}Ca), and 50-70 μg microsomal protein. Total volume 2 ml. Incubation time 5 min. Temperature 20°C. Incubation was stopped by filtration through membrane filters.
○, △, dashed line, no ATP; ●, ▲, solid line, 5 mM ATP (2 experiments).

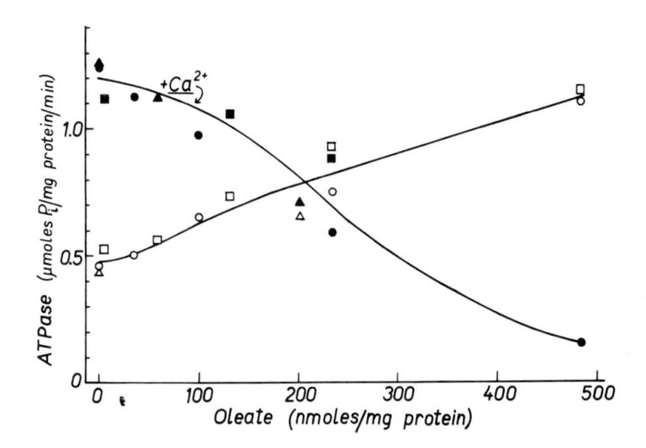

Figure 12. Effect of oleate on ATPase activity of rabbit muscle microsomes.
Incubation medium: 10 mM KCl, 10 mM histidine-HCl buffer, pH 7·2, 5 mM $MgCl_2$, 5 mM oxalate, 5 mM ATP, and 50-70 μg microsomal protein. Other conditions as in Fig. 11.
○, △, □ Controls, ●, ▲, ■ 0·1 mM $CaCl_2$ (3 experiments).

REFERENCES

1. Skrede, S. and Bremer, J., *Acta chem. scand.* **19** (1965) 1995.
2. Pressman, B. C. and Lardy, H. A., *Biochim. biophys. Acta* **21** (1956) 458.
3. Scholefield, P. G., *Can. J. Biochem. Physiol.* **34** (1956) 1227.
4. Lehninger, A. L. and Remmert, L. F., *J. biol. Chem.* **234** (1959) 2459.
5. Hülsmann, W. C., Elliott, W. B. and Slater, E. C., *Biochim. biophys. Acta* **39** (1960) 267.
6. Wojtczak, L. and Wojtczak, A. B., *Biochim. biophys. Acta* **39** (1960) 277.
7. Borst, P., Loos, J. A., Christ, E. J. and Slater, E. C., *Biochim. biophys. Acta* **62** (1962) 509.
8. Joel, C. D., Neaves, W. B. and Rabb, J. M., *Biochem. biophys. Res. Commun.* **29** (1967) 490.
9. Prusiner, S. B., Eisenhardt, R. H., Rylander, E. and Lindberg, O., *Biochem. biophys. Res. Commun.* **30** (1968) 508.
10. Ahmed, K. and Scholefield, P. G., *Nature, Lond.* **186** (1960) 1046.
11. Scholefield, P. G., *Can. J. Biochem. Physiol.* **34** (1956) 1211.
12. Vázquez-Colón, L., Ziegler, F. D. and Elliott, W. B., *Biochemistry, N.Y.* **5** (1966) 1134.
13. Wojtczak, A. B., Ławgińska, E. and Wojtczak, L., *Acta biochim. pol.* **15** (1968) 15.
14. Reshef, L. and Shapiro, B., *Biochim. biophys. Acta* **98** (1965) 73.
15. Avi-Dor, Y., *Biochim. biophys. Acta* **39** (1960) 53.
16. Zborowski, J. and Wojtczak, L., *Biochim. biophys. Acta* **70** (1963) 596.
17. Bos, C. J. and Emmelot, P., *Biochim. biophys. Acta* **64** (1962) 21.
18. Chefurka, W., *Biochemistry, N.Y.* **5** (1966) 3887.
19. Chefurka, W. and Dumas, T., *Biochemistry, N.Y.* **5** (1966) 3904.
20. Cereijo-Santaló, R., *Can. J. Biochem.* **45** (1967) 897.
21. Erecińska, M., Abstracts, 4th FEBS Meeting, Oslo, 1967, p. 94.
22. Siekevitz, P., Löw, H., Ernster, L. and Lindberg, O., *Biochim. biophys. Acta* **29** (1958) 378.
23. Wojtczak, L. and Zaluska, H., *Biochem. biophys. Res. Commun.* **28** (1967) 76.
24. Klingenberg, M. and Pfaff, E., *in* "Regulation of Metabolic Processes in Mitochondria" (edited by J. M. Tager, S. Papa, E. Quagliariello and E. C. Slater), Elsevier, Amsterdam, 1966, B.B.A. Library, Vol. 7, p. 180.
25. Klingenberg, M., Pfaff, E. and Kröger, A., *in* "Rapid Mixing and Sampling Techniques in Biochemistry" (edited by B. Chance), Academic Press, New York, 1964, p. 333.
26. Winkler, H. H., Bygrave, F. L. and Lehninger, A. L., *J. biol. Chem.* **243** (1968) 20.
27. Harris, E. J., van Dam, K. and Pressman, B. C., *Nature, Lond.* **213** (1967) 1126.
28. Quagliariello, E. and Palmieri, F., *Eur. J. Biochem.* **4** (1968) 20.
29. Hildebrand, G. P. and Reilley, Ch. N., *Analyt. Chem.* **29** (1957) 258.
30. Drahota, Z. and Honová, E., *Acta biochim. pol.* **15** (1968) 227.
31. Włodawer, P., Parsons, D. F., Williams, G. R. and Wojtczak, L., *Biochim. biophys. Acta* **128** (1966) 34.
32. Hasselbach, W., *Prog. Biophys. biophys. Chem.* **14** (1964) 167.
33. Hasselbach, W. and Makinose, M., *in* "Biochemistry of Muscle Contraction" (edited by J. Gergeley), Little, Brown and Co., Boston, 1964, p. 247.

FEBS Symposium, Volume 17, 1969, pp. 93-100

Analysis of Phosphorylation of Endogenous ADP and of Translocation Yielding the Overall Reaction of Oxidative Phosphorylation

H. W. HELDT

Lehrstuhl für Physikalische Biochemie der Universität, München, West Germany

Oxidative phosphorylation of mitochondria has been shown to proceed via the phosphorylation of endogenous ADP [1]. The endogenous ATP thus formed is exchanged in a specific reaction against exogenous ADP. This reaction, which has been called adenine nucleotide translocation, was shown to limit the rate of the phosphorylation of exogenous ADP at lower temperatures [2].

It is the purpose of this paper to investigate the quantitative relationship between translocation and phosphorylation. For this the partial reactions are subjected to kinetic analysis. From the kinetic constants thus obtained, a computing model for the overall reaction of oxidative phosphorylation will be proposed.

Analysis of the reaction of endogenous adenine nucleotides

First of all it should be recalled that the interconversion of endogenous AMP to ADP is very slow [2]. In our short-time experiments this reaction may be neglected. Thus the reactive part of the endogenous adenine nucleotides is represented by endogenous ADP + ATP = A (see p. 100) only.

Figure 1 shows the uncoupler-stimulated hydrolysis of endogenous ATP. The logarithmic plot of the ATP decay yields a straight line. This indicates the reaction to be first order.

Similarly, the phosphorylation of endogenous ADP displays first-order kinetics. Figure 2 shows the phosphorylation of endogenous ADP following the addition of oxygen to the anaerobic mitochondria. However, the endogenous ADP is not completely phosphorylated. The reaction stops after a certain ratio ATP/ADP has been established. This kinetic behaviour is characteristic for a reversible reaction consisting of first-order forward and backward reactions (equations 1-3, Table 1). The kinetic constants may be evaluated as follows: The slope of the logarithmic plot yields the sum of $k_1 + k_1'$. If the ratio of

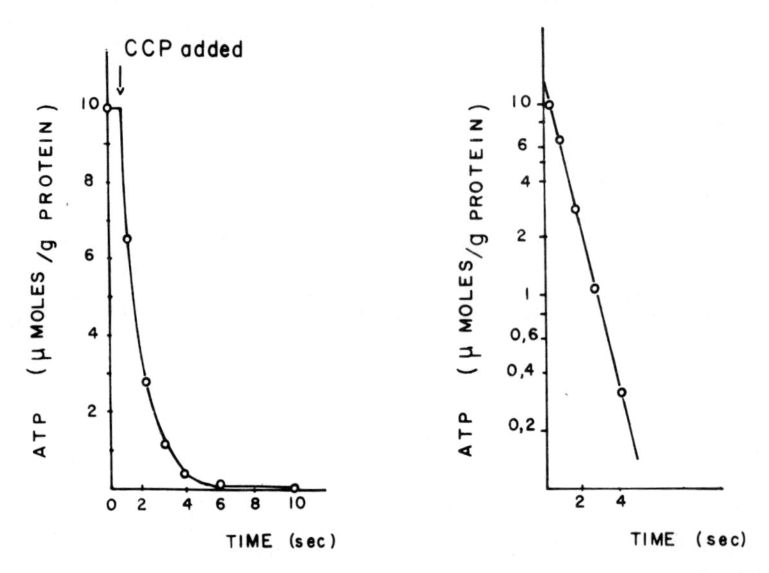

Figure 1. Rat liver mitochondria, temperature 4°C. The hydrolysis was started by addition of 5 μM CCP. For conditions of assay, see [2].

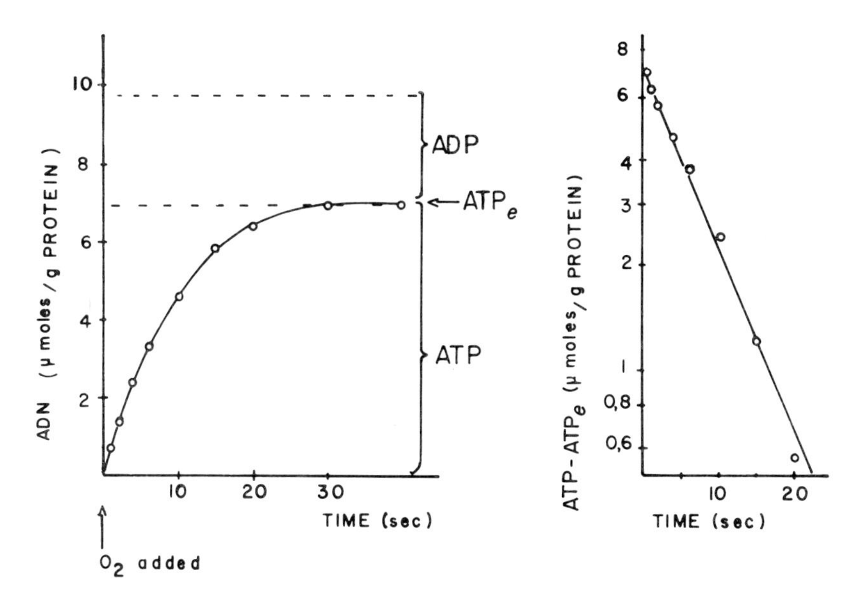

Figure 2. Rat liver mitochondria, temperature 10°C. Substrate 200 μM TMPD + 3 mM ascorbate. For conditions of incubation and assay, see [2].

Table 1

$$\mathrm{ADP} + \mathrm{P}_i \underset{k_1}{\overset{k_1}{\rightleftarrows}} \mathrm{ATP} \tag{1}$$

$$\mathrm{d}\,\frac{\mathrm{ATP}}{\mathrm{d}t} = k_1\mathrm{ADP} - k_1{}'\mathrm{ATP} \tag{2}$$

$$\ln \frac{\mathrm{ATP}_e - \mathrm{ATP}}{\mathrm{ATP}_e} = -\,(k_1 + k_1{}')t \tag{3}$$

in case of equilibrium,

$$\mathrm{d}\,\frac{\mathrm{ATP}}{\mathrm{d}t} = 0 \tag{4}$$

$$\frac{\mathrm{ATP}_e}{\mathrm{ADP}_e} = \frac{k_1}{k_1{}'} \tag{5}$$

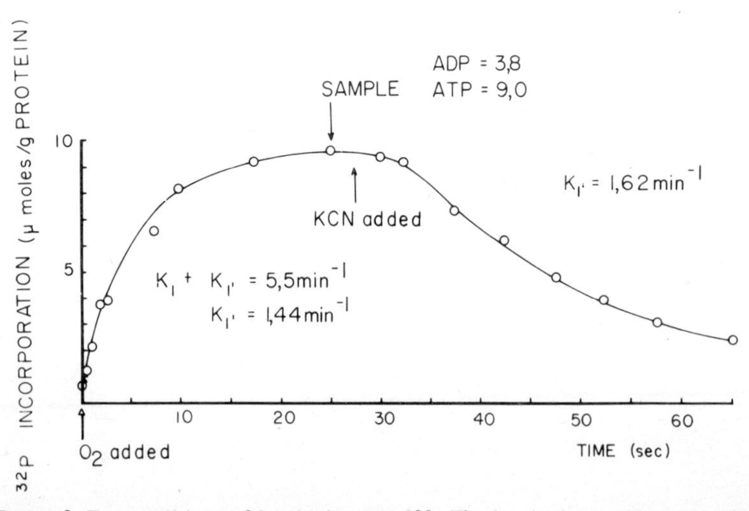

Figure 3. For conditions of incubation, see [2]. The incubation medium contained 8·7 mg mitochondrial protein/ml, 200 μM TMPD, 3 mM ascorbate and 5 μM rotenone. Temperature 10°C. In the sample indicated, ADP and ATP were measured enzymatically. From a logarithmic plot of the phosphorylation curve $k_1 + k_1{}'$ were estimated according to equation 3, Table 1 and $k_1{}'$ according to equation 5, Table 1. From a logarithmic plot of the hydrolysis of endogenous ATP, initiated by addition of 1 mM KCN, $k_1{}'$ was estimated in analogy to equation 2, Table 1.

endogenous ATP/ADP is now assayed, e.g. by enzymatic methods, k_1 and k_1' can be calculated according to equation 5, Table 1.

Figure 3 shows an example for such an evaluation of k_1 and k_1'. To test our considerations, k_1' was assayed in another way. For this the forward reaction of oxidative phosphorylation was eliminated by inhibition of respiration. Thus the addition of KCN to the respiring mitochondria induces the hydrolysis of endogenous ATP. The first-order hydrolysis constant should be equal to k_1'. Indeed, there is rather good agreement between the two values for k_1'.

These results strongly suggest that endogenous ADP and ATP are in full equilibrium. There is no indication for any compartmentation of endogenous ADP as proposed earlier [3]. This finding is important for the observed differences in the controlled state between the ATP/ADP ratios measured in the mitochondria and in the incubation medium [4]. Considering the concentrations of inorganic phosphate, the phosphorylation potential of the ATP/ADP system in the surrounding medium appears to be much higher than in the mitochondria. Our data on the endogenous adenine nucleotides clearly show that these differences are true. The cause for this difference in phosphorylation potential is based on the properties of adenine nucleotide translocation, as will be discussed in another paper of this symposium [5].

Properties of translocation

The exchange of endogenous ADP and ATP follows first-order kinetics [6]. From this, the zero-order reaction constant for translocation (k_2) may be obtained. The adenine nucleotide translocation displays an asymmetrical behaviour. The transport of adenine nucleotides from outside to inside is highly specific for ADP [7]. For the other direction there appears to be no such specificity; endogenous ADP and ATP are equally well transported to the

Table 2

$$j\overset{\text{ADP}}{\longleftarrow} = k_2 \tag{6}$$

$$j\overset{\text{ATP}}{\longrightarrow} + j\overset{\text{ADP}}{\longrightarrow} = k_2 \tag{7}$$

$$\frac{j\overset{\text{ATP}}{\longrightarrow}}{j\overset{\text{ADP}}{\longrightarrow}} = \frac{\text{ATP}}{\text{ADP}} \tag{8}$$

$$\text{ADP} + \text{ATP} = \text{A} \tag{9}$$

$$j\overset{\text{ATP}}{\longrightarrow} = k_2\,\frac{(\text{A} - \text{ADP})}{\text{A}} \tag{10}$$

outside [6]—see equation 8, Table 2. Thus the transport of endogenous ATP from inside to outside appears to be proportional to the share of endogenous ATP in the sum of ADP + ATP.

Calculation of the overall rate of the phosphorylation of exogenous ADP

In the steady state of phosphorylation, the rate of ATP formation inside the mitochondria should be equal to the rate of ATP transport out of the mitochondria (equation 11, Table 3). Thus the steady state concentration of

Table 3

Steady state:

$$\mathrm{d}\frac{\mathrm{ATP}}{\mathrm{d}t} = j^{\mathrm{ATP}}_{\rightarrow} \qquad (11)$$

$$(k_1 + k_1')\mathrm{ADP} - k_1'\mathrm{A} = k_2 \frac{(\mathrm{A} - \mathrm{ADP})}{\mathrm{A}} \qquad (12)$$

$$\mathrm{ADP} = \frac{k_1'\mathrm{A}^2 + k_2\mathrm{A}}{(k_1 + k_1')\mathrm{A} + k_2} \qquad (13)$$

$$\mathrm{d}\frac{\mathrm{ATP}}{\mathrm{d}t} = \frac{k_1 \cdot k_2}{k_1 + k_1' + \dfrac{k_2}{\mathrm{A}}} \qquad (14)$$

endogenous ADP may be calculated according to equation 13, Table 3. The steady state concentration of endogenous ADP is now introduced into equation 2, Table 1, yielding the steady state rate of phosphorylation of exogenous ADP (equation 14, Table 3). Equations 15 and 16 (Fig. 4) have been derived by integration of equations 2 (Table 1) and 10 (Table 2). In this way the entire time course of phosphorylation of endogenous and exogenous ADP may be computed.

Figure 4 shows the validity of our computation. From the experimental data the numeric values for k_1, k_1', k_2 and A were assayed as described before. The dotted line represents the computed time course according to equations 15 and 16. There is good agreement between the experimental and the theoretical results, indicating the validity of our calculation model.

Both curves show a lag phase of phosphorylation after ADP addition. Such a lag phase has been observed earlier [2]. From our model the cause of this lag phase is well explained; it reflects the time in which a new steady state of the endogenous ADP is reached.

$$ATP_{endog} = \frac{k_1\,A}{k_1 + k_1'} \cdot \left(1 - e^{-(k_1 + k_1')\cdot t}\right) \quad [15]$$

$$ATP_{tot} = \frac{A}{\frac{k_1'}{k_1} + 1} + \frac{k_1 \cdot k_2 \cdot t}{\left(k_1 + k_1' + \frac{k_2}{A}\right)} - \frac{k_1 \cdot k_2}{\left(k_1 + k_1' + \frac{k_2}{A}\right)^2}\left(1 - e^{-(k_1 + k_1' + \frac{k_2}{A})\cdot t}\right) \quad [16]$$

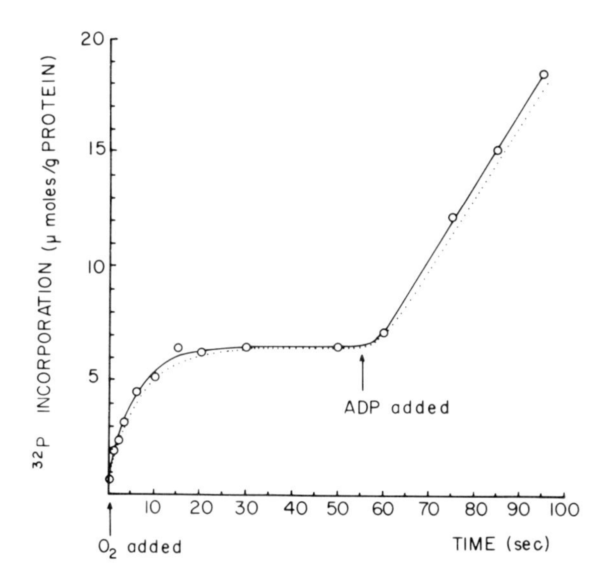

Figure 4. For conditions of incubation, see [2]. The incubation medium contained 5·4 mg mitochondrial protein/ml, 200 μM TMPD, 3 mM ascorbate and 5 μM rotenone. Temperature 10°C. The mitochondria were preincubated (0°C, 30 min) with [^{14}C] ADP ($6{\cdot}5 \times 10^{-3}$ μC/mg protein) for parallel assay of translocation. At the time indicated, 0·5 mM ADP was added. The dotted line represents the computed time course of phosphorylation using equations 15 and 16. Values employed: $k_1 + k_1' = 8{\cdot}1$ min^{-1}, ADP = 2·77, ATP = 6·39 mM/mg prot. (assayed in the sample taken at 50 s); $k_2 = 44{\cdot}2 \times 10^{-9}$ moles min^{-1}.

It appears from our results that it is possible to describe by a comparatively simple model a part of the reaction sequence of oxidative phosphorylation. It should be kept in mind, however, that this might only be a simplification of a more complex system.

ACKNOWLEDGEMENTS

The author is indebted to Professor M. Klingenberg and Dr E. Pfaff for many stimulating discussions and to Dr G. Adam for his mathematical advice, especially for deriving equations 15 and 16. The technical assistance of Miss R. Heeger is gratefully acknowledged. This research has been supported by a grant of the Deutsche Forschungsgemeinschaft.

REFERENCES

1. Heldt, H. W., Jacobs, M. and Klingenberg, M., *Biochem. biophys. Res. Commun.* **18** (1965) 174.
2. Heldt, H. W. and Klingenberg, M., *Eur. J. Biochem.* **4** (1968) 1.
3. Ernster, L., *in* "Round Table Discussion on Mitochondrial Structure and Compartmentation" (edited by J. M. Tager, S. Papa, E. Quagliariello and E. C. Slater), Adriatica Editrice, Bari, 1967, p. 339.
4. Klingenberg, M., Heldt, H. W. and Pfaff, E., *in* "Round Table Discussion on the Energy Level and Metabolic Control in Mitochondria" (edited by J. M. Tager, S. Papa, E. Quagliariello and E. C. Slater), Adriatica Editrice, Bari, 1969, p. 237.
5. Klingenberg, M., Wulf, R., Heldt, H. W. and Pfaff, E., this volume, p. 59.
6. Pfaff, E., Heldt, H. W. and Klingenberg, M., *Eur. J. Biochem.* in press.
7. Pfaff, E. and Klingenberg, M., *Eur. J. Biochem.* **6** (1968) 66.

Note added in proof

The following symbols are employed in the text:

k_1 first order reaction constant of phosphorylation;
k_1' first order reaction constant of ATP hydrolysis;
k_2 activity of translocation;
$j\overset{ADP}{\longleftarrow}$ rate of ADP transport into the matrix;
$j\overset{ATP}{\longrightarrow}$ rate of ATP transport from the matrix to the outside;
A sum of endogenous ADP + ATP.

FEBS Symposium, Volume 17, 1969, pp. 101-109

Interrelations between Diphospho- and Triphospho-pyridine Nucleotides

H. A. KREBS and R. L. VEECH*

Metabolic Research Laboratory,
Nuffield Department of Clinical Medicine,
University of Oxford, England

Lynen [2], Holzer *et al.* [3], and Bücher and Klingenberg [4], have developed the concept that the ratio [free NAD]/[free $NADH_2$] in cells can be determined by measuring the concentrations of the substrates of dehydrogenase systems and applying the equations for the equilibrium constant. If K is known, and if the system is in equilibrium, the ratios of the free pyridine nucleotides can be calculated from the equation

$$\frac{[\text{Oxidized Substrate}]\,[\text{NAD(P)H}_2]}{[\text{Reduced Substrate}]\,[\text{NAD(P)}]} = K$$

or

$$\frac{[\text{NAD(P)}]}{[\text{NAD(P)H}_2]} = \frac{1}{K} \times \frac{[\text{Oxidized Substrate}]}{[\text{Reduced Substrate}]}$$

By this definition the ratio of the free pyridine nucleotides is the product of a ratio of metabolite concentrations times the reciprocal of the equilibrium constant of the dehydrogenase system.

Why is the information on this ratio of interest? Because it tells us something of the thermodynamic characteristics of the dehydrogenase systems. The ratio determines the direction of reversible reactions. In the case of the liver, this applies to the triose phosphate dehydrogenase reaction which in glycolysis and gluconeogenesis operates in opposite directions in the same compartment and is catalysed by the same enzyme. The ratio also determines the extent to which pyridine nucleotides can be effective as reducing agents, for example, in the synthesis of fatty acids or of glutamate. Furthermore, the value of the ratio determines the magnitude of free energy changes of oxido reductions, such as those accompanying the transport of electrons from $NADH_2$ to flavoproteins in the electron transport chain. Unless the free energy change of this reaction is

* U.S. Public Health Postdoctoral Research Fellow.

above a critical minimum, there can be no effective coupling with the synthesis of ATP. If in fact, as recent work of Chance's laboratory suggests [5], the first step of the electron transport chain is the reduction of a flavoprotein which has a very negative redox potential, then electron transport could not be initiated unless the NAD-couple is relatively highly reduced.

Chance [6] has criticized this approach and argues that it is of limited value in the absence of information on the relative concentration of the *bound* pyridine nucleotides. On this point there seems to be a difference of opinion. However important the bound forms of the pyridine nucleotides are for the kinetics of the dehydrogenase systems, they are not directly, (and the operative word is "directly") relevant to the thermodynamic characteristics. The thermodynamic properties of any system depend solely on the concentrations of the starting materials and end products and they are not directly affected by the mechanism of the reactions which link starting materials and end products. The bound forms are part of this mechanism. They are intermediary stages but neither starting material, nor end products.

In a paper published in 1967 [7], we determined the ratios for the NAD-couples in the cytoplasm and in the mitochondria of rat liver after freeze-clamping, using the lactate/pyruvate system as an indicator of the cytoplasmic NAD-couple, and the β-hydroxybutyrate and glutamate dehydrogenase systems as indicators of the mitochondrial NAD-couple. Results obtained from freeze-clamped material may be taken to indicate the *in vivo* situation. The published data show that glutamate dehydrogenase and β-hydroxybutyrate dehydrogenase give virtually the same value for the NAD ratios, which must be taken to indicate that the two enzymes are in equilibrium with the same mitochondrial coenzyme pool, although glutamate dehydrogenase is located in the matrix and β-hydroxybutyrate dehydrogenase in the cristae. The mitochondrial $[NAD]/[NADH_2]$ ratio is about 100 times lower than the cytoplasmic one; in other words, the mitochondrial NAD-couple is in a much more reduced state.

A crucial feature of the evidence is the agreement between the values of the ratios obtained with different dehydrogenase systems. For the cytoplasm it has already been shown by Bücher and his colleagues [4, 8] that lactate, α-glycerophosphate and malate dehydrogenase give the same values under certain conditions. This shows that these dehydrogenase systems are poised at near equilibrium. It is evident that the agreement between the $NAD/NADH_2$ ratio calculated from the glutamate and β-hydroxybutyrate dehydrogenase systems justifies the assumptions on which the method rests, namely the existence of near equilibria.

The aim of the new work was to make analogous measurements for the NADP-couple. Of the five major cytoplasmic NADP-linked dehydrogenases (glucose 6-phosphate dehydrogenase, 6-phosphogluconate dehydrogenase, glutathione reductase, malic enzyme and isocitrate dehydrogenase) only the latter two proved suitable for various reasons.

For the assay of the [NADP]/[$NADPH_2$] ratio in the cytoplasm, the concentration of the substrates of the malic enzyme system and the isocitric dehydrogenase system were therefore measured. In general, there is again good agreement between the values obtained from the two enzyme systems (Table 1).

Table 1. Assay of $\frac{[NADP]}{[NADPH_2]}$ ratio in cytoplasm by different dehydrogenase systems

State of animal	[NADP]/[$NADPH_2$] Derived from malic enzyme system	[NADP]/[$NADPH_2$] Derived from isocitrate dehydrogenase system
Well-fed (12 obs.)	0·0118 ± 0·002	0·0143 ± 0·001
Sucrose-casein diet (8 obs.)	0·0146 ± 0·002	0·0134 ± 0·001

For full experimental details see ref. 1.

Table 2. Ratio of free pyridine nucleotides in rat liver

	Cytoplasm		Mitochondria	
	$\frac{NAD^+}{NADH}$	$\frac{NADP^+}{NADPH}$	$\frac{NAD^+}{NADH}$	$\frac{NADP^+}{NADPH}$
Control	1164	0·014	7·7	12·1
Starved 48 h	564	0·0057	5·5	8·7
Casein-fat diet (3 days)	526	0·0034	4·1	6·5
Casein-sucrose diet (3 days)	1820	0·013	6·7	10·8

For details of the experimental procedure see ref. 1. The mitochondrial NADP-ratios were calculated on the assumption that equilibrium in the NAD- and NADP-systems was established.

They are about 100,000 times lower than the corresponding cytoplasmic NAD/$NADH_2$ ratios (Table 2). That there is a difference of this order of magnitude has already been anticipated by Bücher and Klingenberg [4] and by Bücher and Rüssman [8].

We asked ourselves how the great differences in the redox states of the NAD- and NADP-couples in the cytoplasm may arise. The following considerations throw light on this question. When an NAD-linked and an NADP-linked dehydrogenase system, such as lactate dehydrogenase and glucose 6-phosphate dehydrogenase, are present in a solution, the redox states of the two couples are independent of one another–unless they share a common reactant. The malic enzyme and lactic dehydrogenase share pyruvate as a common reactant, and when equilibrium exists in such a mixture, the differences in the redox states of the NAD- and NADP-couples are no longer independent, but fixed by the equilibrium constants of the two systems and the concentrations of the substrates. The following equations show the relations which hold at equilibrium:

$$[\text{Pyruvate}] = \frac{[\text{Lactate}]\,[\text{NAD}]}{[\text{NADH}_2]} \times K_{\text{Lactate}}$$

$$[\text{Pyruvate}] = \frac{[\text{Malate}]\,[\text{NADP}]}{[\text{CO}_2]\,[\text{NADPH}_2]} \times K_{\text{Malic enzyme}}$$

Combination of the two equilibrium equations and substitution of the values for the equilibrium constants and the experimentally observed substrate concentrations, shows that at equilibrium, differences of the order of 10^5 in the redox states of the NAD- and NADP-couples are to be expected.

$$\frac{[\text{NAD}]/[\text{NADH}_2]}{[\text{NADP}]/[\text{NADPH}_2]} = \frac{[\text{Malate}]}{[\text{Lactate}]\,[\text{CO}_2]} \times \frac{K_{\text{Malic enzyme}}}{K_{\text{Lactate}}}$$

$K_{\text{Malic enzyme}} = 3{\cdot}44 \times 10^{-2}$; $K_{\text{Lactate}} = 1{\cdot}1 \times 10^{-4}$ (pH 7·0);
[Lactate] = 0·83 mM; [Malate] = 0·31 mM; [CO_2] = 1·16 mM.

The value of the above ratio is $0{\cdot}88 \times 10^{-5}$, which is in agreement with the value obtained by independent measurement derived from the isocitrate dehydrogenase system. The implication is that there is nothing miraculous or special in the differences of the redox states of the two pyridine nucleotides, in the sense that no special forces, no spatial separation, and no energy-driven mechanisms are required. The differences of the redox states arise partly from the differences in the equilibrium constants of the two reactions sharing pyruvate and partly from the concentration of metabolites.

Malic enzyme and lactate dehydrogenase are not the only systems in the cytoplasm which establish and maintain the differences in the redox states of the NAD- and NADP-couples. When NAD-linked and NADP-linked dehydrogenases do not share a reactant directly they can be linked by a third reaction. Such a 3-reaction system is provided in the cytoplasm by the NADP-linked isocitric dehydrogenase, malate dehydrogenase and glutamate-oxaloacetate transaminase. The equilibrium equations of the three reactants are as follows:

$$(1)\quad \frac{[\alpha\text{-Oxoglutarate}]\,[\text{CO}_2]\,[\text{NADPH}_2]}{[\text{Isocitrate}]\,[\text{NAD}]} = K_{\text{isocitrate (NADP)}}$$

$$(2)\ \frac{[\alpha\text{-Oxoglutarate}]\,[\text{Aspartate}]}{[\text{Glutamate}]\,[\text{Oxaloacetate}]} = K_{\text{GOT}}$$

$$(3)\ \frac{[\text{Oxaloacetate}]\,[\text{NADH}_2]}{[\text{Malate}]\,[\text{NAD}]} = K_{\text{malate}}$$

Divide (1) by (2) and (3)

$$\frac{[\text{Glutamate}]\,[\text{Malate}]\,[\text{CO}_2]\,[\text{NADPH}_2]\,[\text{NAD}]}{[\text{Isocitrate}]\,[\text{Aspartate}]\,[\text{NADP}]\,[\text{NADH}_2]} = \frac{K_{\text{isocitrate}}}{K_{\text{GOT}} \times K_{\text{malate}}}$$

or

$$\frac{[\text{NAD}]/[\text{NADH}_2]}{[\text{NADP}]/[\text{NADPH}_2]} = \frac{[\text{Isocitrate}]\,[\text{Aspartate}]}{[\text{Glutamate}]\,[\text{Malate}]\,[\text{CO}_2]} = \frac{K_{\text{isocitrate}}}{K_{\text{GOT}} \times K_{\text{malate}}}$$

On the assumption that equilibrium exists, the quantitative relations between the NAD- and NADP-couples can be calculated from the equilibrium constants of the isocitrate and malate dehydrogenases and the glutamate-oxaloacetate transaminase, and the concentrations of the reactants of these three enzymes:

$$\frac{[\text{NAD}]/[\text{NADH}_2]}{[\text{NADP}]/[\text{NADPH}_2]} = \frac{[\text{Isocitrate}]\,[\text{Aspartate}]}{[\text{Glutamate}]\,[\text{Malate}]\,[\text{CO}_2]} \times \frac{K_{\text{isocitrate}}}{K_{\text{GOT}} \times K_{\text{malate}}}$$

$K_{\text{isocitr.}} = 0{\cdot}91\,\text{M}$; $K_{\text{GOT}} = 6{\cdot}7$; $K_{\text{mal.}} = 2{\cdot}78 \times 10^{-5}$ (38°C; pH 7·0).

Well-fed liver:

[Glutamate] 2·4 mM; [Aspartate] 0·74 mM; [Isocitrate] 0·2 mM; [Malate] 0·31 mM; [CO_2] $1{\cdot}2 \times 10^{-3}$M

$$\frac{0{\cdot}02 \times 0{\cdot}74}{2{\cdot}4 \times 0{\cdot}31 \times 1{\cdot}2 \times 10^{-3}} \times \frac{0{\cdot}91}{6{\cdot}7 \times 2{\cdot}78 \times 10^{-5}} = 0{\cdot}82 \times 10^5$$

Again a value of the order of 10^5 is obtained.

There are other reactions which establish links between the redox states of the NAD- and NADP-couples. In fact there is a multiplicity of interlinked reactions which maintain the redox states of the main pyridine nucleotide-linked dehydrogenase systems in the cytoplasm at near equilibrium, and are responsible for the differences in the redox states of the NAD- and NADP-couples.

So far the discussion explains how differences in the redox states between the NAD- and NADP-couples in the cytoplasm are established and maintained. How is the absolute level of the redox state controlled? An attractive answer is the following. Through the glyceraldehyde phosphate dehydrogenase plus 3-phosphoglycerate kinase reaction, a relation is established between the redox state of the NAD-couple and the ATP-ADP-Pi system in the cytoplasm. There are good reasons in support of the assumption that the glyceraldehyde phosphate dehydrogenase and 3-phosphoglycerate kinase systems are in near equilibrium because the capacity of the enzymes is very high. It can be calculated that the rate of flux through these enzyme systems is much lower than the capacity of the enzymes. Furthermore, the fact that in the liver these two reactions readily occur in both directions (depending on whether glycolysis or gluconeogenesis takes place), supports the conclusion that the systems are in near equilibrium. At equilibrium the relation holds:

$$\text{Glyceraldehyde-P} + \text{NAD} + \text{ADP} + \text{P} \rightleftharpoons \text{3-P-G} + \text{NADH}_2 + \text{ATP}$$

$$\frac{[\text{3-P-G}]\,[\text{NADH}_2]\,[\text{ATP}]}{[\text{Glyceraldehyde-P}]\,[\text{NAD}]\,[\text{ADP}]\,[\text{P}]} = K$$

or

$$\frac{[\text{NAD}]}{[\text{NADH}_2]} = \frac{1}{K} \times \frac{[\text{3-P-G}]}{[\text{GAP}]} \times \frac{[\text{ATP}]}{[\text{ADP}]\,[\text{Pi}]}$$

If this relation is combined with the equilibrium equation of the lactate dehydrogenase system another equation is obtained which can be put to the experimental test:

$$\frac{[\text{Pyruvate}]}{[\text{Lactate}]} = K \times \frac{[\text{3-P-G}]}{[\text{GAP}]} \times \frac{[\text{ATP}]}{[\text{ADP}]\,[\text{Pi}]}$$

where

$$K = \frac{K_{\text{lactate}}}{K_{\text{GAPDH}} \times K_{\text{3-P-G-kinase}}}$$

The first to carry out such a test, using mammalian red cells, were Minikami and Yoshikawa [9] and they found that the relation holds. Red cells are devoid of nuclei and mitochondria and are therefore relatively simple systems. In the liver, the direct experimental verification of the relation is at present impracticable because some of the reactants, especially the adenine nucleotides, are unevenly distributed between cell compartments and so far no method for their separate determination in the cytoplasm and mitochondria is available. However, the indirect evidence already referred to—the high activity of the enzyme systems involved and the fact that in the liver carbohydrate metabolism can be rapidly switched over from glycolysis to gluconeogenesis—indicates that the systems are poised at near equilibrium. This would mean that the redox state of the cytoplasmic NAD-couple is linked to the phosphorylation state of the adenine nucleotide system, i.e. the value of the ratio [ATP] / [ADP] [Pi].

The link between the redox state of the cytoplasmic pyridine nucleotide system does not imply a fine control by the adenine nucleotide system of the rates of glycolysis and gluconeogenesis, but it sets a level where both these processes are possible. Both can occur within a relatively wide range of cytoplasmic [NAD]/[NADH] ratios.

As for the redox states of mitochondrial pyridine nucleotides, reference has already been made to the NAD-couple which was determined in earlier work with the help of β-hydroxybutyrate and glutamate dehydrogenases.

So there remains the redox state of the mitochondrial NADP-couple to be considered. The only NADP-linked dehydrogenase of rat liver mitochondria suitable for the assay is the glutamate dehydrogenase, which, in this tissue, is a highly active enzyme. It is known to react *in vitro* at about equal rates with NAD and NADP and it may therefore be expected to maintain equilibrium with both the NAD- and NADP-couples. However, this is a matter of controversy because it has been suggested [10-12] that, in intact mitochondria, glutamate dehydrogenase may react with NADP only. It must be emphasized that this idea is based on experiments with isolated mitochondria and there is no convincing evidence demonstrating that, in the intact liver, glutamate dehydrogenase does not react readily with both the NAD- and NADP-couples. Under innumerable experimental conditions we find that the concentrations of glutamate, α-oxoglutarate, and ammonia in the liver are exactly as expected if they were in equilibrium with the NAD- and NADP-couples as calculated from the β-hydroxybutyrate dehydrogenase system and the known differences in the potentials of the NAD- and NADP-couples. There is a very simple and telling experimental test by which the existence of such equilibria can be checked. By intramuscular injection of ammonium salts, the equilibrium of the glutamate dehydrogenase system in the liver can be upset, and whenever this is done there is an exactly parallel change in the β-hydroxybutyrate system (Table 3). Another way of upsetting equilibria is by anaerobiosis. Again the concentrations of the glutamate dehydrogenase system move precisely parallel to those of the β-hydroxybutyrate dehydrogenase system.

These conclusions have to be reconciled with the existence in mitochondria of an energy-linked transhydrogenase which, as Lee and Ernster [13] have shown, can under certain conditions shift the equilibrium constant of the transhydrogenase system from 1·56 to 0·0022. If this reaction occurred *in vivo*, and if glutamate dehydrogenase readily reacted with NAD and with NADP, (as is the case *in vitro*), concentrations of glutamate, α-oxoglutarate and ammonia would be expected in the tissue which are entirely different from those observed. This discrepancy leads to the conclusion that the energy-linked transhydrogenation occurs in the liver preparations to a significant extent only under test-tube conditions.

The work so far discussed indicates very great differences of the redox states of the NAD-couples of the cytoplasm and of the mitochondria and the question must now be considered how these differences arise and how they are maintained. This problem can be approached in the same manner as that of the origin of the differences between the redox states of the NAD- and NADP-couples in the cytoplasm, i.e. by exploring the principles of the sharing of common reactants between two dehydrogenases located in the two compartments. The subject has been fully discussed in another publication [1].

To sum up, the large differences between the redox state of the NAD- and NADP-couples in the cytoplasm (and between the NAD- and NADP-systems of cytoplasm and mitochondria) can be explained on a simple physico-chemical basis. They are essentially the result of thermodynamic equilibria which are

Table 3. Effect of intramuscular NH_4Cl injection on the concentration of metabolites in rat liver

	Control (mM)	NH_4Cl treatment (mM)
[Glutamate]	2·22	2·70
[α-Oxoglutarate]	0·065	0·035
[NH_4^+]	0·56	1·81
[β-Hydroxybutyrate]	1·98	1·82
[Acetoacetate]	0·71	1·04
$\frac{[\text{Glutamate}]}{[\text{Oxoglutarate}][NH_4^+]}$	62	43 (fall 31%)
$\frac{[\beta\text{-Hydroxybutyrate}]}{[\text{Acetoacetate}]}$	2·8	1·8 (fall 36%)

For details see Williamson *et al.* [7]. The data show that the changes in the redox state of the glutamate- and β-hydroxybutyrate dehydrogenase systems are parallel after NH_4Cl treatment. Both become more oxidized.

determined by the equilibrium constants of a number of highly active, readily reversible enzymes and the concentration of the substrate and products of these enzymes. The crucial feature is the fact that the NAD- and NADP-couples in the two compartments share certain substrates.

This sharing represents links between the two couples in the cytoplasm and between the cytoplasmic and mitochondrial couples. Through these links the order of magnitude of the differences in the redox state is fixed.

The redox states of the pyridine nucleotide couples are also linked to the adenine nucleotide system, i.e. to the ratio [ATP]/[ADP] [Pi] of the cytoplasm, again by enzyme systems which establish equilibria. There is thus a network of equilibria, or more correctly near equilibria, which is ultimately regulated by the degree of phosphorylation of the adenine nucleotide system, i.e. by the respiratory chain. The cytoplasmic adenine nucleotide system is directly linked to the redox state of the cytoplasmic NAD-couple, which in turn is linked to the redox state of the cytoplasmic NADP-couple and that of the two mitochondrial couples. The system of links discussed in this paper is bound to be incomplete because of gaps in the relevant information.

The network of near equilibria in which the pyridine and adenine nucleotides participate is likely to be a fundamental component of the energy-transforming mechanisms in the liver cell. It establishes a basic level of the redox states of the two pyridine nucleotide couples in the two main cell compartments where the energy-transforming mechanisms are located, and it links the redox states to the supply of ATP. It sets the cytoplasmic level of the NAD-couple to be suitable for both glycolysis and gluconeogenesis. It sets the cytoplasmic redox state of the NADP-couple at a much more reduced level so as to be effective in reductive syntheses such as that of fatty acids, the redox potential being about 150 mV more negative than that of the NAD-couple.

This network is one of thermodynamic equilibria whereas living cells, of course, do not represent equilibria but steady states. But the fact that thermodynamic equilibrium of a cell is synonymous with death does not imply that equilibria cannot play an important part in the organization of the chemical cell dynamics. Equilibria form a basic framework upon which virtually irreversible processes are superimposed. These flow through the equilibria, more or less upsetting them and modifying the strict equilibria to a steady state.

It is a flux through the network of equilibria which causes the degradation of food and the synthesis of the cell material–both processes which, as a whole, are irreversible. The framework of equilibria may be looked upon as a component of regulatory mechanisms; as irreversible processes flow through and mildly upset the equilibria, they automatically set in motion steps which re-establish the "resting" state.

REFERENCES

1. Krebs, H. A. and Veech, R. L., *in* "The Energy Level and Metabolic Control in Mitochondria (edited by S. Papa, J. M. Tager, E. Quagliariello and E. C. Slater), Adriatica Editrice, Bari, 1969, p. 329.
2. Lynen, F., quoted in ref. 3.
3. Holzer, H., Holzer, E. and Schultz, G., *Biochem. Z.* **326** (1954) 385; see also Holzer, H., Schultz, G. and Lynen, F., *Biochem. Z.* **328** (1956) 252.
4. Bücher, Th. and Klingenberg, M., *Angew. Chem.* **70** (1958) 552.
5. Chance, B., Ernster, L., Garland, P. B., Lee, C. P., Light, P. A., Ohnishi, T., Ragan, C. I. and Wong, D., *Proc. natn. Acad. Sci. U.S.A.* **57** (1948) 1967.
6. Chance, B. *Adv. Enzyme Regulation* **5** (1967) 435.
7. Williamson, D. H., Lund, P. and Krebs, H. A., *Biochem. J.* **103** (1967) 514.
8. Bücher, Th. and Rüssman, W., *Angew. Chem.* **75** (1964) 881, and International Edition, **3** (1964) 426.
9. Minikami, S. and Yoshikawa, H., *J. Biochem., Tokyo* **59** (1966) 140.
10. Klingenberg, M. and Slenczka, W., *Biochem. Z.* **331** (1959) 486.
11. Klingenberg, M. and Pette, D., *Biochem. biophys. Res. Commun.* **7** (1962) 430.
12. Tager, J. M. and Papa, S., *Biochim. biophys. Acta* **99** (1965) 570.
13. Lee, C. P. and Ernster, L., *Biochim. biophys. Acta* **81** (1964) 187.

FEBS Symposium, Volume 17, 1969, pp. 111-126

Components and Compartments of Mitochondrial Fatty Acid Oxidation

P. B. GARLAND, B. A. HADDOCK and D. W. YATES

Department of Biochemistry,
University of Bristol,
Bristol, England

The reactions and components involved in the β-oxidation of fatty acids are summarized in Fig. 1. The salient features of this scheme are (1) a set of enzymes that catalyse the step-wise degradation of the carbon chain of fatty acyl-CoA esters, (2) acid:CoA ligases that convert a fatty acid into its CoA ester, and (3) intermediate redox carriers, NAD and Fp_{ETF}, that transfer reducing equivalents from the primary dehydrogenases to the respiratory chain.

Although the individual reactions and components of Fig. 1 are relatively well characterized, they do not provide a sufficient basis to adequately explain the experimentally observed behaviour of isolated mitochondria. For instance, the immediate substrates for β-oxidation are fatty acyl-CoA thioesters, and yet such compounds are not readily oxidized when added to intact mitochondria [1]. The discovery that (-)-carnitine (3-hydroxy-4-trimethylammonium-butyrate) stimulated the mitochondrial oxidation of fatty acyl-CoA thioesters led to the recognition of palmityl-CoA carnitine transferase [1, 2] and the scheme of Fritz and Yue [1] (Fig. 2) in which the components of β-oxidation were placed in a compartment separate from an extramitochondrial acid:CoA ligase. Acyl-transfer between the two compartments was considered to occur by the action of two palmityl-carnitine transferases.

1. MITOCHONDRIAL MEMBRANES AND COMPARTMENTS OF FATTY ACID OXIDATION

(i) The membranes

The outer and inner membranes of mitochondria define two spaces, an intermembranous or cristal space, and the space (matrix) surrounded by the

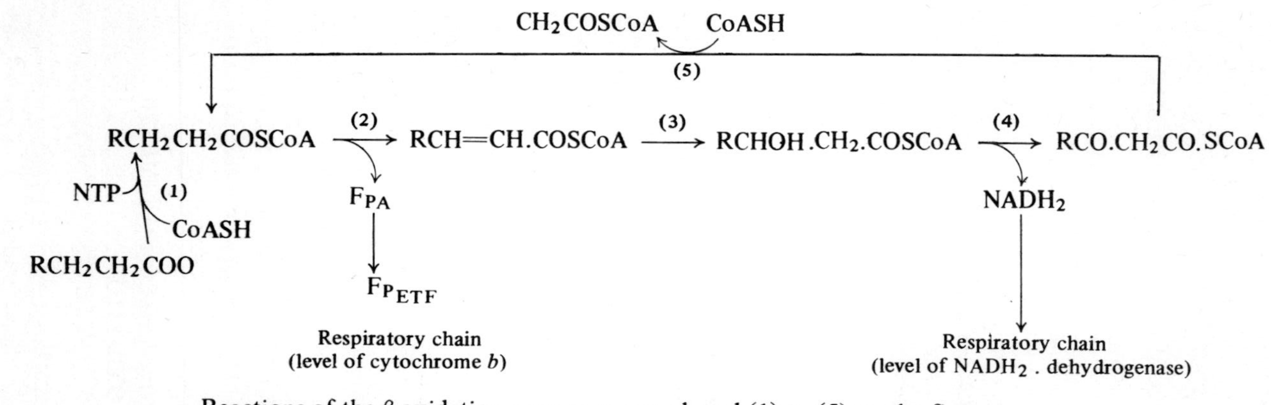

Reactions of the β-oxidation process are numbered (1) to (5) on the figure as follows:

(1) Acid:CoA ligase; enzymes which use either ATP or GTP are known [28]
(2) Acyl-CoA dehydrogenase
(3) Enoyl-CoA hydratase
(4) L-3-Hydroxacyl-CoA dehydrogenase
(5) 3-Ketoacyl-CoA thiolase

Abbreviations on figure:
NTP: nucleoside triphosphate
$NADH_2$: reduced nicotinamide-adenine dinucleotide
Fp_A: flavin of acyl-CoA dehydrogenase
Fp_{ETF}: electron-transferring flavoprotein
CoASH: thiol form of CoA

Figure 1. β-Oxidation of fatty acids.

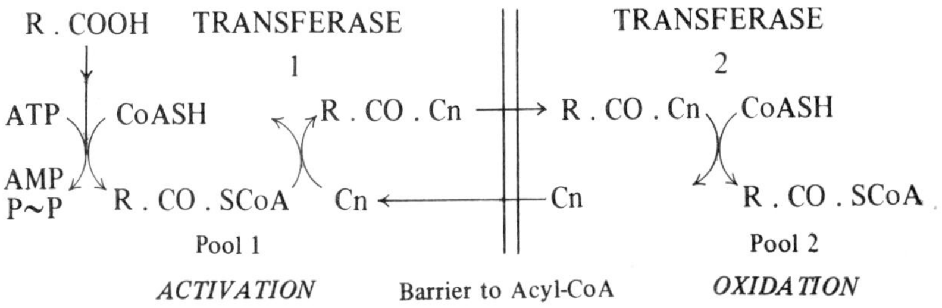

R . COOH TRANSFERASE 1 || TRANSFERASE 2

ATP, CoASH → AMP, P~P — R . CO . SCoA ⇄ R . CO . Cn → R . CO . Cn ⇄ CoASH — R . CO . SCoA

Cn ← Cn

Pool 1 — Pool 2

ACTIVATION — Barrier to Acyl-CoA — *OXIDATION*

Figure 2. Fritz and Yue (1963) scheme for the action of carnitine on fatty acid oxidation.

inner mitochondrial membrane [3]. There are two complementary experimental approaches to determining the topographical relationship of mitochondrial enzymes and functions to the membranes and their defined spaces. One is a structural approach and involves the actual physical separation of the inner and outer mitochondrial membranes from one another, their recognition by electron microscopy, and analysis of the enzymic content. The other approach is operational and starts from the assumption that certain behavioural properties of isolated but intact mitochondria are best explained in terms of compartments and permeability barriers. The permeability barriers are then equated, in the first instance, with one or other of the mitochondrial membranes. It is important to compare the models drawn by these two independent approaches; any severe discrepancy would indicate that one or even both models are incorrect or incomplete. It is also important to stress the assumption underlying the operational approach, namely, that certain phenomena are due to compartments and permeability barriers. This may not be so, and alternative explanations are then preferable.

(ii) Structural studies

Electron microscopy provides information about the structure of the two membranes, and the spaces between them. The primary criterion for discriminating between the two membranes both in intact mitochondria and sub-mitochondrial fractions is based on the distinctive differences of the two membranes as shown by electron microscopy [4-8].

The separation of outer and inner mitochondrial membranes. A number of procedures have been devised for this purpose, and in essence all of them are similar [7-12]. The outer membrane is removed from the inner membrane either by dissolution with detergent, e.g. cholate [9], digitonin [10, 11], or by phospholipase treatment [9], or by rupture and detachment of the outer membrane following mitochondrial swelling [7, 8, 12]. Differential or density gradient centrifugation is then used to separate the membrane fractions which are recognized by electron microscopy. If most of the total mitochondrial activity of a given enzyme is recovered in a certain fraction, and with an increased specific activity over the starting material, then it is reasonably concluded that the enzyme is a component of that fraction. Table 1 summarizes the results of this approach as applied to the enzymes of β-oxidation.

Table 1. Distribution of enzymes of fatty acid oxidation in rat liver mitochondria

Enzyme	Mitochondria (100%)	Outer membrane	Inner membrane	Soluble
Cytochrome a, a_3*	0·065	–	0·109 (91%)	–
Cytochrome b_5*	0·054	0·309 (42%)	0·016 (16%)	–
Monoamine oxidase†	3·49	14·5 (36%)	1·86 (25%)	–
Malate dehydrogenase‡	3·0	–	3·74 (59%)	4·47 (44%)
3-Hydroxyacyl-CoA dehydrogenase†	705	216 (3%)	1050 (71%)	454 (18%)
3-Ketoacyl-CoA thiolase†	212	6 (0·2%)	202 (45%)	356 (47%)
Citrate synthase†	153	16 (0·9%)	205 (63%)	182 (35%)
Palmityl-carnitine transferase†	9·3	–	17·6 (98%)	–
Fp_{ETF}*	0·114	–	0·161 (68%)	0·127 (33%)
Palmitate:CoA ligase (ATP)†	5·8	22·5 (29%)	3·3 (31%)	11·3 (25%)
Decatetraenoate:CoA ligase (ATP)†	0·128	0·216 (17%)	0·071 (24%)	0·173 (39%)
Decatetraenoate:CoA ligase (GTP)†	0·064	<0·02	<0·02	0·103 (36%)

* nmole/mg † nmole/min/mg ‡ μmole/min/mg.
Fractions were prepared essentially as described in ref. 8.

(iii) Operational studies

Permeable spaces. As much as 70% of the total mitochondrial water is permeable to sucrose [13]. The remaining sucrose-impermeable space must therefore account for the osmometric behaviour of mitochondria in sucrose solutions of varying strength [14]. The sucrose-permeable space is also readily penetrated by nucleoside triphosphates [15, 16], NAD [15], CoASH [17] and acetyl-CoA [18], carnitine and acetyl-carnitine [17]. From such studies it is concluded that the mitochondrial water is subdivided into two compartments, one peripheral and permeated by low molecular weight solutes, the other more central, not readily permeated by external solutes, and osmotically active [15]. The structurally defined spaces that are generally equated with these operational spaces are the cristal space (peripheral and sucrose permeable) and the matrix space (central and sucrose impermeable).

The behaviour of intramitochondrial nucleotides. Fatty acid oxidation involves coenzyme A, adenine nucleotides, and NAD. All of these nucleotides are present to a relatively constant extent in isolated mitochondria. Since all of these nucleotides when added can penetrate the sucrose-permeable space, it is likely that they occupy the sucrose-impermeable space when they are present as endogenous components of mitochondria. Studies of the redox state of NAD [19], the phosphorylation state of adenine nucleotides [15], and the acylation state of CoA [20] have directly demonstrated the involvement of these intramitochondrial nucleotides in their metabolic roles. It is difficult to conceive how these nucleotides could be remote from the enzymes for which they act as substrates, and it is more certain that they occupy or border the space occupied by the relevant enzymes.

Latency of mitochondrial enzymes towards their substrate(s). Latency phenomena alone do not yield information on whether the permeability barrier involved is peripheral or relatively more central. However, the interpretation can be developed further in conjunction with measurements of mitochondrial penetration by the substrates. Thus an enzyme is unlikely to be peripherally located in the mitochondrion if it is inactive towards a substrate that penetrates the sucrose-permeable space. Table 2 summarizes the behaviour of a number of enzymes of rat liver mitochondria towards their added substrates.

In the case of the enzymes that are clearly latent or non-latent, a relatively central or peripheral location can be assigned to the enzyme. The acid:CoA ligase and palmityl-carnitine transferase activities exhibit more complex behaviour that depends on the choice and location of substrates, and the simplest explanation for these observations is that there are three acid:CoA ligase activities and two acyl-transferase activities.

The effects of inhibitors: Atractyloside. Intramitochondrial and extramitochondrial adenine nucleotides (ATP and ADP) can exchange rapidly with each other, yet without a change in the total adenine nucleotide content in the intramitochondrial compartment [15]. The exchange mechanism or adenine

Table 2. The latency of some mitochondrial enzymes

Non-latent:
- Adenylate kinase [21]
- Monoamine oxidase [11]
- Cytochrome b_5 reductase [8]

Latent [24]:
- Citrate synthase
- 3-Hydroxyacyl-CoA dehydrogenase
- 3-Ketoacyl-CoA thiolase
- ETF reduction by acyl-CoA
- Oxoglutarate dehydrogenase
- Pyruvate dehydrogenase

Mixed:
- Acid:CoA ligases:
  - Palmitate:CoA ligase (non-latent) [18]
  - Medium-chain-length acid:CoA ligase (latent) [18]
  - Carnitine acyltransferase [22] (80% is latent to palmityl-CoA, but all is accessible to palmityl-carnitine)

An enzyme was classified as non-latent if the activity assayed with extramitochondrial substrates and intact mitochondria was at least 95% of the activity with broken mitochondria. The corresponding value for a latent enzyme was 10%.

nucleotide translocase is inhibited by atractyloside [15, 16, 21, 25, 26]. The sensitivity of acid:CoA ligases towards atractyloside in intact mitochondria under conditions where the only source of ATP is extramitochondrial can be used to locate the acid:CoA ligases in relation to the atractyloside-sensitive adenine nucleotide translocase [21-23, 27]. Experiments illustrating this approach are shown in Figs. 3 and 4 using the non-oxidizable 2,4,6,8-decatetraenoic acid as a substrate for acid:CoA ligase activity [22]. Synthesis of the thioester of this acid is accompanied by the appearance of a new absorption band with a peak at 376 nm. Fig. 3b shows the same difference spectrum for intact mitochondria. The time course of decatetraenoyl-CoA synthesis can be followed using a double-beam spectrophotometer. A convenient wavelength pair is 376-412 nm. Fig. 4a illustrates a decatetraenoate:CoA ligase activity in intact mitochondria (rat liver) that is insensitive to uncoupling agents or atractyloside, but is inhibited by arsenate or phosphate. This activity corresponds to the GTP-linked acid:CoA ligase [28-30]. Fig. 4a also reveals a second activity [29] that is not inhibited by arsenate but that requires added ATP. This activity is inhibited by atractyloside in intact mitochondria but not in disintegrated mitochondria [24]. A third decatetraenoate:CoA ligase activity is also demonstrated in Fig. 4a and is distinguished from the previous two activities by its requirement for added CoASH, Mg^{2+} and ATP, and insensitivity towards atractyloside.

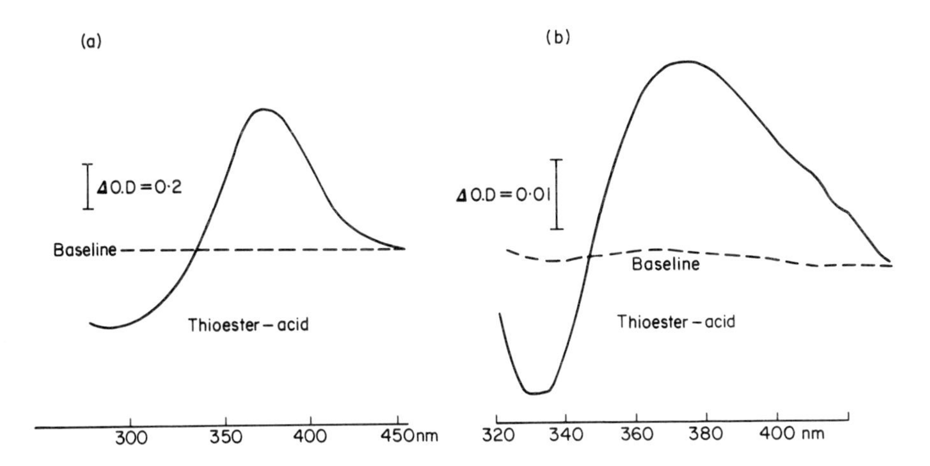

Figure 3. Difference spectra for decatetraenoyl-CoA minus decatetraenoate. (a) Soluble system. (b) Rat liver mitochondria. In (a), spectra were recorded with a Unicam SP800 spectrophotometer. Initially each cuvette contained in a final volume of 2 ml, 200 μmole Tris-HCl pH 7·4, 4 μmole $MgCl_2$, 2 μmole ATP, 50 nmole deca-2,4,6,8-tetraenoic acid, and 0·2 units acid:CoA ligase purified from ox liver mitochondria. After recording the baseline, 20 nmole CoASH were added to the sample cuvette, and the spectrum shown was recorded after the reaction was completed (20 min). In (b), spectra were recorded with a split-beam spectrophotometer. Initially each cuvette contained in a final volume of 2·5 ml, 200 μmole KCl, 50 μmole Tris-HCl pH 7·2, 2·5 μmole EDTA, 2 nmole FCCP, 10 μmole sodium arsenate pH 7·2, and 7 mg mitochondrial protein. After 3 min, 2·5 μg oligomycin and 50 nmole decatetraenoic acid were added to each cuvette. A baseline was recorded, then 2 μmole ATP was added to the test cuvette. After 2 min, the spectrum shown was recorded. In each experiment the optical path length was 1 cm, the temperature 25°C.

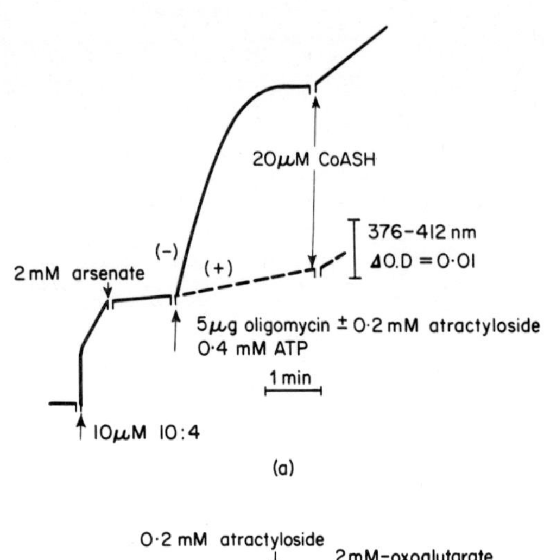

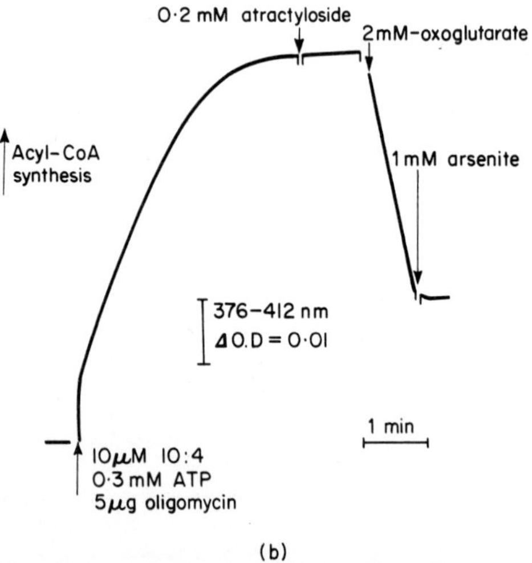

Figure 4. Double-beam spectrophotometric measurements of decatetraenoyl-CoA synthesis by rat liver mitochondria. Initially 2·2 ml of 80 mM-KCl, 20 mM-Tris-HCl pH 7·2, 1 mM-EDTA, 2 mM-$MgCl_2$ at 25°C contained 5 μM-pentachlorophenol and 2·5 mg of mitochondrial protein. Further additions were as shown. 10:4 is the abbreviation for decatetraenoic acid, and the sudden increase in optical density at 376 mμ is a reagent blank due to the acid.

From experiments of this type we concluded that two acid:CoA ligases were demonstrated to have access to intramitochondrial CoASH and Mg^{2+}, and that the two enzymes utilized the same pool of CoASH [22]. Since one of these enzymes is probably GTP-specific, it could be anticipated that succinyl-CoA synthetase would also have access to this pool of CoASH, as would oxoglutarate dehydrogenase. In keeping with this, the addition of oxoglutarate to mitochondria after the acylation of intramitochondrial CoA by decatetraenoate caused a rapid deacylation which was inhibited by arsenite (Fig. 4b). The actions involved are shown in this scheme:

NAD

oxoglutarate, CoASH ⇌ CO_2, succinyl-CoA ⇌ GTP (+ succinate), GDP + Pi ⇌ R.COOH, R.CO.SCoA

$NADH_2$

oxoglutarate dehydrogenase	succinyl-CoA synthetase	Acid:CoA ligase (GTP)

The effects of inhibitors: 2-Bromostearoyl-CoA. This compound is a potent and only slowly reversible inhibitor of palmityl-carnitine transferase [31]. Tubbs and Chase reported [31] that 2-bromostearoyl-CoA inhibited the carnitine-dependent oxidation of palmityl-CoA by rat liver mitochondria, but not the oxidation of palmityl-carnitine. In conclusion they agreed that there were both latent and non-latent acyl-transferase activities [17], though in this case the latency was towards the substrate analogue. Further support for this conclusion is provided by the experiment of Fig. 5, where acyl-transferase activity was directly assayed with a fluorimetric technique.

Pools of coenzyme A. The experiments of Fig. 4 indicated that oxoglutarate dehydrogenase, succinyl-CoA synthetase, a GTP-specific acid-CoA ligase and an ATP-specific acid:CoA ligase all have access to a common pool of CoASH. We have previously presented evidence to show that pyruvate and oxoglutarate dehydrogenases share a common pool of CoASH [32]. Furthermore, since the acetyl-CoA formed from either pyruvate or fatty acid oxidation can be converted to either acetoacetate or citrate, depending on the experimental conditions [32, 33], there is no reason to believe that any of these CoA-specific pathways are in separate compartments from each other. The problem of carnitine-stimulated palmitate oxidation will be discussed below.

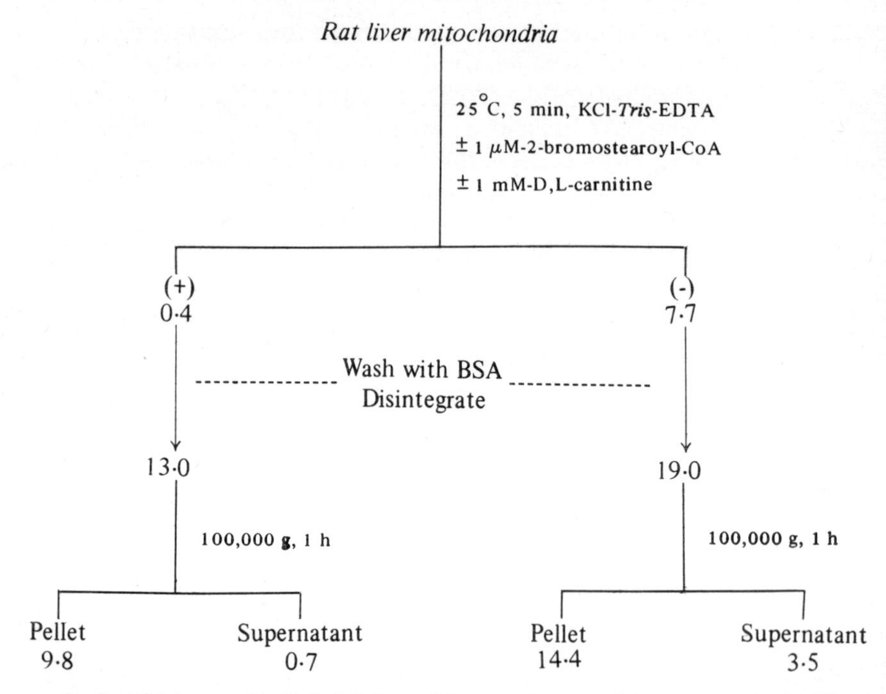

Figure 5. Latency of palmityl-CoA:carnitine transferase activity towards the inhibitor, 2-bromostearoyl-CoA. Assays were performed fluorimetrically [17], and the rates are expressed as nmole/min/mg of initial mitochondrial protein. Disintegration of mitochondria was effected with ultrasonic vibration. Exposure of mitochondria to ultrasonic vibration followed by 2-bromostearoyl-CoA resulted in loss of over 90% of the palmityl-CoA:carnitine transferase activity. 2-Bromostearoyl-CoA was a gift from Dr P. K. Tubbs.

2. CORRELATION OF STRUCTURAL AND OPERATIONAL STUDIES

All of the data for rat liver mitochondria presented or referred to above can be accommodated in the scheme of Fig. 6. The salient features of this scheme are:

(a) An outer mitochondrial membrane that is permeable to low molecular weight solutes, and is the site of an ATP-specific acid:CoA ligase that shows highest activity with long chain fatty acids. The acyl-CoA produced by this enzyme is derived from added CoA *in vitro* and is not oxidized unless carnitine is added.

(b) An inner mitochondrial membrane that is impermeable to nucleotides and carnitine, and contains the respiratory chain. The major part of the palmityl-carnitine transferase is firmly attached to this membrane, and is considered to be orientated in such a manner that it catalyses a vectorial reaction [17]:

$$\text{palmityl-carnitine}_{out} + \text{CoASH}_{in} \rightleftharpoons \text{palmityl-CoA}_{in} + \text{carnitine}_{out}$$

(c) A minor part of the palmityl-carnitine transference activity occupying a position that is functionally external to the inner membrane, and catalysing a non-vectorial reaction between added acyl-CoA and carnitine.

(d) A mitochondrial matrix containing intramitochondrial CoA, and adenine and pyridine nucleotides. The degree of binding of these nucleotides to the inner mitochondrial membrane *in situ* remains to be determined. In the case of coenzyme A, less than 10% of the total mitochondrial CoA is associated with the membrane fraction obtained by ultrasonic disintegration and centrifugation [24].

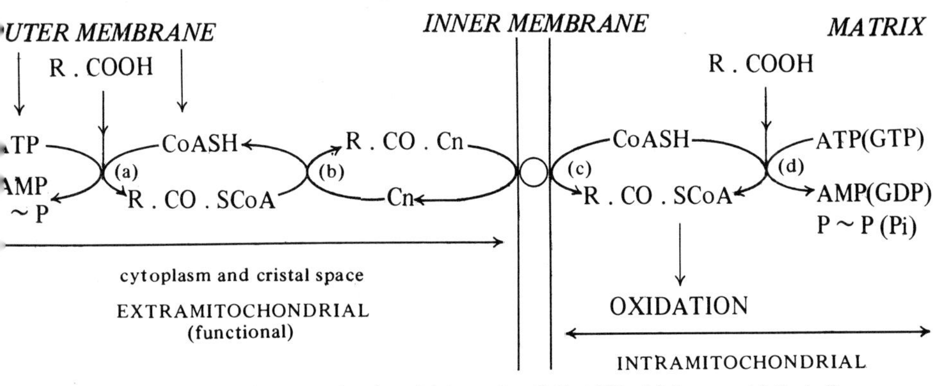

Figure 6. Intra- and extra-mitochondrial pools of CoASH. (a) is an acid:CoA ligase attached to the outer membrane, (b) is the non-latent palmityl-CoA:carnitine acyltransferase, (c) is the latent acyltransferase, (d) is both the ATP- and GTP-specific acid:CoA ligases that utilize intramitochondrial CoASH. Since there appears to be free diffusion of low molecular weight solutes across the outer membrane, the cytoplasmic and cristal spaces are drawn as a single functional "extramitochondrial" space. On this definition, only the nucleotides confined by the inner membrane are truly intramitochondrial.

(e) The ATP- and GTP-specific acid:CoA ligases, possibly duplicated by the chain length specificity in the case of the ATP-specific enzyme, are present in the matrix space.

(f) The remaining enzymes of β-oxidation that are readily solubilized by ultrasonic disintegration are also in the matrix space. These include Fp_{ETF} and the other coenzyme A-specific enyzmes such as pyruvate and oxoglutarate dehydrogenases, citrate synthase, succinyl-CoA synthetase, and the enzymes of acetoacetate synthesis.

3. THE STIMULATION OF PALMITATE OXIDATION BY CARNITINE

The scheme of Fig. 6 provides an adequate basis for the observation that fatty acyl-CoA synthesized outside the inner membrane is oxidized only in the presence of carnitine or structural damage. This scheme explicitly provides for the carnitine-independent oxidation of fatty acyl-CoA synthesized within the matrix, and this certainly occurs in the case of medium chain length fatty acids (invariably) and of long chain length fatty acids (variably). The experimental conditions for observing carnitine-independent oxidation of palmitate by rat liver mitochondria are such that the energy for palmitate activation is generated intramitochondrially [18, 23, 24, 34], and Van den Bergh [27] has proposed that the activating system involved in the carnitine-independent route of palmitate oxidation has a relatively high K_m for ATP. By contrast, the K_m of the acid:CoA ligase involved in the carnitine-dependent oxidation of palmitate would be

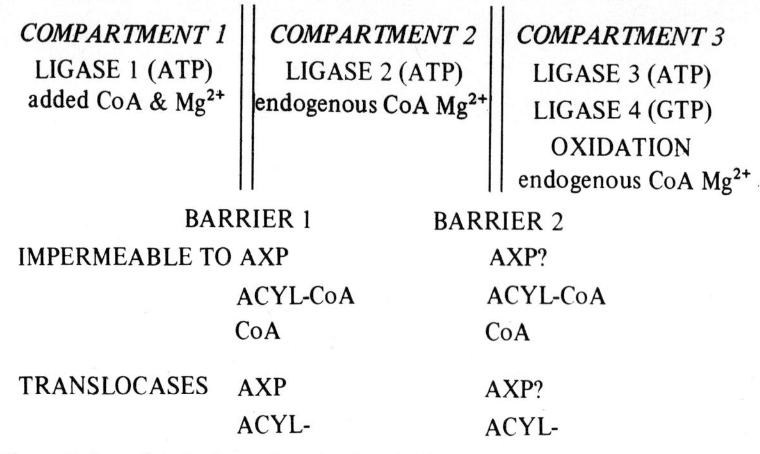

Figure 7. Lamellated scheme for mitochondrial compartments. This scheme is essentially that of Van den Bergh [27] and Chappell and Crofts [21], except that the latter authors postulated only ligases 2 and 3. The "ligases" refer to acid:CoA ligases of long chain length specificity, AXP stands for adenine nucleotides, and the translocases are the adenine nucleotide translocase and, for acyl groups, palmityl-CoA:carnitine acyltiansferase. Barriers 1 and 2 are permeability barriers for the compounds listed.

relatively lower. However, the problem still remains that if the carnitine-dependent route of palmitate oxidation proceeds with CoASH as the initial acyl-acceptor, then it is necessary to postulate that there are two pools of intramitochondrial CoASH between which acyl transfer is mediated by palmityl-carnitine transferase [21, 23, 27]. No matter how the necessarily associated permeability barriers are drawn, a considerable number of other (unwarranted) hypotheses must be made. For instance, the onion-like lamellated schemes that have been proposed imply a multiplicity (four) of palmityl-carnitine transferases and conceivably a similar doubling of other translocating systems (Fig. 7).

The arbitrarily erected permeability barriers between the two proposed pools of intramitochondrial CoASH can either be vaguely specified to the extent of being meaningless, or be so decorated with secondary hypotheses that they become unacceptable. In this respect our own previously proposed scheme [23] is no better than any others [21, 27]. In view of these difficulties, it is appropriate to reconsider the initial assumption, namely, that all palmitate oxidation proceeds through palmitate:CoA ligases.

AN ALTERNATIVE MECHANISM

We propose that the carnitine dependent route of palmitate oxidation proceeds through an acceptor AH, according to the reactions:

$$\text{AH + Palmitate + ATP} \leftrightarrow \text{Palmityl-A + AMP + P} \sim \text{P} \quad \ldots (1)$$

$$\text{Palmityl-A carnitine} \rightleftharpoons \text{Palmityl-carnitine + AH} \quad \ldots (2)$$

(Palmityl-A:carnitine acyltransferase)

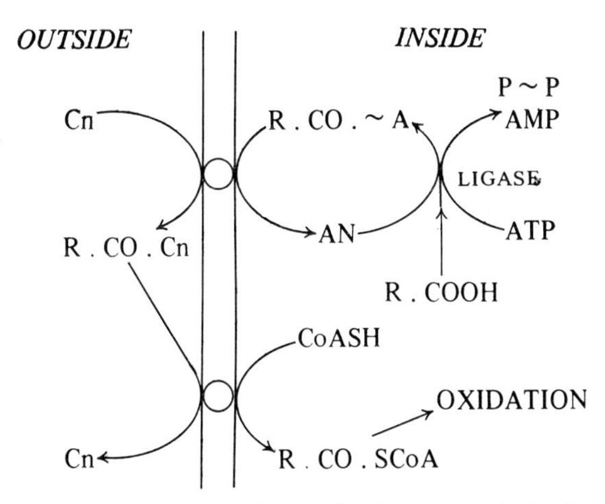

Figure 8. Mechanism for the stimulation of palmitate oxidation by carnitine. This mechanism is proposed for the case where neither CoASH or Mg^{2+} are added, and "activation" occurs via an intramitochondrial pathway (acid:AH ligase). When the ATP is provided externally, atractyloside inhibits this route of palmitate activation.

AH is specified as being a compound other than free CoASH; an enzyme bound form of CoA would perhaps be a candidate. In any case, palmityl-A would be a substrate for palmityl-CoA carnitine transferase but not for palmityl-CoA dehydrogenase. The reactions would readily fit into the scheme of Fig. 6, as shown in Fig. 8.

This scheme has the advantage over others in (1) having fewer assumptions, (2) being more explicit, and (3) being amenable to experimental test. The sum of reactions (1) and (2) is (3), an overall palmitate:carnitine ligase activity that is ATP specific.

$$\text{Palmitate + ATP + carnitine} \leftrightarrow \text{Palmityl-carnitine + AMP + P} \sim \text{P} \quad \ldots (3)$$

This reaction can be assayed as a CoA independent conversion of $[C^{14}]$carnitine into palmityl-$[^{14}C]$carnitine. The rates of reaction (3) under a variety of experimental conditions are shown in Table 3 for ultrasonically disintegrated mitochondria from rat liver. Three conclusions emerge from these data: (1) The rates of palmityl-carnitine synthesis are reasonably close to the rates required (1-2 nmole/min/mg) for the rate of carnitine stimulated palmitate oxidation in

Table 3. Palmityl-carnitine synthesis by rat liver mitochondria

Fraction	Palmityl-carnitine synthesis m μmole/mg protein/min	Total endogenous CoA in assay M $\times 10^{-7}$
Sonically disintegrated mitochondria	1·9	8
Supernatant after 105,000 g for 40 min	1·4	12
Pellet after 105,000 g for 40 min	0·6	2

Rat liver mitochondria were disintegrated by exposure to ultrasonic vibration, and separated into a high-speed supernatant and a pellet fraction. Each fraction (approximately 3·0 mg protein/ml) was incubated at 30°C in 80 mM-KCl; 20 mM-Tris-chloride, pH 7·2; 1 mM-EDTA; 5 μg oligomycin/ml; 0·1 μg rotenone/ml; 5 mg bovine plasma albumin/ml; 50 μM-palmitate; 1 mM-D,L-carnitine, ^{14}C-labelled with a specific activity of 750 d.p.m./mμmole. The reaction was started with 1 mM-ATP, and samples were removed, quenched, and assayed for palmityl-$[^{14}C]$carnitine according to ref. 40. Total CoA refers to the CoA assayed with oxoglutarate dehydrogenase after treatment of the sample with alkali and thiol; the total therefore includes free CoASH, short and long chain acyl-CoA and CoA disulphides [20].

rat liver mitochondria; (2) there are low concentrations of CoA (assayed as total CoA) in all samples; and (3) the maximal concentration of palmityl-CoA that could have accumulated in the assays was 0·2 to 1·2 μM, according to the fraction (Table 3). The K_m of the palmityl-CoA:carnitine acyl transferase for palmityl-CoA is 30 μM (calf liver [38]) or about 20 μM for rat liver (see Fig. 1 of ref. 39). Under the conditions used in Table 3, the rates of palmityl-carnitine synthesis from added palmityl-CoA (0·2-1·2 μM) were an order of magnitude lower than the rate measured with reaction (3) itself. It is therefore unlikely that palmityl-CoA was an intermediate in reaction (3).

CONCLUSIONS

The original scheme of Fritz and Yue [1] for the relationship between the components and compartments of mitochondrial fatty acid oxidation has been extended in greater detail to allow for recent advances. The modified scheme is entirely consistent with virtually all experimental data on the operational behaviour of rat liver mitochondria oxidizing fatty acids. The concept of two pools of mitochondrial CoA separated by a permeability barrier is discussed and discarded in favour of an alternative and experimentally supportable hypothesis involving a newly described enzyme, palmitate:AH ligase. AH is defined as protein-bound acyl acceptor distinct from free CoASH. There are numerous precedents for acyl-acceptors of this type [35-37].

ACKNOWLEDGEMENTS

These studies were supported by the Medical Research Council and The Royal Society. B.A.H. is in receipt of a Science Research Council Studentship.

REFERENCES

1. Fritz, I. B. and Yue, K. T. N., *Adv. Lipid Res.* **4** (1963) 279.
2. Bremer, J., *J. biol. Chem.* **238** (1963) 2774.
3. Palade, G. E., *Anat. Rec.* **114** (1952) 427.
4. Fernández-Morán, H., *Circulation* **26** (1962) 1039.
5. Fernández-Morán, H., Oda, T., Blair, P. V. and Green, D. E., *J. Cell Biol.* **22** (1964) 63.
6. Parsons, D. F., *Science, N.Y.* **140** (1963) 985.
7. Parsons, D. F., Williams, G. R. and Chance, B., *Ann. N.Y. Acad. Sci.* **137** (1966) 643.
8. Sottocasa, G. L., Kuylenstierna, B., Ernster, L. and Bergstrand, A., *J. Cell Biol.* 32 (1967) 415.
9. Allmann, D. W. and Bachmann, E., *Meth. Enzym* **10** (1967) 438.
10. Lévy, M., Toury, R. and André, J., *C. r. Séanc. Soc. Biol.* **262** (1966) 595.
11. Schnaitman, C., Erwin, V. G. and Greenawalt, J. W., *J. Cell Biol.* **32** (1967) 719.
12. Caplan, A. I. and Greenawalt, J. W., *J. Cell Biol.* **31** (1966) 455.
13. Werkheiser, W. C. and Bartley, W., *Biochem. J.* **66** (1957) 79.
14. Tedeschi, H. and Harris, D. L., *Archs Biochem. Biophys.* **58** (1955) 52.
15. Klingenberg, M. and Pfaff, E., *in* "Regulation of Metabolic Processes in Mitochondria" (edited by J. M. Tager, S. Papa, E. Quagliariello and E. C. Slater), Elsevier, Amsterdam, 1966, p. 180.
16. Brierley, G. P. and Green, D. E., *Proc. natn. Acad. Sci. U.S.A.* **53** (1965) 73.
17. Yates, D. W. and Garland, P. B., *Biochem. biophys. Res. Commun.* **23** (1966) 460.
18. Yates, D. W., Ph.D. Thesis, University of Bristol, 1968.

19. Chance, B. and Williams, G. R., *Adv. Enzym.* **17** (1956) 65.
20. Garland, P. B., Shepherd, D. and Yates, D. W., *Biochem. J.* **97** (1965) 587.
21. Chappell, J. B. and Crofts, A. R., *Biochem. J.* **95** (1965) 707.
22. Yates, D. W. and Garland, P. B., *Biochem. J.* **102** (1967) 40P.
23. Garland, P. B. and Yates, D. W., *in* "Round Table Discussion on Mitochondrial Structure and Compartmentation" (edited by E. Quagliariello, S. Papa, E. C. Slater and J. M. Tager) Adriatica Editrice, Bari, 1967, p. 385.
24. Haddock, B. A. and Garland, P. B., unpublished work.
25. Bruni, A., Luciani, S. and Contessa, A. R., *Nature, Lond.* **201** (1964) 219.
26. Kemp, A., Jr. and Slater, E. C., *Biochim. biophys. Acta* **92** (1964) 178.
27. Bergh, S. G. Van den, *in* "Round Table Discussion on Mitochondrial Structure and Compartmentation" (edited by E. Quagliariello, S. Papa, E. C. Slater and J. M. Tager), Adriatica Editrice, Bari, 1967, p. 400.
28. Rossi, C. R. and Gibson, D. M., *J. biol. Chem.* **239** (1964) 1694.
29. Bergh, S. G. Van den, *Biochim. biophys. Acta* **98** (1965) 442.
30. Yates, D. W., Shepherd, D. and Garland, P. B., *Nature, Lond.* **209** (1966) 1213.
31. Tubbs, P. K. and Chase, J. F. A., *Abstr., 4th FEBS Meeting,* Oslo, 1967, p. 315.
32. Nicholls, D. G., Shepherd, D. and Garland, P. B., *Biochem, J.* **103** (1967) 677.
33. Garland, P. B., Shepherd, D. and Nicholls, D. G., *in* "Round Table Discussion on Mitochondrial Structure and Compartmentation" (edited by E. Quagliariello, S. Papa, E. C. Slater and J. M. Tager), Adriatica Editrice, Bari, 1967, p. 424.
34. Rossi, C. R., Galzinga, L. and Gibson, D. M., *in* "Regulation of Metabolic Processes in Mitochondria" (edited by J. M. Tager, S. Papa, E. Quagliariello and E. C. Slater), Elsevier, Amsterdam, 1966, p. 143.
35. Rao, G. A. and Johnston, J. M., *Biochim. biophys. Acta* **114** (1967) 25.
36. Bar-Tana, J. and Shapiro, B., *Biochem. J.* **93** (1964) 533.
37. Marjerus, P. W., Alberts, A. W. and Vagelos, P. R., *J. biol. Chem.* **240** (1965) 4723.
38. Norum, K. R., *Biochim. biophys. Acta* **89** (1964) 95.
39. Bremer, J. and Norum, K. R., *Eur. J. Biochem.* **1** (1967) 427.
40. Bremer, J., *J. biol. Chem.* **238** (1963) 2774.

FEBS Symposium, Volume 17, 1969, pp. 127-135

The Effect of Propionyl-CoA and of the Acetyl-CoA/CoA Ratio on the Formation of Acetoacetate in Rat Liver Mitochondria

J. BREMER and M. AAS

Institute of Clinical Biochemistry, University of Oslo, Rikshospitalet, Oslo, Norway

Recently we have found that propionylcarnitine is present in relatively large amounts in the liver of fed rats [1]. When the rats are fasted or fat-fed the propionylcarnitine disappears almost completely, while the level of acetylcarnitine and long-chain acylcarnitine increases [2, 3].

In the present paper we report some studies on the effect of propionyl-CoA and of the acetyl-CoA/CoA ratio on the formation of acetoacetate in isolated rat liver mitochondria.

MATERIALS AND METHODS

The carnitine derivatives and rat liver mitochondria were prepared as previously described [2, 4, 5].

Acetyl-CoA:carnitine *O*-acetyltransferase (carnitine acetyl-transferase) (EC 2.3.1.7.) and CoA were obtained from Boehringer & Soehne, Mannheim, Germany.

Disrupted mitochondria were prepared by treating an ice-cold mitochondrial suspension in 0·15M KCl with ultrasonic vibration for 3 x 10 s in a Branson sonifier with standard tip at maximal yield (approximately 5 A).

The oxygen uptake experiments were performed at 25°C essentially as described by Chappell [6].

In the acetoacetate formation experiments with disrupted mitochondria, the vessels were preincubated for at least 5 min, to secure equilibrium concentrations of CoA and CoA esters, before the reaction was started by addition of the suspension of disrupted mitochondria.

Acetoacetate was determined according to Walker [7]. Acetoacetylcarnitine does not disturb the assay of free acetoacetate with this method. Disappearance of (-)palmityl[C^3H_3]carnitine was determined with *n*-butanol extraction as previously described [5].

RESULTS

Acetoacetate formation in intact mitochondria

Table 1 shows that propionyl-CoA precursors inhibit the formation of acetoacetate from caprinylcarnitine, caprinate, and pyruvate. The strongest effect was obtained with α-ketobutyrate which also inhibited the respiration.

Table 2 shows that the utilization of palmitylcarnitine was inhibited by α-ketobutyrate and propionate.

The reduced acetoacetate formation in these experiments cannot be explained by an increased formation of β-hydroxybutyrate as the oxygen uptake rates were also decreased by α-ketobutyrate.

The effect of bicarbonate was tested since the first step in the metabolism of propionyl-CoA is a carboxylation to methylmalonyl-CoA. Tables 1 and 2 show that bicarbonate always stimulated the rate of oxygen uptake and the rate of palmitylcarnitine disappearance was also increased.

Caprinylcarnitine was more efficient as acetoacetate precursor than was palmitylcarnitine. Pyruvate was the poorest precursor. In agreement with previous observations the formation of acetoacetate from pyruvate was most easily suppressed [8].

Acetoacetate formation in disrupted mitochondria

According to recent studies, acetoacetate is formed by the following sequence of reactions [9-11]:

Acetyl-CoA → Acetoacetyl-CoA → β-hydroxy-β-methylglutaryl-CoA →
acetoacetate + acetyl-CoA.

In this sequence, the formation of hydroxymethylglutaryl-CoA seems to be rate limiting [11]. The equilibrium of the first step in the sequence is strongly displaced to the left ($K = 2 \times 10^{-5}$ at pH 8·1) [12].

To test the relative importance of acetyl-CoA, propionyl-CoA, and free CoA in the mitochondria we have developed a test system where the relative concentrations of these metabolites can be varied freely within wide limits. Both acetyl- and propionylcarnitine react freely with carnitine acetyltransferase [2]. By using varying proportions of acetylcarnitine, propionylcarnitine, carnitine, and catalytic amounts of CoA in the presence of excess carnitine acetyltransferase, the relative amounts of the corresponding CoA derivatives in the reaction mixture can be calculated when the equilibrium constants are known.

Figure 1 shows the determination of the equilibrium constant

$$K = \frac{[\text{acetylcarnitine}]\,[\text{CoA}]}{[\text{acyl-CoA}]\,[\text{carnitine}]}$$

with acetylcarnitine and propionylcarnitine with a method similar to that used

Table 1. Effect of propionyl-CoA precursors on acetoacetate formation and respiration rate in liver mitochondria

Experiment No.	Additions	Acetoacetate (μmoles/6 min)		Oxygen uptake (μatoms/min)
I	Caprinylcarnitine	0·65	0·63	0·21
	Caprinylcarnitine + α-ketobutyrate	0·34	0·35	0·12
	Caprinylcarnitine + α-ketobutyrate + bicarbonate	0·35	0·39	0·16
	Caprinylcarnitine + propionylcarnitine	0·54	0·34 (?)	0·22
	Palmitylcarnitine	0·44	0·45	0·17
	Palmitylcarnitine + α-ketobutyrate	0·23	0·23	0·10
	Palmitylcarnitine + α-ketobutyrate + bicarbonate	0·29	0·28	0·16
	Palmitylcarnitine + propionylcarnitine	0·39	0·39	0·17
II	Caprinylcarnitine	0·57	0·57	0·32
	Pyruvate	0·31	0·31	0·15
	Pyruvate + α-ketobutyrate	0·10	0·09	0·15
	Pyruvate + α-ketobutyrate + bicarbonate	0·07	0·07	0·20
	Pyruvate + propionylcarnitine	0·26	0·26	0·19
III	Caprinylcarnitine	0·53	0·56	0·25
	Caprinate	0·43	0·42	0·22
	Caprinate + α-ketobutyrate	0·21	0·20	0·16
	Caprinate + propionate	0·28	0·27	0·21
	Caprinate + propionate + bicarbonate	0·23	0·25	0·26

Acetoacetate formation and respiration rate were measured in separate incubations performed with the same mitochondria preparation. Additions: Rat liver mitochondria, approximately 4 mg of protein; TES buffer (pH 7·3), 40 μmoles; ADP, 10 μmoles; phosphate, 10 μmoles. The substrates were added in the following amounts: Caprinylcarnitine, 0·5 μmole; palmitylcarnitine, 0·2 μmole; caprinate, 0·5 μmole; pyruvate, 5 μmoles; α-ketobutyrate, 5 μmoles; propionate, 5 μmoles; propionylcarnitine, 5 μmoles; $NaHCO_3$, 10 μmoles. The mitochondria were preincubated with the propionyl-CoA precursors for approximately 4 min. Acetoacetate formation was measured 6 min after the addition of the acetoacetate precursor. The respiration rates are the rates observed immediately after addition of the acetoacetate precursor. Total volume 2 ml, temperature 30°C.

Table 2. Effect of propionyl-CoA precursors on the oxidation of palmitylcarnitine in rat liver mitochondria

Additions	Acetoacetate (μmoles)		Palmitylcarnitine disappeared (μmoles)	
Palmitylcarnitine	0·17	0·18	0·076	0·069
Palmitylcarnitine + α-ketobutyrate	0·09	0·09	0·031	0·033
Palmitylcarnitine + α-ketobutyrate + bicarbonate	0·12	0·12	0·047	0·041
Palmitylcarnitine + propionate	0·11	0·11	0·065	0·051
Palmitylcarnitine + propionate + bicarbonate	0·13	0·13	0·065	0·075

Additions: Mitochondria, appriximately 2·5 mg of protein; (-)palmityl[C^3H_3] carnitine, 0·2 μmole. Other additions as stated in Table 1. Incubation time, 6 min (no preincubation performed).

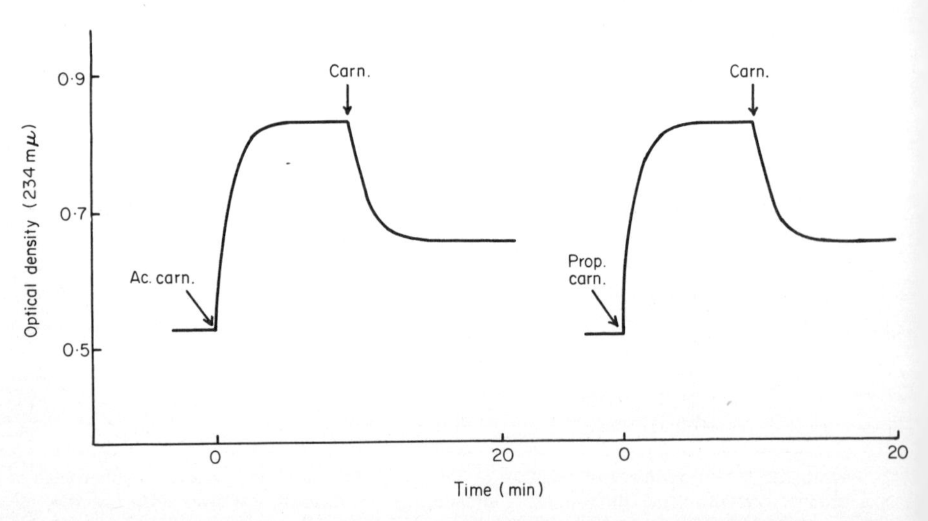

Figure 1. Determination of the equilibrium constant of the reaction acyl-CoA + carnitine ⇌ acylcarnitine + CoA. Approximately 0·1 unit of carnitine acetyltransferase, 0·1 μmole of CoA, and 30 μmoles of Tris buffer (pH 7·5) were mixed in a total volume of 1·5 ml. Acetylcarnitine, propionylcarnitine, and carnitine (5 μmoles in 50 μl) were added as shown. The reaction was followed at 233 mμ in a Zeiss RPQ II reading spectrophotometer at 25°C. Corrections for dilution by the added substrate solutions have been made.

by Norum [13] for the reaction with palmitylcarnitine. With acetylcarnitine we found $K = 1{\cdot}5$ which is in fairly good agreement with measurements by Fritz *et al.* [14] who found $K \approx 1{\cdot}7$. With propionylcarnitine $K = 1{\cdot}3$ and with butyrylcarnitine $K = 1{\cdot}0$. Previously, Norum [13] has found $K = 0{\cdot}5$ with palmitylcarnitine. Thus, the equilibrium of this reaction is displaced slightly to the advantage of acyl-CoA with increasing length of the carboxylic acid.

Since acetoacetylcarnitine also reacts with carnitine acetyltransferase, although more slowly [2, 15], this compound can also be used as substrate in the system.

Table 3. Acetoacetate formation in intact and disrupted mitochondria

Incubation system	Acetoacetate (μmoles/mg of protein)	
Intact mitochondria + caprinylcarnitine + ADP + phosphate	0·185	0·185
Intact mitochondria + caprinylcarnitine + dinitrophenol	0·135	0·15
Disrupted mitochondria + acetylcarnitine	0·105	0·115
Disrupted mitochondria + acetoacetylcarnitine	0·16	0·21

Intact liver mitochondria (4 mg of protein) were incubated with TES buffer (pH 7·3), 40 μmoles; caprinylcarnitine, 0·5 μmole; ADP, 10 μmoles; phosphate, 10 μmoles; or dinitrophenol, 0·25 μmole, in a total volume of 2 ml.

Disrupted mitochondria (2 mg of protein) were incubated with TES buffer (pH 7·3), 20 μmoles; carnitine acetyltransferase, 0·4 unit; CoA, 0·4 μmole; and acetylcarnitine, 10 μmoles; or acetoacetylcarnitine, 5 μmoles, in a total volume of 1 ml. Incubation time 10 min.

Table 3 shows that the rate of acetoacetate formation from acetylcarnitine in disrupted mitochondria was significantly slower than from caprinylcarnitine in a corresponding amount of intact mitochondria. With acetoacetylcarnitine the acetoacetate formation was about equal to that obtained in intact mitochondria.

Table 3 also shows that less acetoacetate was formed when the respiration was stimulated with dinitrophenol than when it was stimulated with ADP and phosphate. The reason for this remains unknown.

Figure 2A shows how the acetyl-CoA/CoA ratio influences the rate of acetoacetate formation. The reaction rate evidently does not depend simply on the acetyl-CoA concentration. In Fig. 2B the same experiment has been redrawn against the ratio $[\text{acetyl-CoA}]^2/[\text{CoA}]$ which theoretically is proportional to the equilibrium concentration of acetoacetyl-CoA in the reaction mixture. The

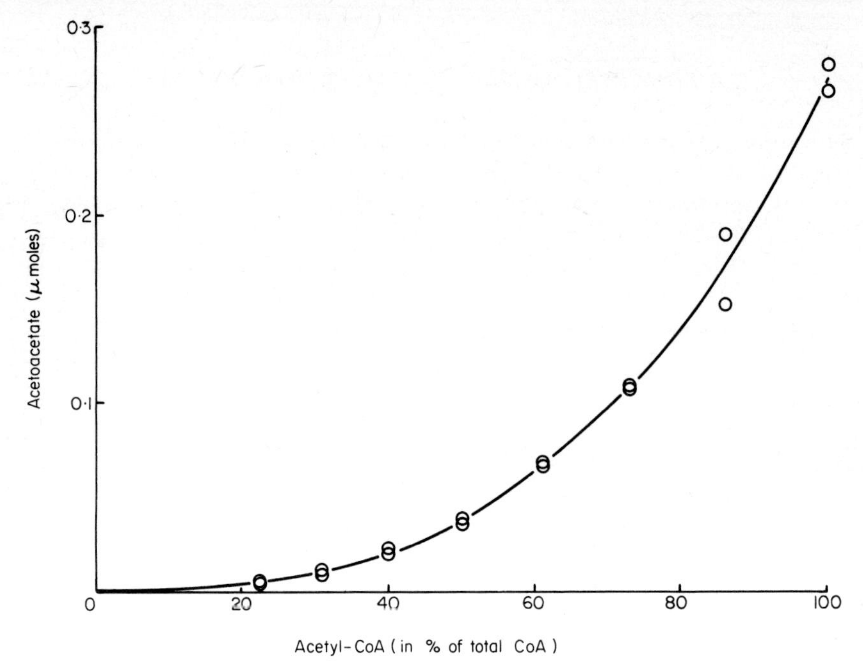

Figure 2A. The effect of the acetyl-CoA/CoA ratio on the rate of acetoacetate formation in disrupted mitochondria (2·7 mg of protein). Additions: TES buffer (pH 7·5), 40 μmoles; carnitine acetyltransferase, 0·2 unit; CoA, 0·2 μmole; KCN, 2·5 μmoles; acetylcarnitine, 5-0 μmoles; carnitine, 0-5 μmoles: Total volume 1 ml, incubation time 10 min. The acetoacetate formation is given in relation to the calculated per cent acetylation of CoA.

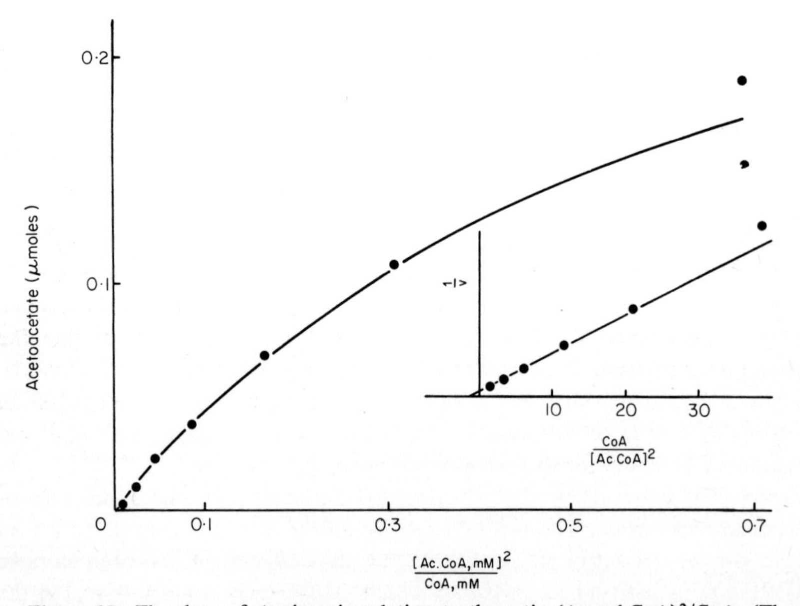

Figure 2B. The data of A given in relation to the ratio $(\text{Acetyl-CoA})^2/\text{CoA}$. (The rate obtained with only acetylcarnitine added is omitted because of the insecure calculation of the acetyl-CoA/CoA ratio). The inserted figure is a Lineweaver-Burk plotting of B.

inserted figure shows a corresponding Lineweaver-Burk plotting. A nearly straight line was obtained. If the equilibrium constant of the acetoacetyl-CoA thiolase reaction is assumed to be 10^{-5} at pH 7·5 [12], the straight line corresponds to a K_M of approximately 7 x 10^{-6}M acetoacetyl-CoA. This value cannot be considered to give more than the order of magnitude, since the concentration of acetyl-CoA increased and the concentration of CoA decreased with the increasing [acetyl-CoA]^2/[CoA ratio].

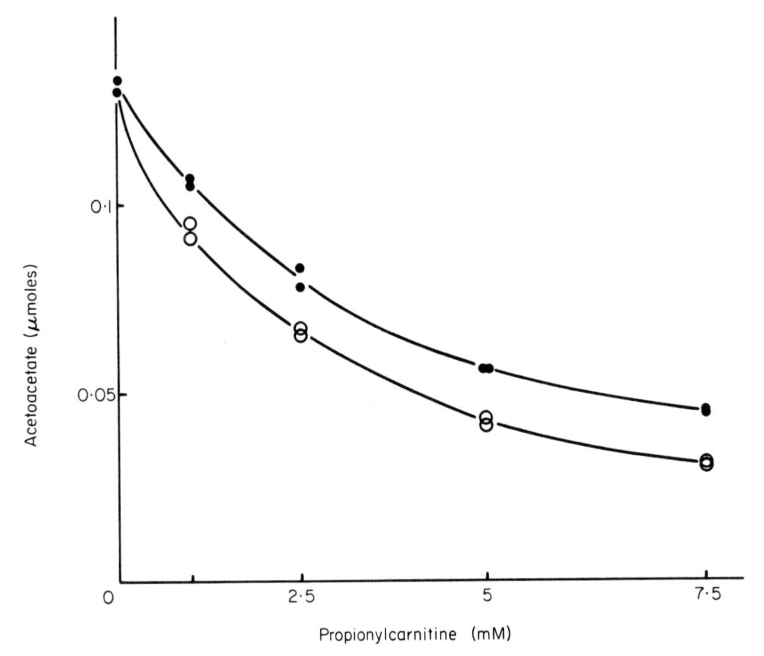

Figure 3. Effect of propionylcarnitine (propionyl-CoA) on the rate of acetoacetate formation in disrupted mitochondria (3·8 mg of protein). Additions: TES buffer (pH 7·5), 40 μmoles; carnitine acetyltransferase, 0·2 unit; KCN, 2·5 μmoles; acetylcarnitine, 2·5 μmoles; propionylcarnitine, 0-7·5 μmoles; CoA, 0·08 μmole (lower curve) or 0·08-0·26 μmole (upper curve). The upper curve represents vessels where the addition of CoA was adjusted to give a constant concentration of acetyl-CoA (approximately 0·075 μmole/ml).

Effect of propionylcarnitine (propionyl-CoA). Figure 3 shows how the acetoacetate production was inhibited by increasing amounts of propionylcarnitine in the reaction mixture. The lower curve in the figure shows the effect when the total amount of CoA in the reaction mixture was kept constant. As a suboptimal amount of CoA was used, the effect evidently might be due to binding of most of the available CoA as propionyl-CoA. However, the inhibiting effect was only to a small extent counteracted by increasing the addition of CoA in such a way that the equilibrium concentration of acetyl-CoA was kept constant (Fig. 3, upper curve). It seems likely, therefore, that propionyl-CoA is an inhibitor of the acetoacetate-forming enzyme system of the mitochondria. At present it is not possible to make conclusions about the mechanism of this inhibition.

DISCUSSION

The reaction

$$2 \text{ acetyl-CoA} \rightleftharpoons \text{acetoacetyl-CoA} + \text{CoA}$$

is strongly displaced to the left ($K = 2 \times 10^{-5}$ at pH 8·1) [12], and therefore it should be expected that the rate of acetoacetate formation depends on the acetyl-CoA/CoA ratio in the mitochondria. The studies presented here support this assumption.

Recently we have found that there are relatively high concentrations of propionylcarnitine in the liver of carbohydrate-fed animals. This metabolite disappears almost completely in fasted and fat-fed animals [1, 2]. It is likely that there is a corresponding variation in the level of propionyl-CoA in the liver. This may be of importance in the regulation of ketogenesis since the present studies indicate that propionyl-CoA inhibits ketogenesis, in part by lowering the amount of CoA available for acetyl- and acetoacetyl-CoA formation, and in part by inhibiting the acetoacetate-forming enzyme system. In this connection, it is noticeable that propionate has a strong effect on acetoacetate formation from pyruvate, while its effect on pyruvate oxidation is moderate [16].

In vivo studies have shown that the acetyl-CoA/CoA and the acetylcarnitine/carnitine ratios are moderately increased in fasted and fat-fed animals as compared to carbohydrate-fed animals [2, 17-19]. The sum of acetyl-CoA + CoA is also higher in fasted animals [19], presumably because a significant fraction of the total CoA is bound as propionyl-CoA (and other CoA esters) in carbohydrate-fed animals. Thus, at least three factors may lower the rate of acetoacetate formation in carbohydrate-fed animals:

1. The acetyl-CoA/CoA ratio is low;
2. Less CoA is available for acetyl-CoA formation;
3. An inhibitor of the acetoacetate-forming enzyme system (propionyl-CoA) is present.

What determines the acetyl-CoA/CoA ratio is still not completely understood. The availability of fatty acids for oxidation is probably important. The level of long-chain acylcarnitine is elevated at least 10 times in fasted, diabetic, and fat-fed animals [2], and long-chain acylcarnitines are oxidized in preference to pyruvate [8]. This may be the reason why it is more difficult to suppress ketogenesis from fatty acids [8].

The present studies show that a high rate of acetoacetate formation is obtained only when the acetyl-CoA/CoA ratio is relatively high. This is in agreement with observations by Garland *et al.* [20] who found that there is very little free CoA left in mitochondria oxidizing palmitylcarnitine to acetoacetate. However, *in vivo* studies have never shown acetylcarnitine/carnitine and acetyl-CoA/CoA ratios in liver tissue approaching those found necessary in the experiments presented here. It seems likely, therefore, that the average acetyl-CoA/CoA ratio of whole liver tissue *in vivo* does not reflect the acetyl-CoA/CoA ratio at the cellular site of acetoacetate formation. The acetyl-CoA formed in the oxidation of fatty acids may have a more direct access to the acetoacetate-forming enzymes, without being in equilibrium with the CoA pool(s) of whole liver.

REFERENCES

1. Böhmer, T. and Bremer, J., *Biochim. biophys. Acta* **152** (1968) 440.
2. Böhmer, T. and Bremer, J., *Biochim. biophys. Acta* **152** (1968) 559.
3. Böhmer, T., *Biochim. biophys. Acta* **144** (1967) 259.
4. Bremer, J., *Biochem. Prep.* **12** (1968) 69.
5. Bremer, J. and Norum, K. R., *J. biol. Chem.* **242** (1967) 1744.
6. Chappell, J. B., *Biochem. J.* **90** (1964) 225.
7. Walker, P. G., *Biochem. J.* **58** (1954) 699.
8. Bremer, J., *Biochim. biophys. Acta* **116** (1966) 1.
9. Lynen, F., Henning, U., Bublitz, C., Sörbo, B. and Kröplin-Rueff, L., *Biochem. Z.* **330** (1958) 269.
10. Caldwell, I. C. and Drummond, G. I., *J. biol. Chem.* **238** (1963) 64.
11. Sauer, F. and Erfle, J. D., *J. biol. Chem.* **241** (1966) 30.
12. Stern, J. R., Coon, M. J. and del Campillo, A., *J. Am. chem. Soc.* **75** (1953) 2277.
13. Norum, K. R., *Biochim. biophys. Acta* **89** (1964) 95.
14. Fritz, I. B., Schultz, S. K. and Srere, P. A., *J. biol. Chem.* **238** (1963) 2509.
15. Bressler, R. and Katz, I. R., *J. clin. Invest.* **44** (1965) 840.
16. Pennington, R. J. and Appleton, J. M., *Biochem. J.*, **69** (1958) 119.
17. Wieland, O. and Weiss, L., *Biochem. biophys. Res. Commun.* **10** (1963) 333.
18. Tubbs, P. K. and Garland, P. B., *Biochem. J.* **93** (1964) 550.
19. Williamson, J. R., Herczog, B., Coles, H. and Danish, R., *Biochem. biophys. Res. Commun.* **24** (1966) 473.
20. Garland, P. B., Shepherd, D. and Yates, D. W., *Biochem. J.* **97** (1965) 587.

FEBS Symposium, Volume 17, 1969, pp. 137-144

Some New Aspects of Fatty Acid Oxidation by Isolated Mitochondria

S. G. Van den BERGH, C. P. MODDER,
J. H. M. SOUVERIJN and H. C. J. M. PIERROT

*Laboratory of Biochemistry, B.C.P. Jansen Institute,
University of Amsterdam, Amsterdam, The Netherlands*

Taking advantage of the vagueness of the title of this paper, we would like to present our most recent results on three separate lines of investigation concerning fatty acid metabolism in mitochondria.

The role of high-energy intermediates of oxidative phosphorylation in fatty acid activation

The possibility that the energy required for the formation of the coenzyme-A esters of fatty acids could be supplied directly by the high-energy intermediates of oxidative phosphorylation was first proposed by Bode and Klingenberg [1]. Evidence that this process occurred in rat liver mitochondria was brought forward by Wojtczak and co-workers [2, 3]. We could demonstrate, however, that the dinitrophenol- and oligomycin-insensitive fatty acid activation observed by these authors was due to substrate-level phosphorylation linked to the α-oxoglutarate oxidation step of the Krebs cycle [4, 5]. If substrate-linked phosphorylation was prevented, fatty acid activation was completely oligomycin-sensitive. On the other hand, fatty acid activation occurring in the presence of oligomycin or dinitrophenol was completely blocked by arsenite, an inhibitor of α-oxoglutarate oxidation. The possibility that high-energy intermediates of oxidative phosphorylation can activate fatty acids without the intervention of nucleoside triphosphates seemed, therefore, ruled out.

More recently, however, the case was reopened by Beattie and Basford [6]. Their results indicated that in brain mitochondria fatty acids are activated by two different mechanisms: an ATP-linked reaction and a system involving high-energy intermediates of oxidative phosphorylation. In order to differentiate the roles of substrate-level phosphorylation and high-energy intermediates in fatty acid activation, these authors also used arsenite. Addition of arsenite to brain mitochondria inhibited fatty acid activation by 15%. The addition of oligomycin to the arsenite-treated system further inhibited fatty acid activation by another 20%. The remaining 65% was considered to be that portion of fatty acid activation involving the high-energy intermediate.

When we started our studies with rat brain mitochondria, we were surprised at their very low rate of fatty acid oxidation. (Of course, we should not have been surprised, since every textbook tells us that the brain does not utilize fatty acids.) The rate of oxygen uptake due to palmitate oxidation is about 0·03 μl of oxygen per h per mg protein at 37°C as compared with over 200 for the substrate pair glutamate + malate, a 7000-fold difference. The only way to measure this very slow oxidation process is by using radioactive substrates.

We then checked the concentration of arsenite required to fully inhibit the oxidation of α-oxoglutarate in brain mitochondria. We found that 1 mM arsenite, the concentration of inhibitor used by Beattie and Basford [6], inhibited by only 65% the production of radioactive CO_2 from 1–^{14}C-labelled palmitate. A similar figure can be calculated from the data of Beattie and Basford [6]. Since a 100% inhibition of radioactive CO_2 production should be found when α-oxoglutarate oxidation is fully blocked, it is clear that from experiments with 1 mM arsenite no conclusions can be drawn about the role of substrate-level phosphorylation in fatty acid activation.

In Table 1 an experiment is shown in which arsenite was used at a concentration of 10 mM. Radioactive palmitate was oxidized by rat brain mitochondria in the presence of a low concentration of malate. After the incubation, radioactivity was counted both in the CO_2 produced and in water-soluble degradation products, mostly Krebs-cycle intermediates. The first thing to note is that with 10 mM arsenite the production of labelled CO_2 is completely blocked. Therefore, α-oxoglutarate oxidation is completely inhibited and so is the substrate-linked phosphorylation. This complete inhibition of substrate-linked phosphorylation leads to a 60% inhibition of fatty acid activation. Inhibition of respiratory-chain phosphorylation by oligomycin brings about a 35% inhibition of fatty acid activation. However, when both phosphorylation processes are inhibited by the combined action of arsenite and dinitrophenol, or arsenite and oligomycin, a 90% inhibition or even higher can be observed.

To satisfy ourselves, we wanted to show that even the remaining 10% of fatty acid activation was not energized by high-energy intermediates. It occurred to us that the amount of fatty acid that was broken down during the experiment was probably less than the high-energy phosphate content of the freshly isolated mitochondria, so that a considerable part of the fatty acid could have been activated by nucleoside triphosphates initially present. In Table 2 an experiment is shown in which the mitochondria were depleted of their high-energy content by preincubation for 5 min with 10 μM dicoumarol. After the preincubation the dicoumarol was removed by addition of bovine serum albumin. The control

Table 1. Oxidation of [1−^{14}C] palmitate by rat brain mitochondria

Conditions	$^{14}CO_2$	Water soluble ^{14}C	Total ^{14}C	Inhibition %
	dpm			
Control	875	1545	2420	–
+ dinitrophenol (0·1 mM)	555	525	1080	55
+ arsenite (10 mM)	0	975	975	60
+ oligomycin (5 μg)	795	790	1585	35
+ oligomycin + dinitrophenol	1075	805	1880	22
+ arsenite + dinitrophenol	0	95	95	96
+ arsenite + oligomycin	0	245	245	90

Rat brain mitochondria (5-6 mg protein) were incubated for 60 min at 37°C in 1 ml of a medium containing 0·02 μmole [1−^{14}C] palmitic acid (192,185 dpm), 0·05 mM L-malate, 30 mM potassium phosphate buffer (pH 7·5), 50 mM sucrose, 15 mM KCl, 2 mM EDTA, 5 mM $MgCl_2$ and 50 mM Tris-HCl buffer (pH 7·5). The reaction was stopped by addition of 0·1 ml 45% $HClO_4$ after which the incubation was continued for another 60 min. Radioactivity was counted in the CO , absorbed in hyamine and in the aqueous medium after three extractions with 2·5 ml *n*-pentane.

Table 2. Effect of preincubation with dicoumarol

Conditions	$^{14}CO_2$	Water soluble ^{14}C	Total ^{14}C	Inhibition %
	dpm			
Not preincubated				
Control	1334	1985	3319	–
+ dinitrophenol + arsenite	0	246	246	93
+ oligomycin + arsenite	0	390	390	88
Preincubated				
Control	1170	1920	3090	–
+ dinitrophenol + arsenite	0	54	54	98
+ oligomycin + arsenite	0	165	165	95

The mitochondria were preincubated at room temperature in the medium described in Table 1, to which 10 μM dicoumarol had been added (palmitate and oligomycin were absent during the preincubation). After 5 min, 1 mg defatted bovine serum albumin was added and oligomycin, if indicated. After another minute the reaction was started by addition of the labelled palmitate.

values of the preincubated mitochondria are not very different from those of the untreated mitochondria, but the inhibition by dinitrophenol plus arsenite, or by oligomycin plus arsenite, approaches 100% even more closely in the preincubated mitochondria.

From these experiments we conclude that also in brain mitochondria fatty acid activation is energized by nucleoside triphosphate and that high-energy intermediates are unable to activate fatty acids directly.

Effects of oleate on liver mitochondria

Wojtczak and Zaluska [7] have described experiments indicating that oleate is an inhibitor of the translocation of adenine nucleotides through mitochondrial membranes; in other words, that oleate has an atractyloside-like effect on rat liver mitochondria. In these experiments, discussed earlier in this Symposium by Dr. Wojtczak, mitochondria were incubated with ^{14}C-labelled ATP or ADP for various periods of time, following which the translocation of adenine nucleotides was stopped by addition of atractyloside. The total radioactivity of the mitochondria, corresponding to the amount of adenine nucleotide which entered the mitochondria during the incubation, was then measured.

In these experiments, which we have confirmed, oleate inhibited the rate of penetration of adenine nucleotides into the mitochondria. We have also obtained the kinetic data, shown by Dr. Wojtczak earlier in this Symposium.

Thus, we subscribe to the conclusion of Wojtczak and Zaluska that oleate, apart from its uncoupling effect on oxidative phosphorylation, exerts an inhibitory effect on adenine nucleotide translocation in mitochondria. We disagree, however, with their conclusion that this atractyloside-like effect of oleate should also explain the inhibition by oleate of the dinitrophenol-induced ATPase of the mitochondria. In our opinion, this inhibition is brought about by a third effect of oleate, which we shall indicate as the oligomycin-like effect of oleate.

Figure 1 shows the effect of increasing concentrations of oleate on the dinitrophenol-stimulated ATPase activity of rat liver mitochondria in the presence of added magnesium ions. Contrary to the data of Wojtczak and Zaluska [7] we find a maximal inhibition at oleate concentrations around 200 nmoles per mg protein. This concentration is about 10 times higher than the concentration at which oleate has its maximal inhibitory effect on adenine nucleotide translocation; it is 5 times higher than the concentration of oleate which fully uncouples oxidative phosphorylation [5]. However, at this very same concentration of 200 nmoles per mg protein, oleate has another effect, which to date has never been fully understood. This effect is shown in Fig. 2, in which the rate of oleate oxidation is plotted against the oleate concentration. At a concentration of 40 nmoles per mg protein, respiratory chain phosphorylation is completely uncoupled and, as with all other fatty acids, above this "critical

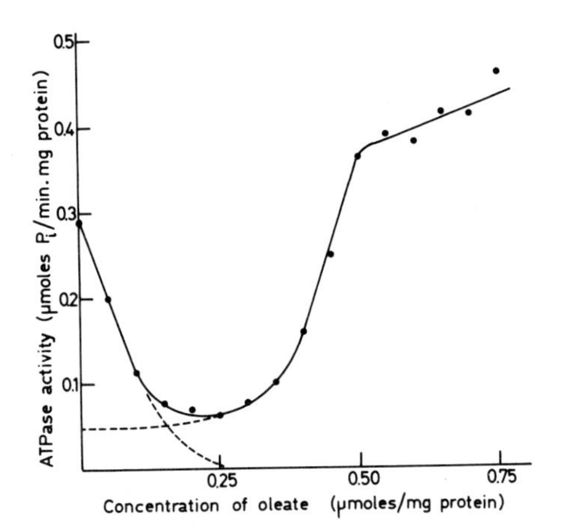

Figure 1. Effect of oleate on the dinitrophenol-induced ATPase of rat liver mitochondria. Mitochondria (0·15 mg protein) were incubated for 15 min at 25°C in 1·5 ml of a medium containing 2 mM ATP, 0·1 mM dinitrophenol, 1·5 mM $MgCl_2$, 0·5 mM EDTA, 50 mM Tris-HCl buffer (pH 7·5), 50 mM sucrose, 75 mM KCl and varying concentrations of oleate. The reaction was stopped by addition of 1·5 ml 10% trichloroacetic acid and the amount of phosphate liberated was determined in the deproteinized reaction medium. The dotted lines represent the ATPase activity in the absence of dinitrophenol and Mg^{2+} ions, respectively.

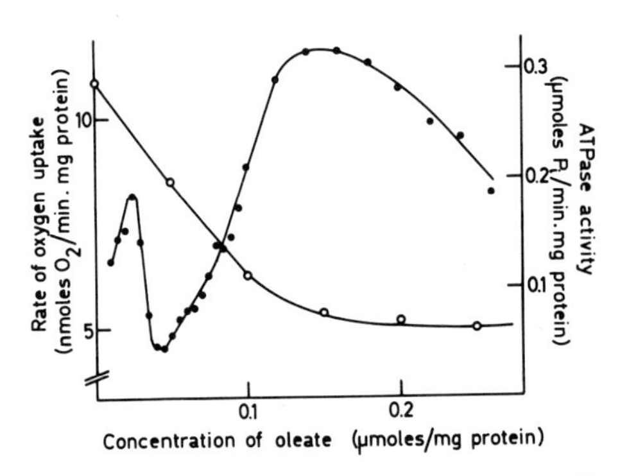

Figure 2. Effects of oleate on its rate of oxidation and on the ATPase activity of rat liver mitochondria. ○——○, ATPase activity, as in Fig. 1. ●——●, Rate of oleate oxidation as measured at 25°C with a vibrating platinum electrode in 2 ml of a medium containing 25 mM potassium phosphate buffer (pH 7·5), 1·5 mM ADP, 50 mM sucrose, 50 mM Tris-HCl buffer (pH 7·5), 15 mM KCl, 2 mM EDTA, 5 mM $MgCl_2$ and varying concentrations of oleate.

ratio" of fatty acid to protein oleate oxidation is inhibited [5]. Unlike other fatty acids, however, oleate oxidation is stimulated again by a further increase in its concentration. This curve with two peaks was found independently by Wojtczak and co-workers [3] and by ourselves [5] and shown by both groups at the 1965 Bari Symposium.

As we have already indicated, oleate oxidation in the second peak is completely dependent on substrate-level phosphorylation. Since this oxidation occurs in the presence of high phosphate concentrations, it cannot be mediated by the GTP-linked thiokinase [8] as we originally thought, but it must be activated by ATP formed in substrate-level phosphorylation, as suggested already by Drahota and Honová [9]. At concentrations of oleate around 50 nmoles per mg protein this substrate-level ATP is obviously not available for fatty acid activation, probably since it is split by the oleate-induced ATPase; addition of oligomycin greatly stimulates oleate oxidation at this concentration of oleate. With increasing oleate concentrations the addition of oligomycin becomes redundant since oleate itself inhibits the ATPase and the substrate-level ATP is now available for oleate activation. In Figure 2 the inhibitory effect of oleate on the dinitrophenol-induced ATPase, shown already in Fig. 1, is plotted again. It can be seen that the concentration of oleate giving maximal inhibition of the ATPase coincides with that giving an optimal rate of oleate oxidation in the second peak. If oleate at a concentration of 200 nmoles per mg protein inhibits the hydrolysis of both externally added ATP and of the ATP formed endogenously in substrate-level phosphorylation, it is clear that this inhibition cannot be exerted at the level of the adenine nucleotide translocase, but that the site of inhibition must be between endogenous ATP and the respiratory chain, i.e. in the same region where oligomycin inhibits energy transfer. This oligomycin-like effect of oleate is most simply explained by a removal of the intramitochondrial magnesium, as suggested already in 1962 by Borst *et al.* [10].

Summarizing, we distinguish four effects of oleate on the energy-conserving system of rat liver mitochondria:

1. An atractyloside-like effect which is optimal at 25 nmoles per mg protein;
2. An uncoupling effect leading to a complete uncoupling of respiratory-chain phosphorylation at 40 nmoles per mg protein;
3. An oligomycin-like effect which is optimal at concentrations around 200 nmoles per mg protein;
4. Even higher concentrations of oleate cause major structural alterations of the mitochondria.

Intramitochondrial localization of acyl-CoA synthetases

Earlier experiments on the effects of carnitine and atractyloside on fatty acid oxidation by isolated mitochondria, summarized so elegantly by Dr Garland earlier in this Symposium, have led us to propose that ATP-linked fatty acid activation occurs at three sites in rat liver mitochondria, separated by the atractyloside- and the carnitine-barriers [11, 12]. Since both these barriers are thought to be located in the mitochondrial inner membrane, the simplest assumption, to our mind, is that ATP-linked thiokinases occur in the matrix, inner membrane, and outer membrane. To obtain direct evidence in favour of this triple location of ATP-linked fatty acid activation, rat liver mitochondria were fractionated by the digitonin method of Schnaitman *et al.* [13] and the ATP-linked thiokinase activity was measured in the various submitochondrial fractions.

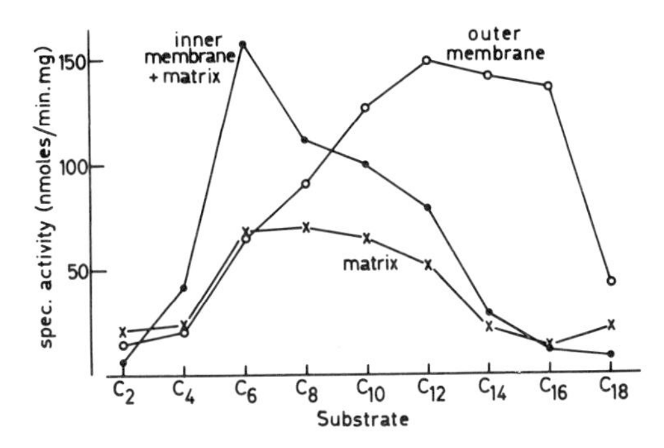

Figure 3. Acyl-CoA synthetase activity in submitochondrial fractions. Rat liver mitochondria were fractionated by the method of Schnaitman *et al.* [13]. Acyl-CoA synthetase activity was assayed at 38°C and at pH 8 by the method of Ellman [14]. Fatty acid substrates were added at 2 mM.

In Figure 3 the results of such an experiment are shown for all even-numbered, saturated, straight-chain fatty acids from acetate through stearate. The outer membrane has the highest specific activity with the longer-chain fatty acids, whereas the inner membrane shows a sharp but very reproducible peak of activity with hexanoate. The matrix has a less pronounced specificity pattern but its activity cannot be explained by contamination with membrane material since this fraction was extremely pure. It contained no measurable cytochrome oxidase (the inner-membrane marker enzyme) and the specific activity of the outer-membrane marker, the rotenone-insensitive NADH-cytochrome *c* reductase, was only 3% of that in the outer membrane fraction. The thiokinase activity in the matrix cannot be explained by the presence of the GTP-linked thiokinase [8] since the observed activity was completely insensitive to fluoride ions.

We therefore regard these data as solid experimental support for our proposal that ATP-linked fatty acid activation occurs at three separate sites in rat liver mitochondria.

ACKNOWLEDGEMENTS

The authors gratefully acknowledge the help and advice of Dr. A. Kemp and the expert technical assistance of Mrs. Y. J. Agterberg.

REFERENCES

1. Bode, C. and Klingenberg, M., *Biochim. biophys. Acta* **84** (1964) 93.
2. Wojtczak, L., Zaluska, H. and Drahota, Z., *Biochim. biophys. Acta* **98** (1965) 8.
3. Wojtczak, L., Drahota, Z., Zaluska, H. and Zborowski, J., *in* "Regulation of Metabolic Processes in Mitochondria" (edited by J. M. Tager, S. Papa, E. Quagliariello, and E. C. Slater), Elsevier, Amsterdam, 1966, p. 134.
4. Van den Bergh, S. G., *Biochim. biophys. Acta* **96** (1965) 517.
5. Van den Bergh, S. G., *in* "Regulation of Metabolic Processes in Mitochondria" (edited by J. M. Tager S. Papa, E. Quagliariello, and E. C. Slater), Elsevier, Amsterdam, 1966, p. 125.
6. Beattie, D. S. and Basford, R. E., *J. biol. Chem.* **241** (1966) 1412.
7. Wojtczak, L. and Zaluska, H., *Biochem. biophys. Res. Commun.* **28** (1967) 76.
8. Galzinga, L., Rossi, C. R., Sartorelli, L. and Gibson, D. M., *J. biol. Chem.* **242** (1967) 2111.
9. Drahota, Z. and Honová, E., *Acta biochim. pol.* **15** (1968) 227.
10. Borst, P., Loos, J. A., Christ, E. J. and Slater, E. C., *Biochim. biophys. Acta* **62** (1962) 509.
11. Van den Bergh, S. G., *in* "Round Table Discussion on Mitochondrial Structure and Compartmentation" (edited by E. Quagliariello, S. Papa, E. C. Slater, and J. M. Tager), Adriatica Editrice, Bari, 1967, p. 400.
12. Van den Bergh, S. G., *Abstr., 4th FEBS Meeting,* Universitets-forlaget, Oslo, 1967, Abstr. 193, p. 49.
13. Schnaitman, C., Erwin, V. G. and Greenawalt, J. W., *J. Cell Biol.* **32** (1967) 719.
14. Ellman, G. L., *Archs Biochem. Biophys.* **82** (1959) 70.

FEBS Symposium, Volume 17, 1969, pp. 145-152

Oxidative Phosphorylation and Compartmentation of Fatty Acid Activation in Mitochondria from Brown Adipose Tissue

K. J. HITTELMAN*, B. CANNON and O. LINDBERG

Wenner-Gren Institute, University of Stockholm, Stockholm, Sweden

Mitochondria isolated from brown adipose tissue by the regular sucrose procedure are not able to carry out oxidative phosphorylation if tested with the normal hexokinase-glucose system [1, 2]. If, however, they are supplemented with massive concentrations of albumin and nucleotides during the incubation, they exhibit more or less normal oxidative phosphorylation and respiratory control [3, 4]. These properties of sucrose-isolated brown fat mitochondria are not a consequence of the high triglyceride content of this particular organ, as other tissues mixed with brown fat during the isolation of mitochondria yield normal mitochondria.

The experiments described here were performed with rat brown fat mitochondria isolated by the usual method employing 0·25M sucrose. The mitochondrial protein concentration was 0·64 mg/ml of reaction mixture in all experiments. The reaction mixture contained *N*-tris(hydroxymethyl) methyl-2-aminoethane sulphonic acid (TES) buffer, pH 7·4, 8 mM; KCl, 50 mM; $MgCl_2$, 6 mM; EDTA, 1 mM; KH_2PO_4, pH 7·4, 14 mM. The total volume was 2·75 ml and the temperature 22°C. Oxygen consumption was determined with a Clark-type O_2 electrode coupled to a sensitive electronic differentiator [5, 6], so that both total O_2 consumption and the rate of O_2 consumption could be monitored. In the figures presented below, the upper of the two traces represents the total O_2 consumption, while the lower represents the O_2 consumption rate.

As first described by Drahota *et al.* [7], brown fat mitochondria supplemented by carnitine and ATP† give a sudden burst of respiration. In the

* U.S.P.H.S. Post-doctoral Fellow

† Abbreviations: ADP, adenosine 5′-diphosphate; ATP, adenosine 5′-triphosphate; Carn, DL-carnitine; CoASH, coenzyme A, reduced; FCCP, carbonyl cyanide *p*-fluoromethoxyphenylhydrazone; FFA, free fatty acid; α-GP, α-glycerophosphate; GTP, guanosine 5′-triphosphate; Hex, hexokinase; α-KG, α-ketoglutarate; P_i, inorganic phosphate.

experiment shown in Fig. 1, ATP was added prior to carnitine. If the order of additions is reversed, the increase of respiration follows upon ATP addition. The *rate* of oxidation of the endogenous substrate is stimulated by added CoASH but the total amount oxidized is not increased. We agree with Drahota *et al.* [7] on the interpretation that the endogenous substrate oxidized in this reaction is some free fatty acid or fatty acids.

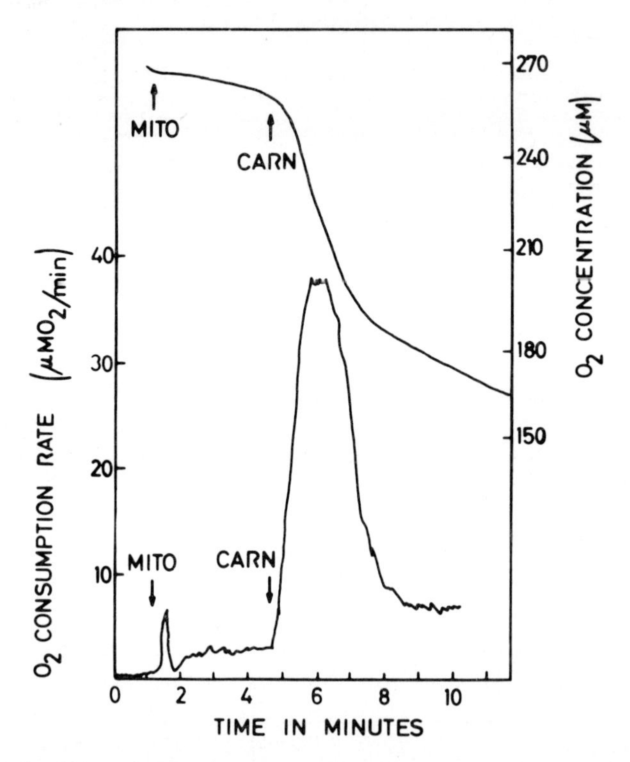

Figure 1. Stimulation of endogenous respiration of brown fat mitochondria by addition of DL-carnitine (3·6 mM). ATP (0·36 mM) and CoASH (5×10^{-6}M) present in the reaction mixture at time of addition of mitochondria.

The well known capacity of free fatty acids to uncouple oxidative phosphorylation raises the question of whether it is these agents that bring about the uncoupling of BAT mitochondria. Figure 2 demonstrates the uncoupled nature of sucrose-isolated brown fat mitochondria. With α-glycerophosphate as substrate, neither oligomycin nor FCCP can influence the respiration. The same holds true for succinate as substrate. If, however, the mitochondria are first treated with ATP and carnitine, and after the evoked respiration has ceased substrate is added, the substrate-induced respiration is inhibited by oligomycin and stimulated by FCCP. This pattern, typical for coupled oxidative phosphorylation, is demonstrated in Fig. 3.

An interruption of the carnitine + ATP-induced respiration can be brought about by the addition of hexokinase + glucose which rapidly consumes the added ATP. If the carnitine + ATP-induced respiration is interrupted in this manner before all the endogenous substrate is oxidized, a less marked effect of the inhibitors is obtained when substrate is added.

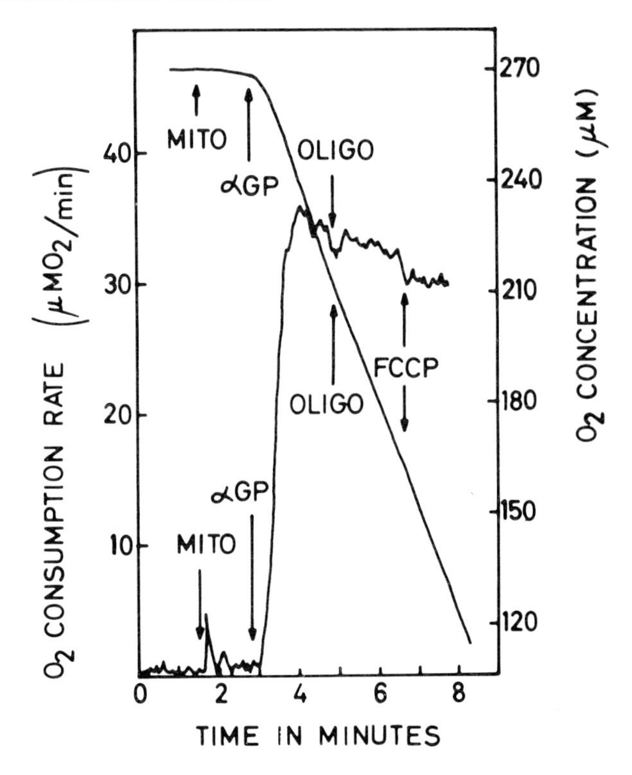

Figure 2. Insensitivity of oxidation of α-glycerophosphate by brown fat mitochondria to oligomycin and FCCP. α-Glycerophosphate, 3·6 mM; oligomycin, 1·8 μg/ml; FCCP, 7·3 x 10^{-8}M; ATP, added before mitochondria, 0·36 mM.

The fractionated oxidation of endogenous substrate can also be achieved by graded additions of ATP. Figure 4 shows a series of experiments in which three different concentrations of ATP are used with α-ketoglutarate as the substrate. Oxidative phosphorylation is followed by measuring the esterification of $^{32}P_i$. It emerges that normal P/O ratios are obtained when the carnitine-induced respiration is completed with an excess of ATP. These experiments have been repeated with the Warburg technique.

It seems justified to state that the endogenous substrate of isolated brown fat mitochondria is free fatty acids and that the presence of these in the mitochondrion is directly or indirectly responsible for the uncoupled state of these isolated particles.

It is well known from experiments in other laboratories that the oxidation of added free fatty acids by mitochondria can be blocked by atractyloside [8, 9]

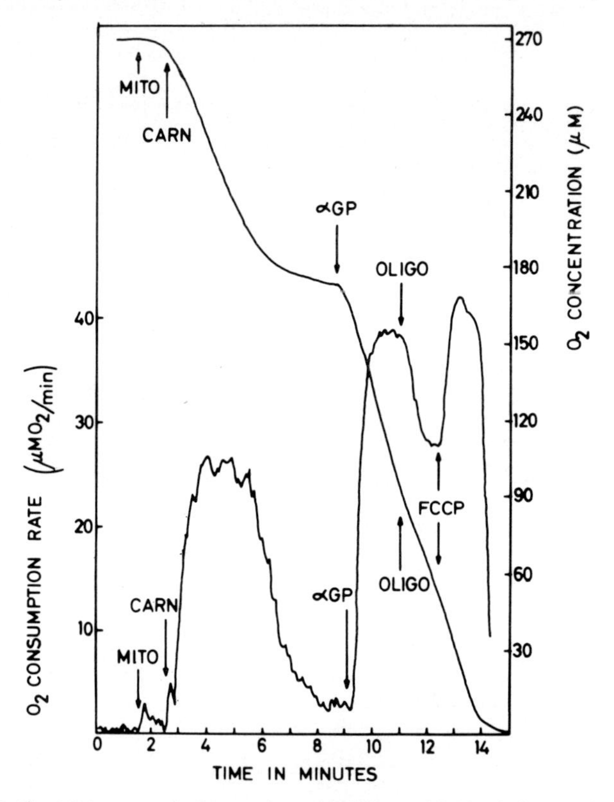

Figure 3. Establishment of oligomycin and FCCP sensitivities in brown fat mitochondria after completion of carnitine-evoked respiration. Reaction mixture contained ATP (0·36 mM) and CoASH (5×10^{-6}M) at time of addition of mitochondria DL-carnitine, 3·6 mM; other components as in Fig. 2.

and that this effect can be overcome by addition of CoASH [9]. Figure 5 shows that both atractyloside inhibition and its release by CoASH are demonstrable in BAT mitochondria. Thus, the compartmentalized activation of exogenous FFA demonstrated in other mitochondrial systems also applies to the activation of the endogenous substrate of brown fat mitochondria.

How this fatty acid is localized, how it is activated, and if it can uncouple oxidative phosphorylation in brown fat mitochondria without being available to the compartment of the mitochondrion where oxidative phosphorylation occurs are also subjects of our interest.

Yates *et al.* [9] suggest that the two compartments in which fatty acid activation takes place, and which are divided by the atractyloside barrier, contain one ATP- and one GTP-requiring system. As emerges from Fig. 6,

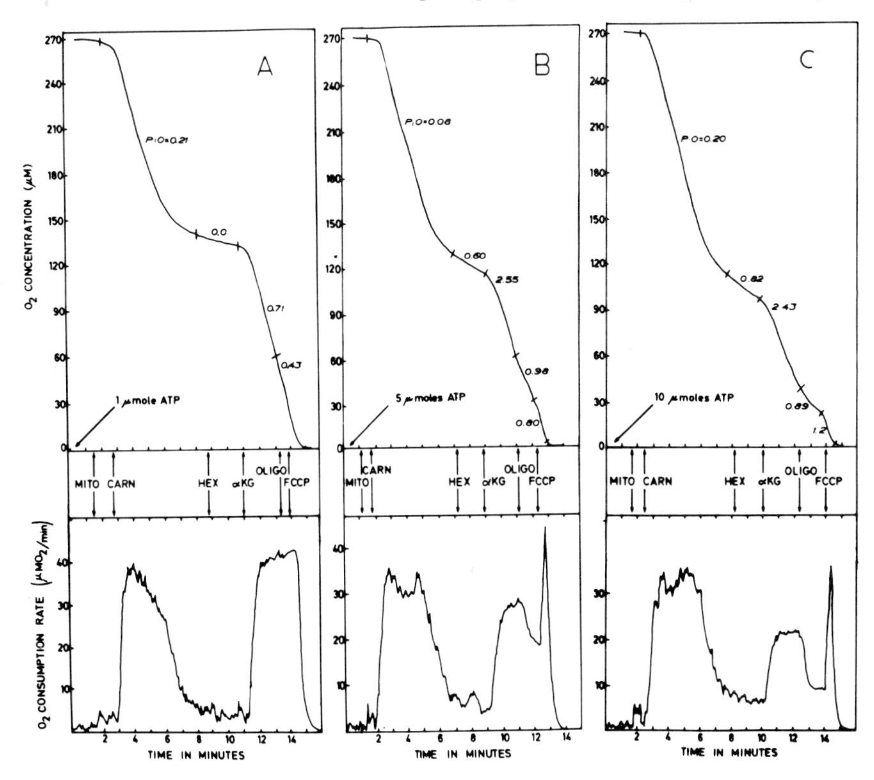

Figure 4. Effects of various concentrations of ATP on the carnitine-evoked respiration and on subsequent P:O ratios and inhibitor sensitivities during α-ketoglutarate oxidation. ATP, in amounts given in the figure, and glucose (17 mM) were present prior to addition of the mitochondria. The figures on the upper curve of each experiment represent the P:O ratios determined over the intervals marked on the curves. α-Ketoglutarate, 3·6 mM; DL-carnitine, 3·6 mM; hexokinase, 14 units; other components as in Fig. 2; the total P_i concentration in these experiments was reduced from 14 mM to 3·5 mM.

however, in our system exogenous GTP is not able to replace ATP as the energy source.

Figure 7 shows an experiment in which brown fat mitochondria are incubated with α-ketoglutarate. No nucleotides are added. The evoked respiration can be inhibited by oligomycin, but this inhibition cannot be released by FCCP. Addition of exogenous ADP increases the respiration. α-Glycerophosphate cannot replace α-ketoglutarate in this respect. This is interpreted in terms of uncoupled electron transport but an obligatorily-coupled substrate-level phosphorylation. Respiration is maintained by the intramitochondrial ATPase which serves to regenerate ADP. This ATPase is blocked by oligomycin. The lack of an FCCP effect indicates the uncoupled nature of the electron transport system. Under the influence of oligomycin the mitochondria may be considered to have

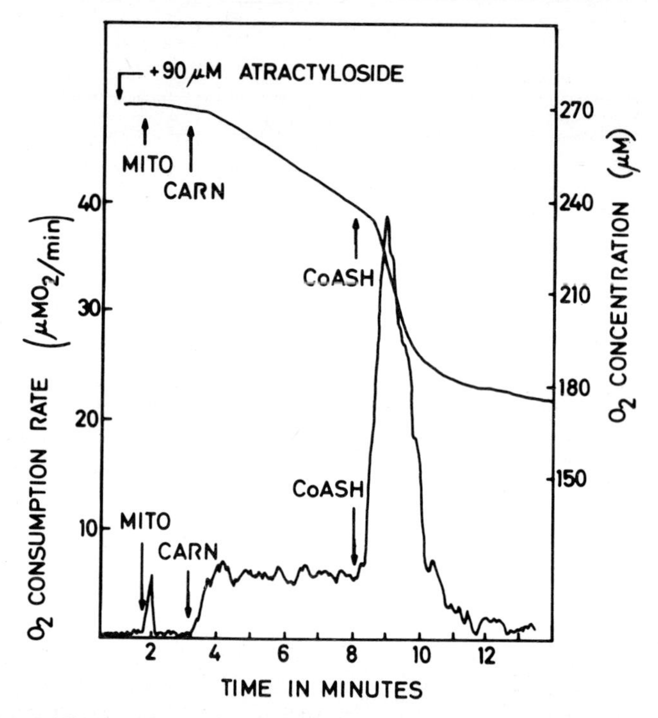

Figure 5. Inhibition of carnitine-evoked respiration by atractyloside (90 μM) and its release by CoASH. ATP (0·36 mM) present prior to addition of mitochondria. DL-carnitine, 3·6 mM; CoASH, 5×10^{-6}M. Compare the O_2 consumption rate after carnitine addition in this experiment with that of Fig. 1.

a high nucleotide triphosphate content, but even so, added carnitine cannot elicit respiration. This experiment demonstrates that internal GTP has no accessibility to the system that can activate the endogenous FFA. Whether or not the ATP formed upon addition of ADP to this system activates fatty acids inside or outside the inner mitochondrial membrane remains in question.

We have suggested earlier that uncoupling plays a role in the mechanism of heat production taking place in the brown adipose tissue. If the phenomena described above are relevant to this problem, the very presence of endogenous free fatty acids in these isolated mitochondria cannot be the whole picture, as

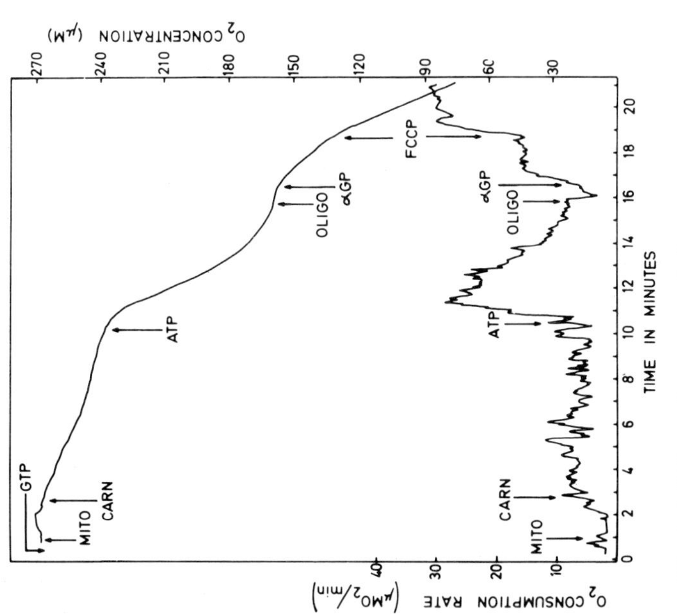

Figure 6. Failure of exogenous GTP to replace ATP as energy source for activation of fatty acids prior to carnitine-evoked respiration. Subsequent addition of ATP, however, results in a burst of endogenous respiration following which the mitochondria show sensitivity to both oligomycin and FCCP. GTP, 5·0 mM; DL-carnitine, 3·6 mM; ATP 5·0 mM; other components as in Fig. 2.

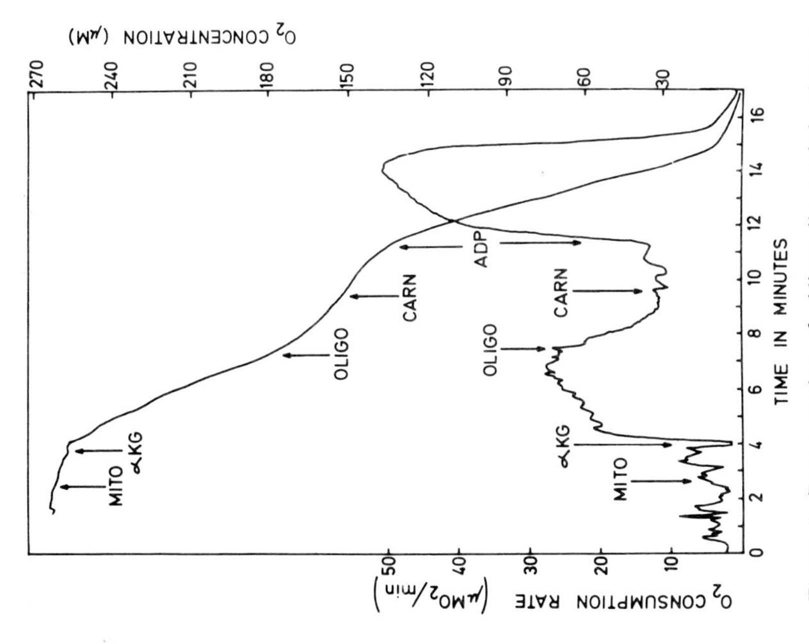

Figure 7. Demonstration of obligatorily-coupled substrate-level phosphorylation and non-participation of endogenous GTP in the fatty acid activating system of brown fat mitochondria. See text for details. α-Ketoglutarate, 3·6 mM; oligomycin, 1·8 μg/ml; DL-carnitine, 3·6 mM; ADP, 5·0 mM.

mitochondria isolated from heart possess the same carnitine + ATP-induced respiration without being uncoupled. That the level of endogenous FFA in isolated brown fat mitochondria, however, may be of some physiological significance is strengthened by the fact that mitochondria isolated from reserpinized rats exhibit a lower level of these fatty acids.

To summarize, we have demonstrated that metabolic depletion of endogenous free fatty acids contained in brown fat mitochondria establishes classical patterns of oxidative phosphorylation. Experiments with atractyloside show that the activation of FFA in brown fat mitochondria may be compartmentalized in a manner similar to that described for other mitochondria. As yet, however, we have not been able to demonstrate a GTP specific fatty acid-activating system, although it is clear that ATP formed from GTP is able to participate in fatty acid activation.

REFERENCES

1. Smith, R. E., Roberts, J. C. and Hittelman, K. J., *Science, N.Y.* **154** (1966) 653.
2. Lindberg, O., de Pierre, J., Rylander, E. and Afzelius, B. A., *J. Cell Biol.* **34** (1967) 293.
3. Joel, C. D., Neaves, W. B. and Rabb, J. M., *Biochem. biophys. Res. Commun.* **29** (1967) 490.
4. Hohorst, J. H. and Rafael, J., *Hoppe-Seyler's Z. physiol. Chem.* **349** (1968) 268.
5. Chappell, C. B. and Crofts, A. R., *Biochem. J.* **95** (1965) 378.
6. Liljesvan, B., personal communication.
7. Drahota, Z., Honová, E. and Hahn, P., *Experientia* **24** (1968) 431.
8. Chappell, C. B. and Crofts, A. R., *Biochem. J.* **95** (1965) 707.
9. Yates, D. W., Shepherd, D. and Garland, P. B., *Nature, Lond.* **209** (1966) 1213.

FEBS Symposium, Volume 17, 1969, pp. 153-160

Fatty Acid Oxidation by Brown Adipose Tissue Mitochondria

Z. DRAHOTA, E. HONOVÁ, P. HAHN, and P. GAZZOTTI

Institute of Physiology, Czechoslovak Academy of Sciences, Prague, Czechoslovakia

In newborn mammals, cold adapted animals, and hibernators, brown adipose tissue has a special function, i.e. thermogenesis [1-3]. It is assumed that the main source for heat formation are fatty acids [4] and that mitochondria in brown adipose tissue are partly or completely uncoupled so that the energy released by substrate oxidation is directly transferred to heat [5, 6]. Brown adipose tissue is well equipped for its thermogenic function. It has high oxidase activity comparable to that of, for example, heart muscle [7] and its cells are filled with mitochondria with a typical arrangement of intramitochondrial cristae [8, 9].

In our experiments we studied the metabolic activity of mitochondria of young 6-30-day-old rats. Also in these newborn rats, just as in cold adapted animals and hibernators, the cells are filled with mitochondria which have the typical structure previously described.

In mitochondria isolated from brown adipose tissue in the usual way in 0·25M sucrose we always obtain uncoupled particles. Phosphorylation coupled to the respiratory chain oxidation may be detected if mitochondria are isolated in the presence of serum albumin [5, 6, 10] or if oxidative phosphorylation is measured in the presence of ATP and carnitine [11]. Uncoupling in mitochondria isolated in the usual way is evidently due to the uncoupling effect of fatty acids. Direct determination of free fatty acids in the isolated brown fat mitochondria revealed that they contain higher amounts of free fatty acids than liver mitochondria [12].

In vivo studies of energy-dependent Ca-ion accumulation in brown adipose tissue and liver mitochondria showed that Ca accumulation is much lower in brown adipose tissue than in liver (not yet published). Experiments with 2,4-dinitrophenol performed with tissue fragments demonstrated that respiration in the presence of glucose and α-ketoglutarate may be activated by 2,4-dinitrophenol, but this is not so in the case of endogenous respiration or in the presence of succinate or α-glycerophosphate. If respiration is activated, for example by nor-epinephrine, 2,4-dinitrophenol has no activating effect on tissue respiration. Thus, *in vivo* the situation in brown adipose tissue is not clear, and we do not yet know exactly whether the mitochondria in the intact tissue are coupled or uncoupled.

Since respiration activated by nor-epinephrine is always accompanied by a high rate of lipolysis [13] it may be expected that in this situation mitochondria are uncoupled. Hence we assumed that, when the respiratory rate in brown adipose tissue is maximal, free fatty acid oxidation must occur at maximal rates even when energy generated by respiratory-chain oxidation cannot be used as a sufficient source for free fatty acid activation. We therefore studied fatty acid oxidation in uncoupled mitochondria, i.e. in mitochondria isolated from brown adipose tissue in the usual way without serum albumin.

Although it is assumed that fatty acids are the main energy source and an important respiratory substrate in brown adipose tissue, their oxidation in isolated mitochondria has not been studied in detail.

First of all we tested how far our mitochondrial preparations are coupled and whether they can perform energy-dependent reactions such as Ca- or valinomycin-induced K-ion accumulation. In contrast to liver mitochondria, in brown fat mitochondria addition of Ca-ions did not activate respiration or H-ion ejection. Energy-dependent valinomycin-induced K-ion accumulation can also not be demonstrated in brown adipose tissue mitochondria. These mitochondria, when compared with liver mitochondria, do not maintain H- and K-ion gradients between extra- and intramitochondrial solutions and their potassium content is 10 times lower than that of liver mitochondria.

It was shown in our previous experiments [12] that the endogenous respiration of mitochondria, isolated in the usual way from brown adipose tissue, may be activated by ATP and carnitine if incubated in simple saline media with malate. When free fatty acids are added in the absence of ATP and carnitine, no stimulation of endogenous respiration occurs. Fatty acids have also no activating effect if respiration is activated by ATP and carnitine [12]. No other substrate such as glutamate, α-ketoglutarate, pyruvate and malate, succinate or α-glycerophosphate raise the respiratory rate as much as ATP and carnitine. Values of the respiratory rate are mostly much lower. The endogenous fatty acid content is apparently sufficient to ensure maximum respiratory rate.

Respiration activated by ATP and carnitine is decreased by bovine serum albumin. Addition of palmitate under the experimental conditions employed activates respiration, but values of ATP-carnitine-activated respiration in the absence of albumin can never be obtained.

In the absence of serum albumin and in the presence of ATP and carnitine, oxidation of added fatty acid cannot be detected by means of an oxygen electrode. The exogenous fatty acids evidently enter the fatty acid pool, and even if they are oxidized no activation of the respiratory rate can be detected. We therefore used [C^{14}]palmitate and measured the activity of evolved CO_2 as affected by the addition of ATP and/or carnitine. We found that production of labelled CO_2 from palmitate is 3 times higher in the presence of ATP and carnitine than in their absence, indicating that exogenous fatty acids may be oxidized by brown adipose tissue mitochondria through an ATP-carnitine-dependent pathway (Fig. 1).

Optimum conditions necessary for the maximum rate of ATP-carnitine-activated respiration were tested. It was shown that:

1. 0·5-2 mM carnitine and 0·5-4 mM ATP were necessary (Fig. 2);
2. in the absence of inorganic phosphate the respiratory rate was reduced by 50% (Fig. 3);
3. addition of EDTA completely inhibited ATP-carnitine-activated respiration, and sufficient amounts of Mg-ion returned respiration to the original rate (Fig. 4);
4. an intact structure was necessary, and addition of the non-ionic detergent Lubrol completely inhibited ATP-carnitine-activated respiration (Fig. 5).

The oxidation of endogenous fatty acids by uncoupled brown fat mitochondria requires carnitine and exogenous ATP. This indicates that both exogenous and endogenous fatty acids are activated at the mitochondrial surface, i.e. outside the carnitine barrier, and that fatty acids cannot be activated in the intramitochondrial space to which fatty acids have free access.

One of the reasons for this is the fact that in contrast to liver mitochondria, atractyloside-sensitive transport of nucleotides [14] through the brown fat mitochondrial membrane is minimal. In liver mitochondria isolated from 15-day-old rats atractyloside-sensitive transport of [^{14}C]ATP is inhibited by oleate and this inhibition is ineffective in the presence of albumin. In brown adipose tissue, mitochondrial atractyloside-sensitive transport of ATP cannot be demonstrated even in the presence of albumin. This might explain the fact that even in the presence of exogenous ATP, fatty acids cannot be oxidized (activated) behind the carnitine barrier.

A further source for intramitochondrial activation of fatty acids might be substrate level phosphorylation. Experiments described up to now were performed in the presence of 14 mM phosphate which inhibits GTP-dependent fatty acid activation [15]. Hence we compared oxidation of [^{14}C]palmitate in a medium with 1 mM phosphate, in the presence of 2,4-dinitrophenol in liver and brown adipose tissue mitochondria. In the presence of ATP and carnitine, the amount of palmitate oxidized is about the same in both mitochondrial preparations. If, however, only carnitine and α-ketoglutarate or only the sparker are present, then the oxidation of fatty acid in brown adipose tissue mitochondria is much lower (Fig. 6).

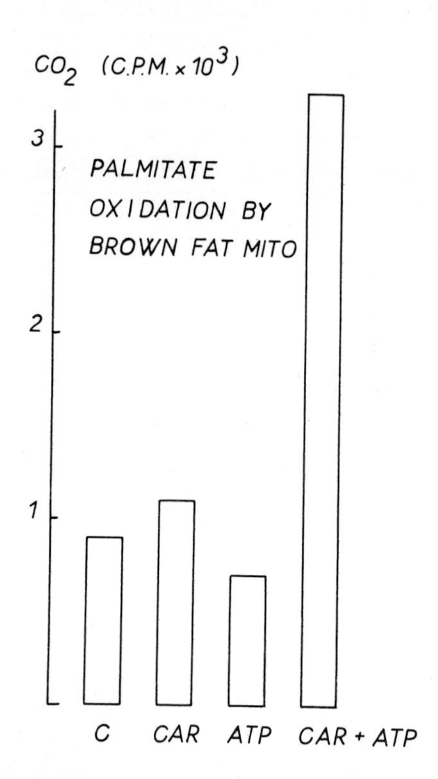

Figure 1. Palmitate oxidation by brown adipose tissue mitochondria from 15-day-old rats. The system contained in 2·2 ml: 5·0 mg of mitochondrial protein, KCl, 118 mM; Tris-Cl pH 7·4, 9 mM; $MgCl_2$, 2·7 mM; phosphate pH 7·4, 0·9 mM; sucrose, 23 mM; malate, 4·5 mM; 2,4-dinitrophenol, 0·1 mM; palmitate (1-C-14), 0·23 mM. Where indicated adenosine triphosphate (ATP) 4·5 mM, and carnitine (CAR) 4·5 mM, were added. C means control. Mitochondria incubated for 30 min at 22°C. Results expressed as c.p.m. x 10^3/mg protein in CO_2.

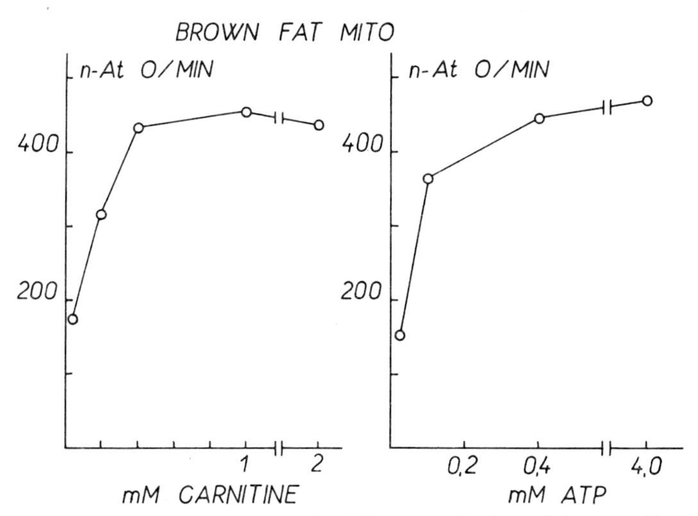

Figure 2. The effect of carnitine and ATP on respiration of brown adipose tissue mitochondria from 10-day-old rats. Respiration measured by Clark oxygen electrode at 30°C. The system contained in 2 ml: 1·5 mg of mitochondrial protein, KCl, 80 mM; Tris-Cl pH 7·4, 20 mM; $MgCl_2$, 6 mM; phosphate pH 7·4, 14 mM; K-malate, 3·3 mM; sucrose, 12·5 mM. Results expressed as nano-atoms of oxygen/min.

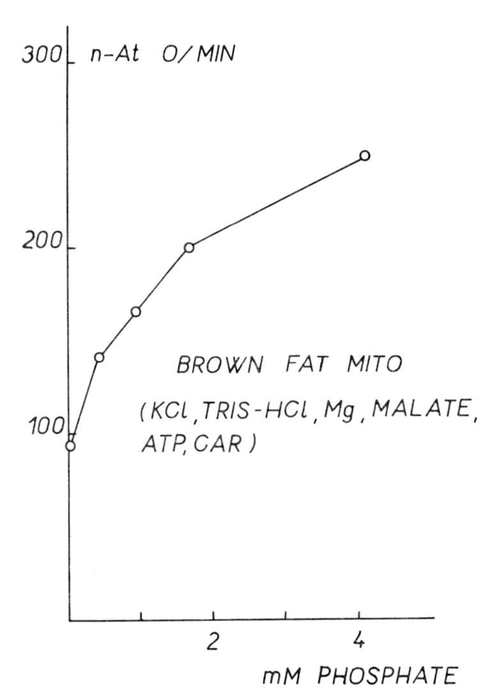

Figure 3. The effect of phosphate concentration on respiration of brown adipose tissue mitochondria from 10-day-old rats. Experimental conditions as in Fig. 2. ATP, 2 mM, carnitine, 1 mM.

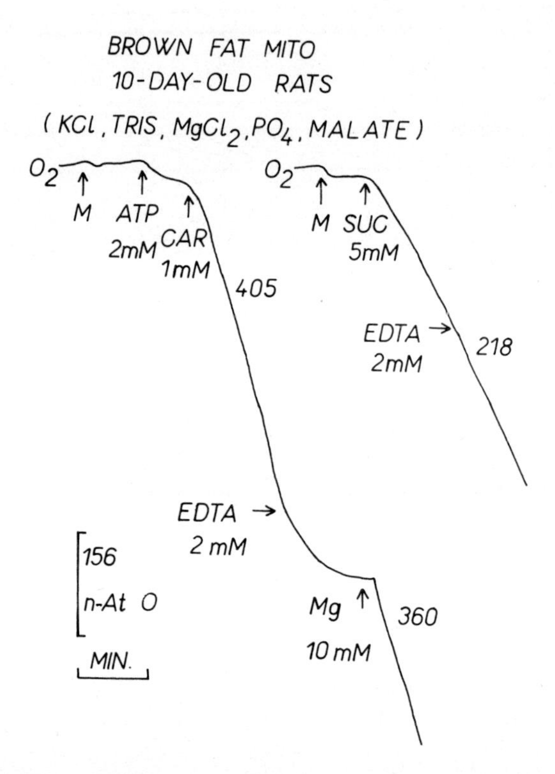

Figure 4. The effect of EDTA on the respiration of brown adipose tissue mitochondria from 10-day-old rats. Experimental conditions as in Fig. 2. $MgCl_2$ was 2 mM. Mitochondria (M), adenosine triphosphate (ATP), carnitine (CAR), ethylene diamine tetracetate (EDTA), $MgCl_2$ (Mg) and succinate (SUC) added where indicated. The numerals indicate respiration expressed as nano-atoms oxygen/min.

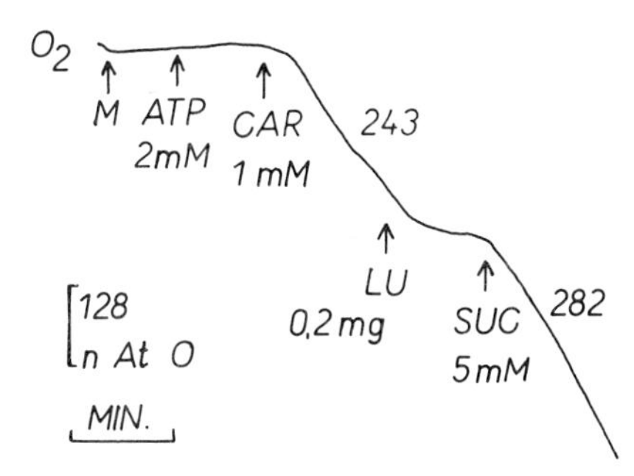

Figure 5. The effect of Lubrol on respiration of brown adipose tissue mitochondria from 10-day-old rats. Experimental conditions as in Fig. 2. Malate was not used in this system. Mitochondria (M), adenosine triphosphate (ATP), carnitine (CAR), detergent Lubrol (LU) and succinate (SUC) added where indicated. The numerals indicate respiration expressed as nano-atoms of oxygen/min.

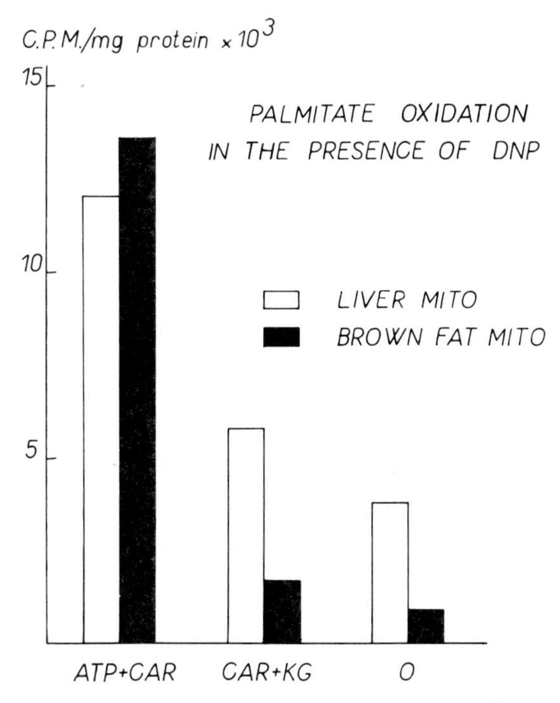

Figure 6. Oxidation of palmitate by liver (white columns) and brown adipose tissue (black columns) mitochondria from 15-day-old rats. Experimental conditions as in Fig. 1. Where indicated adenosine triphosphate (ATP), 4·5 mM, carnitine (CAR), 4·5 mM, and α-ketoglutarate (KG), 4·5 mM, were added. 0 means no additions. Results expressed as c.p.m. x 10^3/mg protein in CO^2 and trichloracetic acid soluble fraction.

All these data suggest that in brown adipose tissue ATP formed by glycolysis is probably the main energy source for activation of fatty acids under conditions of enhanced lipolysis (conditions of maximum respiratory activity). It would appear that fatty acids released by lipolysis are adsorbed onto the mitochondria–the affinity of mitochondria from brown adipose tissue for fatty acids is 3 times greater than that of liver mitochondria [16]–and in such a situation the uncoupling of mitochondria by fatty acids would ensure maximum respiratory rate and direct transformation of energy released by oxidation to heat, and thus the main physiological role of brown adipose tissue, the thermogenesis, would be ensured.

REFERENCES

1. Smith, R. E. and Hock, R. J., *Science, N.Y.* **140** (1963) 199.
2. Ball, E. G., *Ann. N.Y. Acad. Sci.* **131** (1966) 225.
3. Dawkins, M. J. R. and Hull, D., *J. Physiol., Lond.* **172** (1964) 216.
4. Lindberg, O., de Pierre, J., Rylander, E. and Afzelius, B. A., *J. Cell Biol.* **34** (1967) 293
5. Hohorst, J. H. and Stratmann D., *in* "Proceedings of the 4th meeting of the Federation of European Biochemical Societies, Oslo, 1967", p. 109.
6. Guillory, R. J. and Racker, E., *Biochim. biophys. Acta* **153** (1968) 490.
7. Joel, C. D., *in* "Handbook of Physiology, Section 5, Adipose Tissue" (edited by A. E. Renold and G. F. Cahil) American Physiological Society, Washington, 1965, p. 59.
8. Napolitano, L., *in* "Handbook of Physiology, Section 5, Adipose Tissue" (edited by A. E. Renold and C. F. Cahil) American Physiological Society, Washington, 1965, p. 109.
9. Fein, J., Reed, N. and Saperstein, R., *J. biol. Chem.* **242** (1967) 1887.
10. Alridge W. N. and Street B. W., *Biochem. J.* **107** (1968) 315.
11. Lindberg, O., Hittelman, K. J. and Cannon, B., this book, p. 145.
12. Drahota, Z., Honová, E. and Hahn, P., *Experientia* **24** (1968) 431.
13. Joel, C. D., *J. biol. Chem.* **241** (1966) 814.
14. Wojtczak, L. and Zaluska, H., *Biochim. biophys. Res. Commun.* **28** (1967) 76.
15. Van den Bergh, S. G., *Biochim. biophys. Acta* **98** (1965) 442.
16. Erazímová, J. and Drahota, Z., *Physiologia bohemoslov.* (1969) in press.

FEBS Symposium, Volume 17, 1969, pp. 161-177

Amino Acid Incorporation by Isolated Mitochondria: Automated Characterization of Yeast Respiratory Particles Labelled *in vitro*

D. B. ROODYN and L. A. GRIVELL

Department of Biochemistry
University College, London, England

For some reason, research on mitochondria has always engendered a great deal of heat. We might say that it sometimes becomes uncoupled by emotion! The field of research into amino acid incorporation by isolated mitochondria is quite typical in this respect, and in the ten years since the pioneer paper of McLean *et al.* [1] the topic has suffered from two major sources of confusion. The first was the claim that cytochrome *c* is synthesized *in vitro* by isolated mitochondria from rat liver and calf heart [2, 3]. However, when this claim was examined in more detail, it was found that there was, in fact, negligible incorporation into cytochrome *c in vitro* [4, 5]. Since we also found in our laboratory that there was negligible radioactivity in malate dehydrogenase (and, incidentally, catalase which subsequently was shown to be a peroxisomal enzyme), and could also find no evidence for labelling of any other soluble proteins, the potency of the mitochondrial system was put in some doubt. Fortunately, much evidence has accumulated since those now distant days to indicate that we should not expect labelling of soluble proteins, because most, if not all, are synthesized in non-mitochondrial cytoplasmic sites. The evidence for this view is discussed more fully in a recent monograph [6]. It is therefore perhaps rather fortunate that we did not expend too much energy trying to "improve" the counts in soluble proteins.

The second confusing element is the spectre of bacterial contamination, made much more sinister by the many similarities between bacterial and mitochondrial protein synthesis [6]. Unfortunately, it is not easy to obtain rat liver mitochondria completely free of bacteria. From the earliest days of the problem we observed a variable number of variable organisms in our preparations, ranging from about 1000 to 50,000 organisms per ml of reaction mixture. However, there was absolutely no correlation between the incorporation activity and the number of bacteria [7]. For example, a preparation with 10^3 micro-organisms was found to be more active than one isolated soon after with 5×10^4. Lamborg and Zamecnik [8] had shown that at least 10^5 *E. coli* cells per ml were needed to obtain measurable counts, even with exponentially growing cells, fully adapted to the medium. Although they are often quoted to the contrary, it can be argued that experiments in which artificial mixtures of bacteria and inactive mitochondria are incubated with radioactive amino acids [9, 10] confirm that at least 10^5 cells are needed before there is significant interference in the mitochondrial assay, and even then the artificial system has a completely different progress curve to the usual incorporation system [10].

The lack of interference by bacterial contamination in our studies was confirmed over the years by the elucidation of a large number of properties of the system which were incompatible with the view that bacteria could be contributing significantly [11]. To summarize them briefly: The incorporation is energy dependent, has a plateau-shaped progress curve, is specific for certain oxidizable substrates, is negligible in the presence of glucose but absence of succinate, requires added Mg^{2+} ions, adenine nucleotides, P_i, is sensitive to osmotic shock, mechanical damage, lysis by low concentrations of detergent, and incubation for short periods of time in the absence of substrate, is affected by the physiological and hormonal state of the animal, shows no correlation with bacterial counts, and follows mitochondria precisely on sucrose gradients.

Fortunately there are now a healthy number of reports of incorporation by virtually sterile mitochondria (see ref. 6 for a summary of these) and indeed with recent discoveries of the presence of ribosomes, DNA, RNA polymerase, DNA polymerase, tRNA and activating enzymes in mitochondria (again, see ref. 6 for a summary) it would have been somewhat of a disaster if the germ-free mitochondria had been totally inactive! The results below with yeast mitochondria fully confirm that mitochondria isolated under axenic conditions are capable of incorporating amino acids into protein. The general properties of the amino acid incorporation system as they are now known have been summarized recently [6] and are given in Fig. 1. It can be seen that the major incorporation *in vitro* appears to be into insoluble proteins in the inner mitochondrial membrane. Although this diagram looks rather complex, in fact it reveals a large number of questions about the actual mechanism of the incorporation process, and its relation to general mitochondrial function. In addition, the diagram only shows the mitochondrion at one point of time. It is now abundantly clear that the chondriome may undergo dramatic changes during the cell cycle, and it is most dangerous to regard it as a static structure.

For these reasons, and to be free once and for all of the anxieties of bacterial contamination, we decided to work with the amino acid incorporation system of yeast mitochondria [12]. The system is under active study at the present moment, and in this paper we present as an interim report some experiments on the characterization of the incorporating particles by some recently developed techniques of enzyme automation.

The method of fractionation is given in Fig. 2, and is a modification of the techniques of Ohnishi *et al.* [13] and Duell *et al.* [14]. One modification in the scheme we have used is to layer the crude mitochondrial pellet over 1·5M sucrose, followed by a high speed centrifugation. The mitochondria form a layer over the dense sucrose, and are found to be free of any contaminating residual whole cells or protoplasts (it is also likely that this modification is effective in

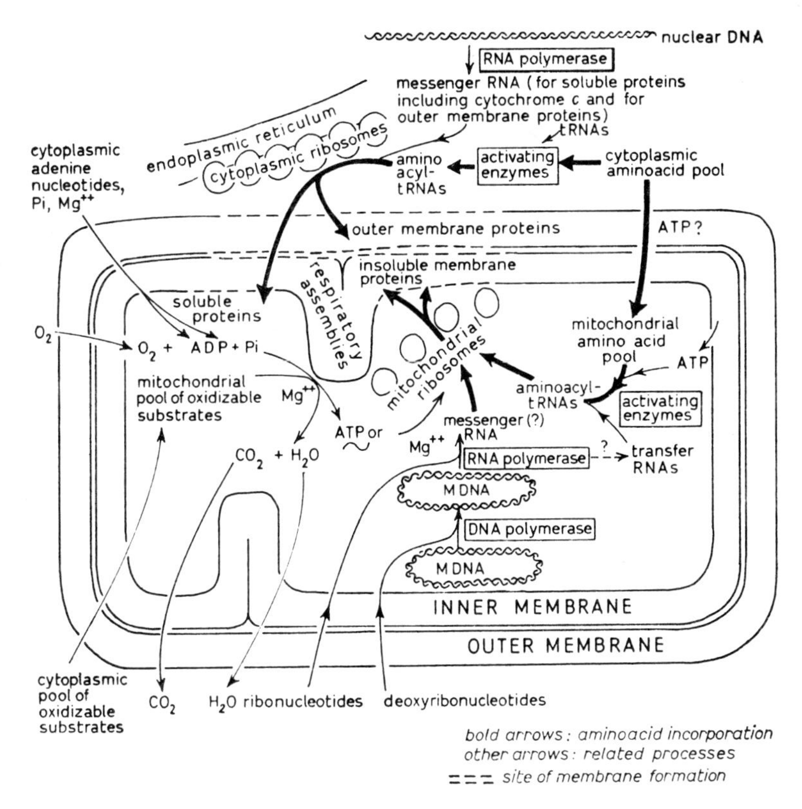

Figure 1. Probable pathways for synthesis of mitochondrial proteins (from ref. 6).

removing contaminant bacteria). In order to reduce the number of experimental variables we have used as simple an incorporation medium as is consistent with reasonable activity [15]. The medium is given in Table 1, and we have used the general techniques of incubation and radioactive assay described previously [16]. Active preparations were obtained that contained negligible numbers of bacteria or yeast cells. The results are given in Table 2 (see ref. 15). The incorporation is dependent on the presence of ADP or ATP, an oxidizable substrate and Mg^{2+} ions. It is greatly inhibited by chloramphenicol, but is not

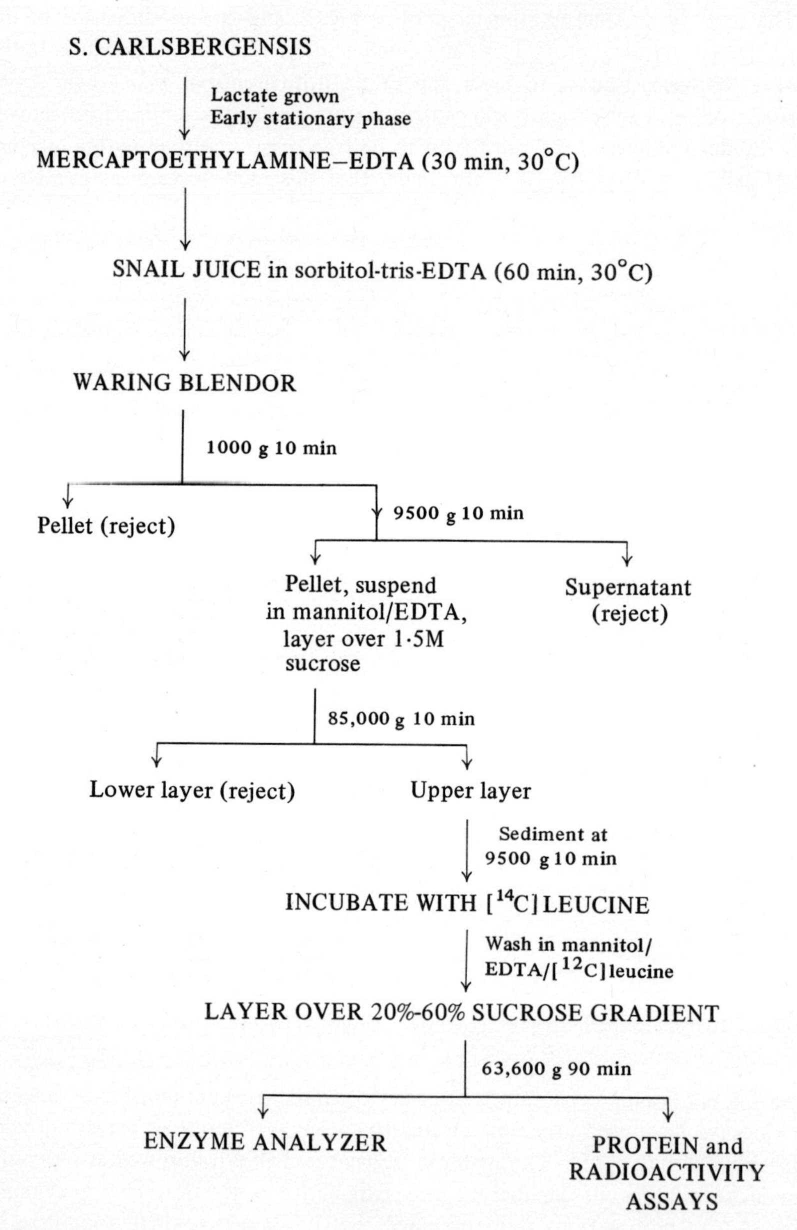

Figure 2. Scheme for isolation and analysis of yeast mitochondria (for details see ref. 15).

Table 1. Simplified amino acid incorporation medium for yeast mitochondria

sucrose	100 mM
KCl	100 mM
$MgCl_2$	10 mM
KH_2PO_4	10 mM
K succinate	10 mM
ADP or ATP	2 mM
[^{14}C] amino acid	
pH 6·7 with tris base	
30° air 1·5 mm fluid depth	
80 cycles/min	0·1 μc/ml

Table 2. Amino acid incorporation by yeast mitochondria

Radioactive amino acid	Expt. no.	Addition	c.p.m./mg	Colonies/plate bacteria	Colonies/plate yeast
U-[^{14}C] leucine	1	–	1351	0	0
(165 mC/mmole)	2	–	1605	0	0
	3	–	752	5	10
		CAP (10 μg/ml)	96	5	10
		CAP (50 μg/ml)	64	5	10
		RNAse (40 μg/ml)	884	5	10
1-[^{14}C] leucine	1	–	1068	0	0
(34·2 mC/mmole)	2	–	2915	0	0
	3	–	1934	1	0
	4	–	695	1	0
	5	–	936	1	1
	6	–	2117	1	0
		–	2548	1	0
		–	2410	2	0
		–	2265	2	0
	7	–	1817	3	0
	8	–	973	7	2
	9	–	2355	4	2
	10	–	778	6	0
		RNAse (50 μg/ml)	706	6	0

Incubation conditions as in Table 1; 0·1 ml reaction mixture plated; 0·6-2·0 mg protein/ml; 30 min incubation. Data from ref. 15. CAP: chloramphenicol.

affected by ribonuclease (Fig. 3). Thus, in the properties so far examined, it closely resembles the rat liver system [6].

Although there has been increased interest in yeast mitochondria in recent years, there is still much to be done in the characterization of these particles. Over the last few years, we have been developing automated systems for the study of yeast enzymes, and it was therefore of interest to apply these to the study of yeast mitochondria. In the technique of "multiple enzyme analysis" [17, 18], groups of enzymes are estimated in the same homogenate or sub-cellular fraction, using some common factor in the assay (e.g. release of P_i, change in E_{340} for NAD(P)-linked enzymes or fall in E_{420} for ferricyanide reductases). In order to obtain an automatic measure of the enzyme activity at

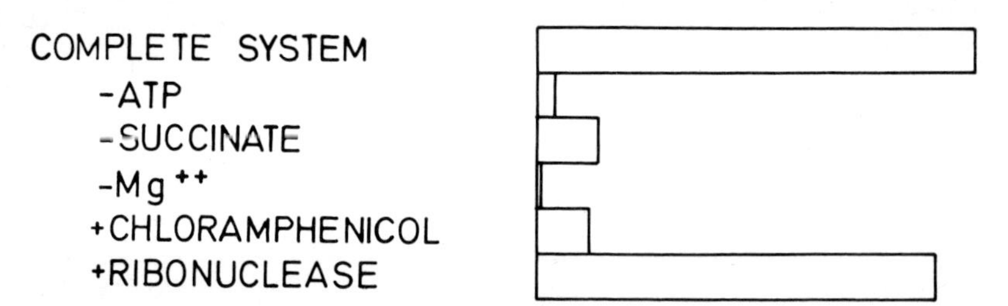

Figure 3. Properties of mitochondrial amino acid incorporation system from *S. carlsbergensis.* The incorporation system in Table 1 was used with 30 min incubation. Chloramphenicol (50 μg/ml) and ribonuclease (50 μg/ml).

various protein concentrations, stock enzyme is pumped into a suitable diluent and thence into the analyzer, so as to produce an "enzyme gradient". (This system is described more fully in refs. 19, 20.) A system for performing this type of assay, with two times of incubation, is given in Fig. 4, and was applied to the study of several ferricyanide (FC) reductases*, which have been used as conveniently cheap model assay systems. The results of such an experiment are given in Fig. 5. It can be seen that there is measurable NADPH-, succinate- and D(-) lactate-FC reductase activity in the mitochondrial fraction, but the two most active enzymes are the L(+)lactate- and NADH-FC reductases.

The advantage of the technique of multiple enzyme analysis is that it is easy to obtain enzyme patterns or "profiles" (cf. Ref. 21). We had done this previously for a number of strains of yeast, and the results for the respiratory competent strains [22] are compared with the present values for yeast mitochondria in Fig. 6. It can be seen that the mitochondrial profile shows relative enrichment of the succinate-FC reductase and a lower ratio of NADH- to L(+)lactate-FC reductase. In terms of activity per mg protein, all the mitochondrial activities are higher, except for the D(-)lactate-FC reductase.

* abbreviation: FC = K ferricyanide

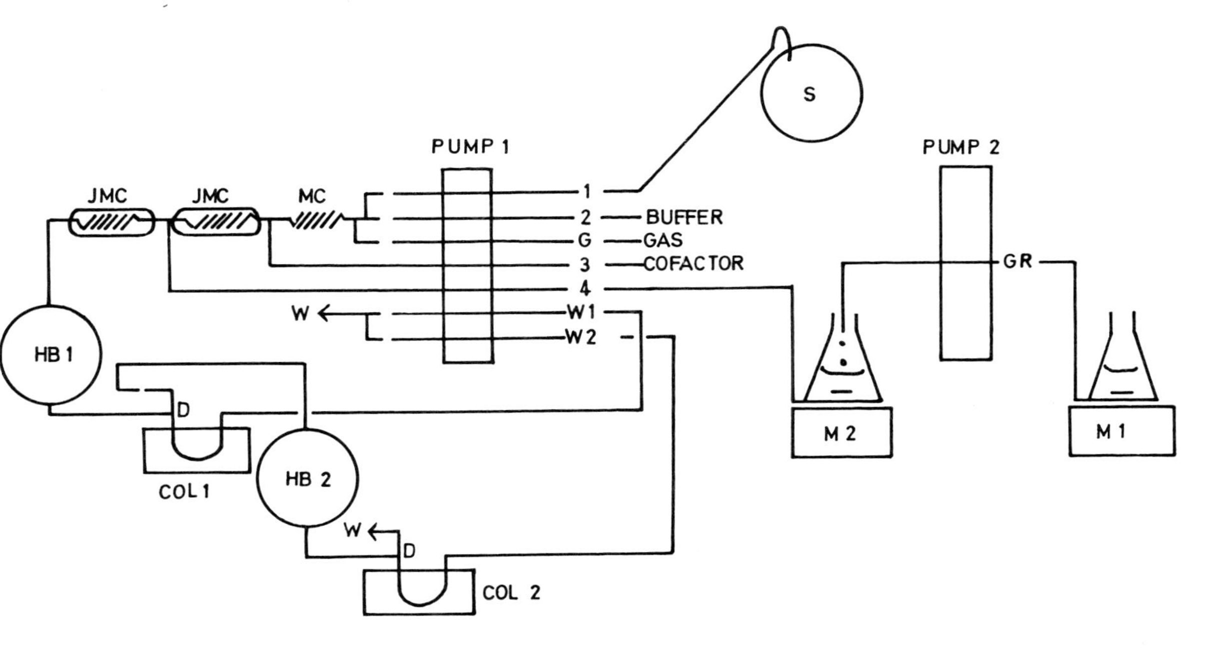

Figure 4. Autoanalyzer with "enzyme gradient" and two incubation times. COL:colorimeter, D:debubbler, GR:gradient pump tube, HB:heating bath, JMC:jacketed mixing coil, M:magnetic stirrer, MC:mixing coil, S:sampler, W:waste. Pumping rates for lines 1,2, G, 3, 4, Wl, W2 and GR were, respectively, 1·0, 3·4, 1·6, 1·2, 1·2, 2·0, 2·9 and 0·6 ml/min.

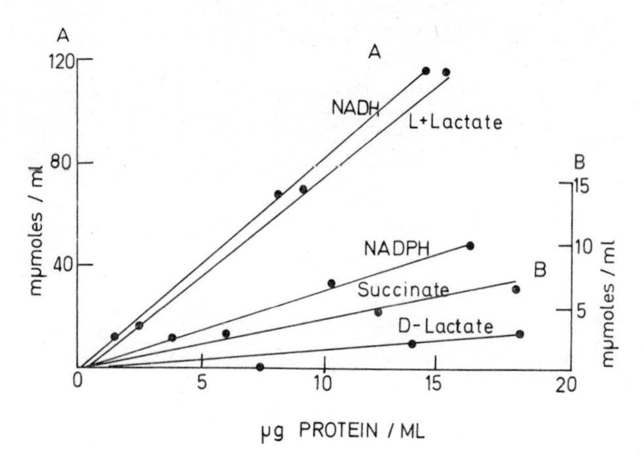

Figure 5. Ferricyanide reductase activities in yeast mitochondrial fraction. System in Fig. 4 used with an "enzyme gradient" of mitochondrial protein. Sampler contained three repetitive cycles of five substrates, 37°C, O_2-free nitrogen, 40 samples/h with water wash between samples. Final concentrations in reaction mixture: (pH 7·4) 25 mM K phosphate, 0·25 mM $K_3Fe(CN)_6$, 0·11M mannitol; substrate concentrations: 1·4 mM succinate, D(-)lactate or L(+)lactate, or 0·3 mM NADH or NADPH. Results given for incubation time of 8·6 min. (Scale A for NADH-, L(+)lactate-FC reductases, scale B for the other enzymes.)

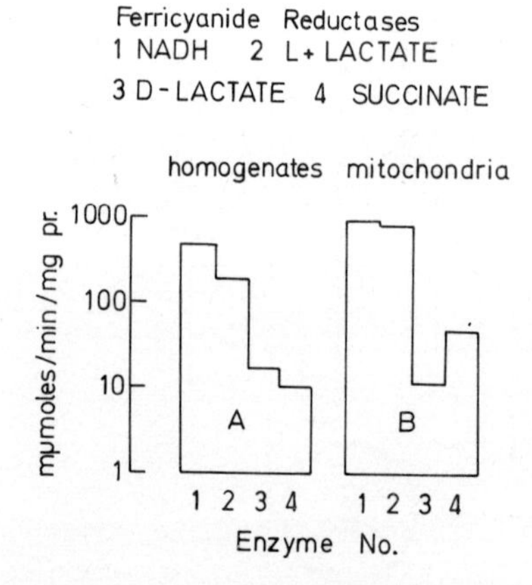

Figure 6. Enzyme pattern for respiratory competent yeasts, and yeast mitochondria. A:average of results with 8 strains of respiratory competent yeast (from ref. 22), B:average of results with 3 mitochondrial fractions.

Table 3. FORTRAN V program for enzyme analysis (I)

INTRODUCTION
Textual comments.
MOLAR CONVERSION FACTOR CALCULATIONS
Deviations from Lambert-Beer Law; interpolation with non-linear calibration curves.
INCUBATION TIME AND CHART SPEED CALCULATIONS
Calculation of retention times, chart speeds.
LINE VOLUME CALCULATIONS
Calculation of flow rates for each reagent. Warning of deviation from linearity.
LINE REAGENT AND REACTION MIXTURE CALCULATION
Calculation of final concentration of each reagent. Required line reagents and dilution of stock reagents.
STOCK SOLUTIONS
Calculation of weights of solid.
Preparation of weighing sheets.
GRADIENT CALIBRATION
Calculation of gradient dilution factors.
Graph of gradient dilution, Factors against reading number.

Table 4. FORTRAN V program for enzyme analysis (II)

ASSAY CONDITIONS
Textual comments.
PROTEIN CONCENTRATIONS
Calculation, tabulation and graph of protein concentration in reaction mixture: CONSTANT, VARIABLE or GRADIENT options.
SAMPLE PATTERN
Calculation of substrate concentrations.
Tabulation of substrates with initial and final concentrations.
RESULTS
Calculation, tabulation and graph of μmoles/ml, μmoles/ml/min and μmoles/mg. protein/min. NO PROTEIN, CONSTANT, VARIABLE or GRADIENT options.
Calculation of c/c_{av} in sucrose gradients. Checks that results are within calibration range. Textual comments.
PLOT
Graph of supplied values of X against Y with textual comments.
COMMENT
Listing of input data, references or other textual comment.

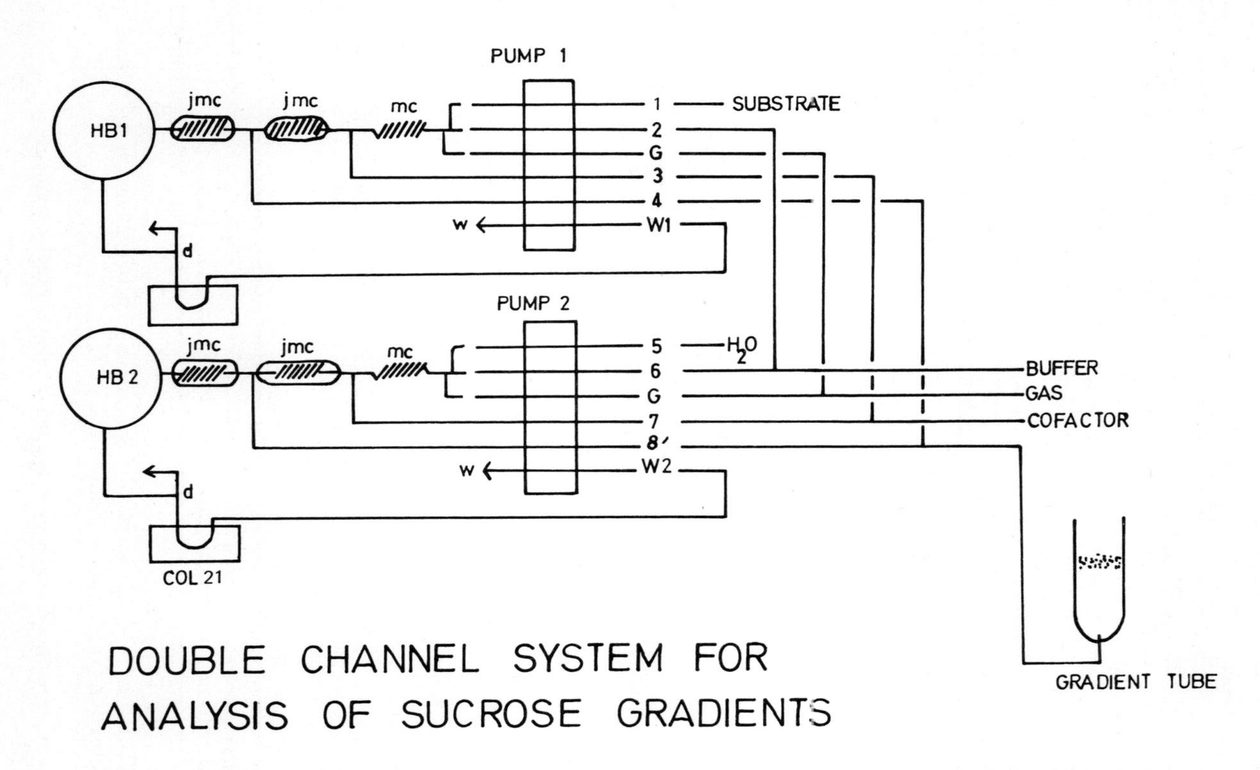

Figure 7. Autoanalyzer system for analysis of sucrose gradients. Symbols as in Fig. 4. Pumping rates for lines 1, 2, G, 3, 4, W1, 5, 6, G, 7, 8 and W2 were respectively, 0·6, 2·0, 1·2, 0·6, 0·6, 2·5, 0·6, 2·0, 1·2, 0·6, 0·6 and 2·5 ml/min.

One of the results of using automated assay procedures is a great increase in the amount of data that requires processing. Various computer programs have been written in our laboratory to overcome this problem [23] and the program currently in use is briefly summarized in Tables 3 and 4. Table 3 describes various calibrations and calculations that are common to a series of experiments, and Table 4 shows the computations that may be performed on individual experiments in the series. At the moment our results are recorded on pen recorders and then punched manually on IBM cards, for batch processing by the ATLAS computer. However, it is clear that the system would be greatly improved by on-line computing.

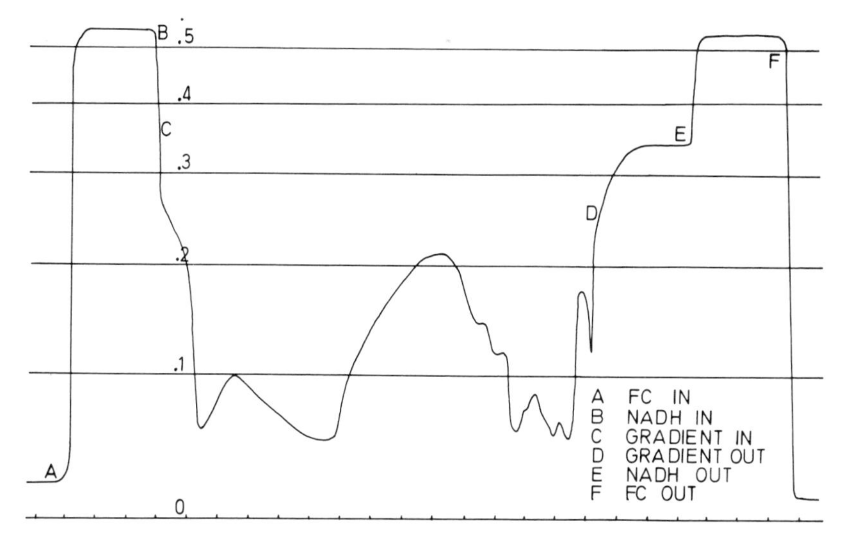

Figure 8. Actual trace in analysis of NADH-FC reductase activity in sucrose gradient. Assay at 30°C, under O_2-free nitrogen. Final concentrations in reaction mixture: 25 mM K phosphate, 0·4 mM $K_3Fe(CN)_6$, 0·05% Triton-X-100, 0·36 mM NADH, pH 7·4. Incubation time 6·4 min. Yeast mitochondria prepared as in Fig. 2, with no incubation with radioactive amino acid. Optical density given by horizontal lines.

Having detected certain enzymes in the mitochondrial fraction, we then analyzed these fractions on sucrose gradients. In the first series of experiments we used a dual channel analyzer, and pumped the contents of the gradient directly into the system (cf. Ref. 24). The arrangement of the two channels may be varied. For example, in Fig. 7 we performed the enzyme assay in channel 1 and measured the blank in the absence of substrate in channel 2. In other experiments we have measured two enzymes simultaneously in the same gradient tube, by having one enzyme in each channel. (One could also measure one enzyme at two different protein concentrations or incubation times, and preliminary experiments have shown that the system may also be adapted for multiple enzyme analysis.) An actual trace in the analysis of a sucrose gradient is given in Fig. 8, which also shows the sequence of operations in the analysis. The distributions of NADH-, L(+)lactate- and succinate-FC reductases along the gradient are given in Fig. 9, together with the "E_{420}". This is the increment in absorption at 420 mμ that is observed when the particles are pumped through the system in the absence of substrate, but the presence of all the other reagents.

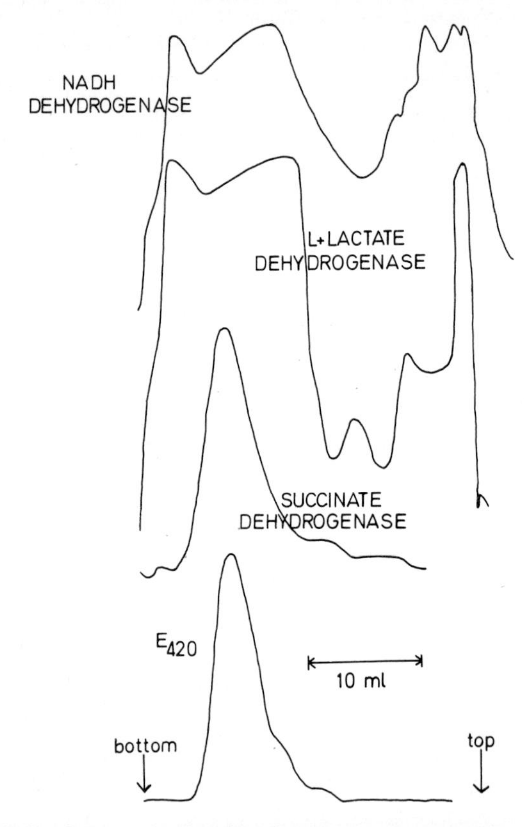

Figure 9. Analysis of yeast mitochondrial fraction on sucrose gradient. Conditions as in Fig. 8 with 0·36 mM NADH or 1·8 mM L(+) lactate or succinate. E_{420} is the increment in absorption at 420 mμ observed as the gradient is analyzed with water in the substrate line (Fig. 7).

It is seen that the succinate-FC reductase and the E_{420} give reasonably homogeneous and symmetrical peaks. The distribution of the enzyme precisely follows the E_{420} (Fig. 10). Since we may take the succinate dehydrogenase as a marker for yeast mitochondria [25, 26], the E_{420} may well be a suitable mitochondrial marker as well. We are not certain whether the absorption at 420 mμ is due to light scatter, or to the Soret bands of mitochondrial cytochromes.

It is seen in Fig. 12 that the succinate-FC reductase closely follows the total protein in the gradient, so that had only these parameters been measured, one would have concluded that the preparation was very homogeneous.

However, the cell components containing NADH- and L(+)lactate-FC reductase show a very heterogenous distribution along the *same* gradients (Fig. 9). In Fig. 9, the L(+)lactate-FC reductase was actually measured in the same gradient tube as the succinate dehydrogenase, with one enzyme in each channel of the autoanalyzer.

Having established the heterogeneity of the population, it was important to establish which particles were labelled *in vitro*. The yeast mitochondrial fraction was incubated with [^{14}C]leucine and analyzed on a sucrose gradient, using the

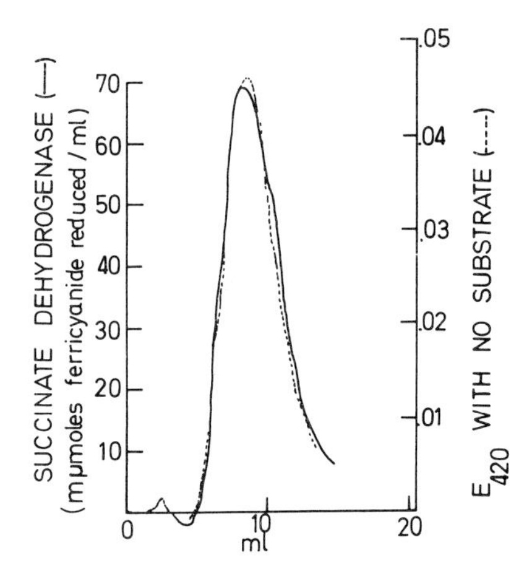

Figure 10. Parallel banding of succinate-FC reductase and E_{420}. Conditions as in Fig. 9.

autoanalyzer system shown in Fig. 11. Fortunately, it was found that the bulk of the radioactivity sedimented as a symmetrical band at precisely the same position as similar shaped bands of total protein and succinate dehydrogenase (Fig. 12). However, the complex pattern with L(+)-lactate-FC reductase was still observed, except that some of the minor peaks near the top of the gradient were absent. (This was possibly due to the pre-incubation of the mitochondria before analysis on the gradient.) There was no indication of radioactivity in the lighter lactate-dehydrogenase particles. It is therefore reasonable to conclude that the only components in the mitochondrial fraction labelled *in vitro* are mitochondria, although it is probable that non-mitochondrial elements are present. The reaction mixture and mitochondrial fraction used in these sucrose gradient analyses was found to be free of bacterial contamination.

Using a sorbitol gradient, Jayaraman *et al.* [27] found that mitochondria banded at a density of 1·175, in reasonable agreement with our results for fresh

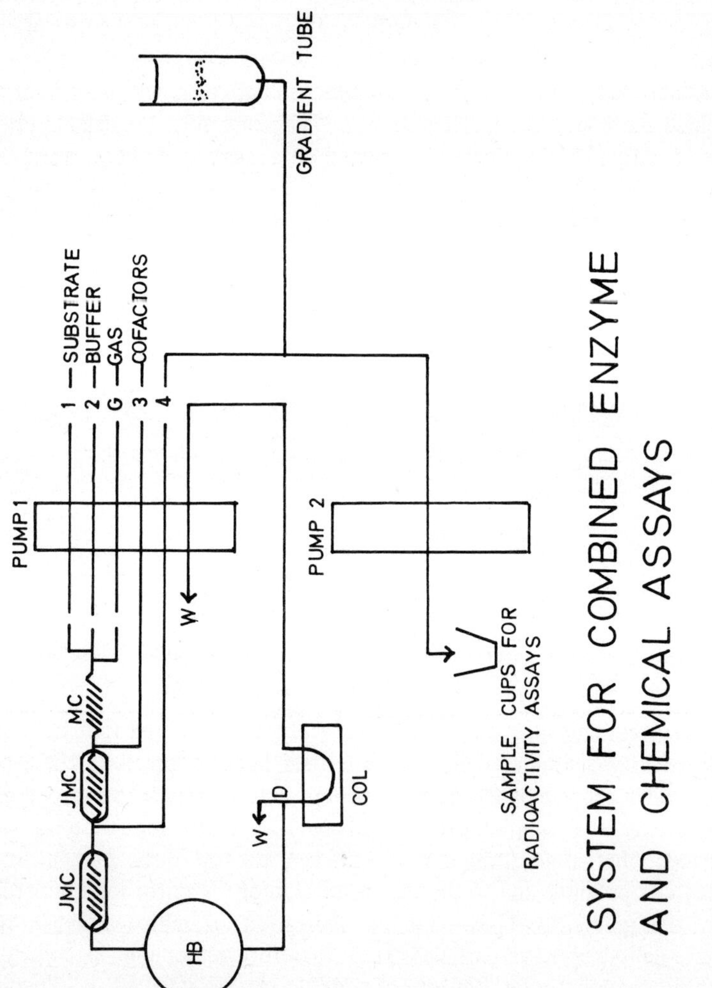

Figure 11. Autoanalyzer system for combined chemical and enzyme assays. Symbols as in Fig. 4. Pumping rates for tubes 1, 2, G, 3 and 4 as in Fig. 7. Line in pump 2 had a flow rate of 0·6 ml/min.

mitochondria. It is interesting that they obtained a reasonably symmetrical peak for NADH oxidase in their gradient. This would indicate that the heterogeneous distribution we observe with NADH-FC reductase is due to the primary dehydrogenase, not the oxidase system. Since respiratory particles from yeast also give homogeneous bands with NADH-cytochrome *c* reductase [28] we have to distinguish between results obtained with different electron acceptors.

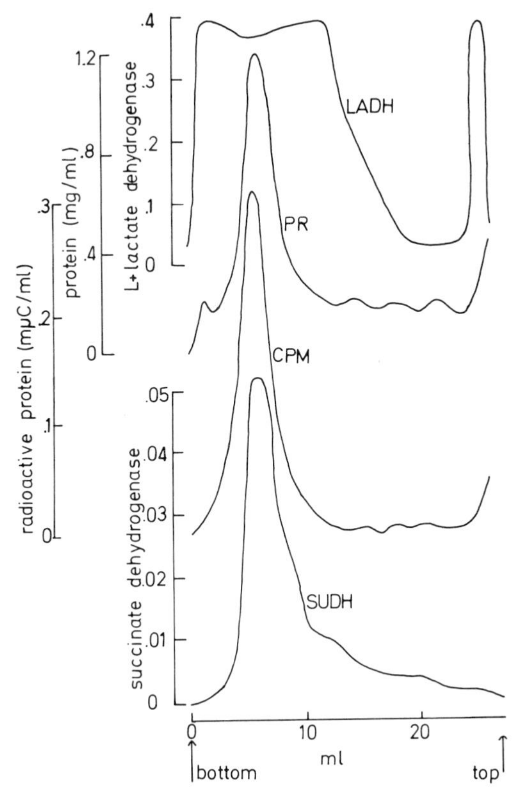

Figure 12. Sucrose gradient analysis of yeast mitochondrial fraction labelled *in vitro* with [^{14}C]leucine. Labelled mitochondria prepared as in Fig. 2 and Table 2. L(+)-lactate- and succinate-FC reductases assayed as in Fig. 9. Values for enzyme activities in μmoles/ml reaction mixture; values for protein and radioactivity in amount/ml gradient.

Analysis of whole homogenates of yeast with glycerol gradients [29] showed that succinate dehydrogenase gave a symmetrical peak banding at a density of 1·18. Schatz *et al.* [30] isolated respiratory particles from mechanically disrupted yeast and found that succinate-cytochrome *c* reductase banded on a sucrose gradient at densities of 1·18-1·19 and at densities of 1·15-1·16 on Urograffin gradients. The protein profile they obtained was less homogeneous than our results (Fig. 12) and they suggested that mitochondrial precursors were present, as well as double membranes rich in ATPase.

The lactate dehydrogenase system of yeast is complex, and the activity obtained depends greatly on the electron acceptor used [31] and on the growth conditions (see ref. 32 of example). The system is summarized and discussed in detail by Labeyrie and Slonimski [33]. Aerobic yeast contains a D(-) lactate dehydrogenase which is particle-bound but does not reduce FC, and an L(+)lactate dehydrogenase (cytochrome b_2) which is partly soluble and partly particle-bound, and which does reduce FC. Our results add another dimension of complexity to the system, since they suggest the possible presence of several types of particle containing L(+) lactate dehydrogenase. The results with NADH-FC reductase are difficult to interpret at the moment. Yeast mitochondria rapidly oxidized exogenous NADH possibly through a non-phosphorylating by-pass of the respiratory chain [34]. The NADH to cytochrome *b* region is insensitive to rotenone, and Site I phosphorylation is probably absent in isolated yeast mitochondria [35, 36]. Our results establish the need to discover whether all the enzyme systems in yeast mitochondrial fractions that oxidize NADH are truly mitochondrial or whether some are present in non-mitochondrial contaminants.

To summarize, it is clear that isolated yeast mitochondria incorporate radioactive amino acids into their protein. The incorporation system has similar properties to the rat liver system, and the radioactive particles have identical buoyant densities to yeast mitochondria as followed by succinate dehydrogenase. However, although the bulk of the protein in the fractions sediments as a single band, analysis of NADH- and L(+)lactate-FC reductases indicate that yeast mitochondrial fractions are probably not homogeneous, and contain several minor non-mitochondrial components that do not contribute greatly to the total protein content. Finally, the results also confirm the many previous findings that amino acid incorporation observed *in vitro* with isolated mitochondria is not due to bacterial contamination.

REFERENCES

1. McLean J. R., Cohn, G. L., Brandt, I. K. and Simpson, M. V., *J. biol. Chem.* **233** (1958) 657.
2. Bates, H. M. and Simpson, M. V., *Biochim. biophys. Acta* **32** (1959) 597.
3. Bates, H. M., Craddock, V. M. and Simpson, M. V., *J. Am. chem. Soc.* **80** (1958) 1000.
4. Simpson, M. V., Skinner, D. M. and Lucas, J. M., *J. biol. Chem.* **236** (1961) PC 81.
5. Roodyn, D. B., Suttie, J. W. and Work, T. S., *Biochem. J.* **83** (1962) 29.
6. Roodyn, D. B. and Wilkie, D., "The Biogenesis of Mitochondria" Methuen, London, 1968.

7. Roodyn, D. B., Reis, P. J. and Work, T. S., *Biochem. J.* **80** (1961) 9.
8. Lamborg, M. R. and Zamecnik, P. C., *Biochim. biophys. Acta* 42 (1960) 206.
9. Beattie, D. S., Basford, R. E. and Koritz, S. B., *J. biol. Chem.* **242** (1967) 3366.
10. Sandell, S., Löw, H. and der Decken, A., *Biochem. J.* **104** (1967) 575.
11. Roodyn, D. B., *in* "Regulation of Metabolic Processes in Mitochondria" (edited by J. M. Tager, S. Papa, E. Quagliariello and E. C. Slater), Elsevier, Amsterdam, 1966, pp. 562-563.
12. Wintersberger, E., *Biochem. Z.* **341** (1965) 409.
13. Ohnishi, T., Kawaguchi, K. and Hagihara, B., *J. biol. Chem.* **241** (1966) 1797.
14. Duell, E. A., Inoue, S. and Utter, M. F., *J. Bact.* **88** (1964) 1762.
15. Grivell, L. A., *Biochem. J.* **105** (1967) 44c.
16. Roodyn, D. B., Freeman, K. B. and Tata, J. R., *Biochem. J.* **94** (1965) 628.
17. Roodyn, D. B., *in* "Abstracts Technicon International Symposium", London, 1964, p. 11.
18. Roodyn, D. B., *in* "Automation in Analytical Chemistry: Technicon Symposia 1967" Mediad Press, New York, 1968, p. 233.
19. Roodyn, D. B., *in* "Automation in Analytical Chemistry: Technicon Symposia 1965" Mediad Press, New York, 1966, pp. 593-594.
20. Roodyn, D. B., *Nature, Lond.* **216** (1967) 1033.
21. Pette, D., Klingenberg, M. and Bücher, Th., *Biochem. biophys. Res. Commun.* 7 (1962) 425.
22. Roodyn, D. B. and Wilkie, D., *Biochem. J.* **103** (1967) 3c.
23. Roodyn, D. B. and Maroudas, N. G., *Analyt. Biochem.* (1968) 496-505.
24. Schuell, H. and Anderson, N. G., *J. Cell Biol.* **21** (1964) 309.
25. Rossi, C., Hauber, J. and Singer, T. P., *Nature, Lond.* **204** (1964) 167.
26. Lukins, H. B., Tham, S. H., Wallace, P. G. and Linnane, A. W., *Biochem. biophys. Res. Commun.* 23 (1966) 363.
27. Jayaraman, J., Cotman, C., Mahler, H. R. and Sharp, C. W., *Archs Biochem. Biophys.* **116** (1966) 224.
28. Mahler, H. R., Mackler, B., Grandchamp, S. and Slonimski, P. P., *Biochemistry* 3 (1964) 668.
29. Schatz, G., *Biochem. biophys. Res. Commun.* 12 (1963) 448.
30. Schatz, G., Haslbrunner, E. and Tuppy, H., *Biochem. biophys. Res. Commun.* **15** (1964) 147.
31. Singer, T. P., Gregolin, C. and Cremona, T., *in* "Control Mechanisms in Respiration and Fermentation" (edited by B. Wright), Ronald Press Co., New York, 1963, pp. 47-79.
32. Somlo, M., *Bull. Soc. Chim. biol.* **48** (1966) 247.
33. Labeyrie, F. and Slonimski, P. P., *Bull. Soc. Chim. biol.* **46** (1964) 1793.
34. Stekhoven, F. M. A. H. S. and Beunen, C. Th. M., *Archs Biochem. Biophys.* **115** (1966) 555.
35. Ohnishi, T., Sottocasa, G. L. and Ernster, L., *Bull. Soc. Chim. biol.* **48** (1966) 1189.
36. Schatz, G., Racker, E., Tyler, D. D., Gonze, J. and Estabrook, R. W., *Biochem. biophys. Res. Commun.* 22 (1966) 585.

FEBS Symposium, Volume 17, 1969, pp. 179-188

Biosynthesis of Mitochondrial Enzymes

B. KADENBACH

Physiologisch-Chemisches Institut der Universität München,
München, Germany

INTRODUCTION

The biogenesis of mitochondria represents a complex process, which involves a controlled synthesis of proteins and some phospholipids inside the pre-existing organelle and a concomitant synthesis of proteins and phospholipids at the endoplasmic reticulum, followed by a transfer into the intact mitochondrium. Cytochrome *c* was shown to be synthesized at the endoplasmic reticulum [1, 2], and in a second step, to be transferred into the mitochondria [3, 4]. The present paper describes quantitative experiments on the kinetics of labelling of cytochrome *c* from mitochondria and microsomes with [^{14}C] L-lysine *in vivo.* The data allow the determination of the true microsomal pool-size for cytochrome *c* and the turnover-time and half-life of cytochrome *c* as well.

RESULTS

With a new method for isolating cytochrome *c* from cell fractions we were able to isolate, purify and estimate very small amounts of cytochrome *c* quantitatively [2]. The principle of the method is shown in Fig. 1. This chromatographic method involves the combined use of TEAE- and CM-cellulose columns [3]. Cytochrome *c* is separated from all other proteins under physiological conditions and in a very short time. The whole procedure from killing the rat until elution of purified cytochrome *c* can be done in less than 12 h. Our method separates isocytochromes *c,* described by Flatmark in detail [5, 6] as shown in Fig. 2.

Using this chromatographic method, cytochrome *c* was extracted from rat liver homogenates and from isolated mitochondria and microsomes. Table 1 represents the data from different experiments. The content of cytochrome *c* in mitochondria agrees well with data in the literature, estimated by spectroscopic measurements [7-9]. From the relation between cytochrome *c* content in total

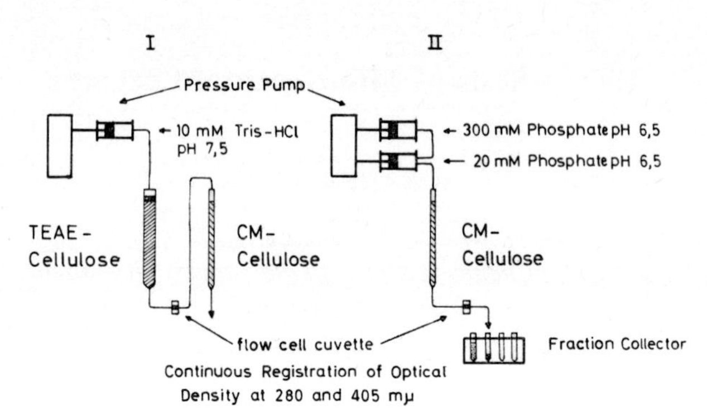

Figure 1. Schematic presentation of the principle of chromatographic isolation of cytochrome *c*. From ref. 2.

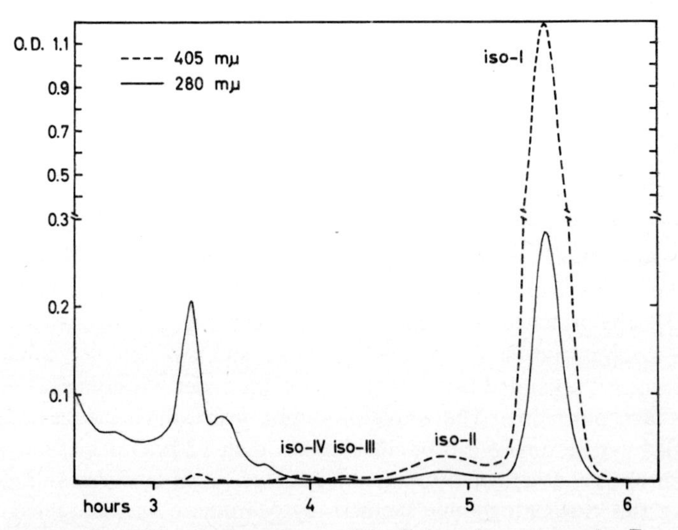

Figure 2. Separation of isocytochromes *c* by CM-cellulose chromatography. From ref. 2.

Table 1. Cytochrome *c* content of total rat liver and of isolated microsomes and mitochondria. From ref. 2.

Experiment No.	Homogenate mμ moles/g wet weight	Microsomes	Mitochondria
		mμ moles / g protein	
1	12·5	10·5	220*
2	11·9	11·7	202*
3	12·4	10·2	228*
4	11·8	12·6	220*
Average	12·1	11·2	217

* Corrected for microsomal contamination by the activity of NADPH-cytochrome *c* reductase.

rat liver homogenate and in isolated mitochondria the amount of mitochondria in rat liver can be calculated. From the data it follows that 29% of total liver protein is mitochondrial protein. This calculation is based on the assumption that all cytochrome *c* is localized within the mitochondria. However, a relatively large and constant amount of cytochrome *c* was found in the microsomal fraction. To calculate the percentage of total cytochrome *c* recovered in the microsomal fraction, the content of microsomes in rat liver was estimated by measuring the activity of NADPH-cytochrome *c* reductase in the total rat liver homogenate and in isolated microsomal fraction (Table 2). NADPH-cytochrome *c* reductase is assumed to be exclusively bound to the endoplasmic reticulum [10]. Since about 23% of total protein of the rat liver is recovered in the microsomal fraction it is calculated that 4% of total cytochrome *c* is bound to the microsomal fraction. This is less than the amount found by González-Cadavid and Campbell, who extracted 11% of the total in their standard microsomal fraction [11]. This may be due to a different homogenization procedure, since it is a well known fact that cytochrome *c* is redistributed during homogenization of tissues [12, 13].

Table 2. Content of microsomal protein in rat liver, estimated by the activity of NADPH-cytochrome *c* reductase in rat liver microsomes and total homogenate

Experiment	Microsomes	Homogenate	% Microsomes
	μ moles/min /g protein		
1	280	62·3	22·2
2	283	63·6	22·4
3	248	56·2	22·6
4	204	48·0	23·5
Average			22·6

In earlier publications we have shown that ^{14}C-labelled microsomes, obtained either from rat liver slices, which were incubated with [^{14}C] L-lysine *in vitro* [3], or from the livers of rats, which were injected with [^{14}C] L-lysine 20 min prior to death [4], transfer labelled cytochrome *c* into unlabelled isolated mitochondria. González-Cadavid and Campbell [1] showed from the kinetics of labelling of microsomal and mitochondrial cytochrome *c*, after injection of [^{14}C] lysine *in vivo*, that microsomal cytochrome *c* is labelled first and to a higher extent than the mitochondrial one. We have done similar experiments, which are shown in Fig. 3. Obviously the microsomal cytochrome *c* is labelled prior to the mitochondrial one and to a higher degree.

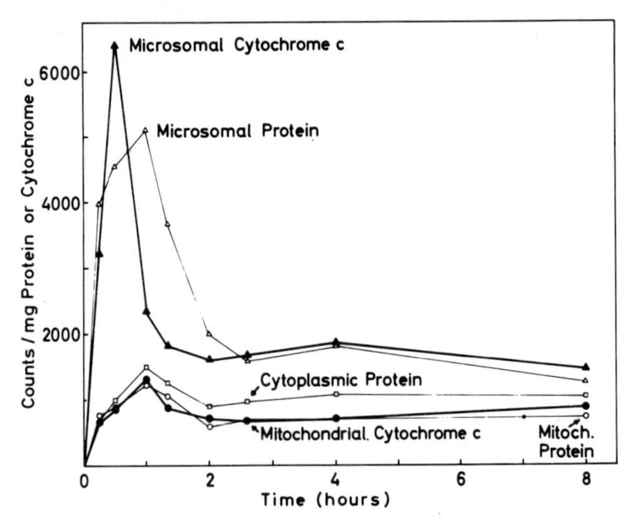

Figure 3. Kinetics of incorporation of [^{14}C] lysine into cytochrome *c* and protein of rat liver microsomes and mitochondria *in vivo*. Starved rats were injected intraperitoneally with 5 μc [^{14}C] L-lysine/100 g rat weight. From ref. 2.

To check the identity of both the mitochondrial and the microsomal cytochrome *c*, a cochromatogram was carried out. *In vivo* ^{14}C-labelled mitochondrial and unlabelled microsomal cytochrome *c* were adsorbed on a CM-cellulose column and eluted together. Figure 4 shows that along the entire peak of cytochrome *c* the specific radioactivity was constant. This result rather indicates that both cytochromes are identical.

In further experiments we tried to analyse the kinetics of labelling cytochrome *c* quantitatively. In Fig. 5 the kinetics of incorporation of [^{14}C] L-lysine into cytochrome *c* and total protein from rat liver mitochondria and microsomes are shown. In addition, the activity of lysine, which was calculated from the radioactivity of the deproteinized supernatant of a rat liver homogenate, assuming a free lysine content of 0·55 μmoles/g wet weight [14] is given in the figure. From the two curves, the time dependency of the specific activity of free [^{14}C]lysine, and of microsomal [^{14}C]cytochrome *c*, it follows that there is a precursor-product relationship. We calculated the turnover rate and the turnover time of the microsomal cytochrome *c* according to the method of Zilversmit *et al.* [15].

We made the following assumptions:

1. The non-precipitable radioactivity in the liver was only due to [^{14}C]L-lysine. This assumption appears reasonable, since we used only data, obtained not later than 20 min after injection.
2. There was full equilibration of labelled lysine with the pool of free lysine in the rat liver.

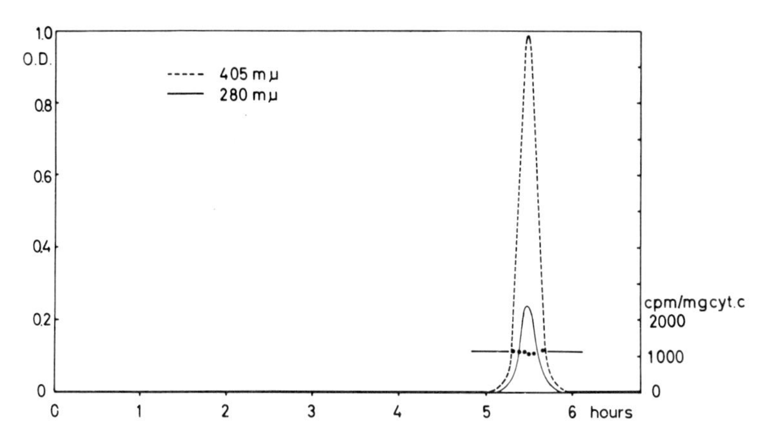

Figure 4. Cochromatography of ^{14}C-labelled cytochrome *c* from mitochondria with unlabelled cytochrome *c* from microsomes. From ref. 2.

3. The rate of synthesis of cytochrome *c* at the endoplasmic reticulum is constant and equals the rate of transfer into the mitochondria and the rate of degradation of cytochrome *c*.
4. One mole of rat liver cytochrome *c* contains 18 moles of lysine.

To compare the specific radioactivities we define the specific activity of free L-lysine: $A = \dfrac{\text{counts/min}}{18 \text{ moles L-lysine}}$ and of microsomal cytochrome *c*:

$$B = \frac{\text{counts/min}}{\text{mole microsomal cytochrome } c}.$$

It is

$$\frac{dB}{dt} = \frac{V}{N}(A(t) - B) \quad (1)$$

where V is the rate of synthesis and N the pool-size of microsomal cytochrome c. Integration of (1) gives

$$B(t_2) - B(t_1) = \frac{V}{N} \int_{t_1}^{t_2} A(t)\mathrm{d}t - \int_{t_1}^{t_2} B(t)\mathrm{d}t \qquad (2)$$

From equation (1) it follows that at the maximum of the curve for microsomal cytochrome c we have: $\mathrm{d}B/\mathrm{d}t = 0$ and $B(t_{\mathrm{max.}}) = A(t_{\mathrm{max.}})$. Therefore the specific activities of free L-lysine and of microsomal cytochrome c must be equal. However, the measured specific activity of microsomal cytochrome c is much lower than that of free lysine. This is obviously due to contamination by mitochondrial cytochrome c, which has a much lower specific activity. These three data allow the calculation of the amount of mitochondrial contamination in the extracted microsomal cytochrome c. The calculation shows that only 1/38th of the "microsomal" cytochrome c is of microsomal origin. The rest has to be considered as mitochondrial cytochrome c. The true pool-size of microsomal cytochrome c then is: $N = 0{\cdot}36$ nmoles/g protein. In other words, only 1/1000th of total rat liver cytochrome c is bound at the endoplasmic reticulum.

To obtain the turnover time of microsomal cytochrome c, we can write: $(V/N) = (1/\tau)$ where $1/\tau$ is the turnover rate and τ the turnover time of microsomal cytochrome c. Using equation (2), τ can be estimated by planimetry of the area between t_2 and t_1 [15]:

$$\tau = \frac{F_1 - F_2}{B(t_2) - B(t_1)} \qquad (3)$$

We have redrawn Fig. 5 by correcting the specific activities of microsomal cytochrome c with the factor of 38 using a molecular weight of 12,340 for cytochrome c, as shown in Fig. 6. The values obtained for τ are 71 min (integrating between 10 and 15 min) and 81 (integrating between 15 and 20 min). This means that in about 75 min the amount of microsomal cytochrome c is synthesized once. Since this is 0·1% of the total cytochrome c, the time required for synthesis of total rat liver cytochrome c is 57,700 min or 40 days. From this turnover time the half-life is calculated to be about 28 days. This value is higher than the one obtained by Fletscher and Sanadi [16], who found 10 days by a different method. Our data are summarized in Table 3.

Since only 2·6% of the isolated "microsomal" cytochrome c is of microsomal origin, this small amount would not be detectable by spectrophotometric methods. In fact the isolated cytochrome c from microsomes was only identified by the absorbance of the contaminated mitochondrial cytochrome c. Therefore we injected ^{59}Fe into rats and as shown in Fig. 7, a similar kinetic of labelling microsomal and mitochondrial cytochrome c is obtained. This seems to indicate that cytochrome c is synthesized at the microsomes *in toto*.

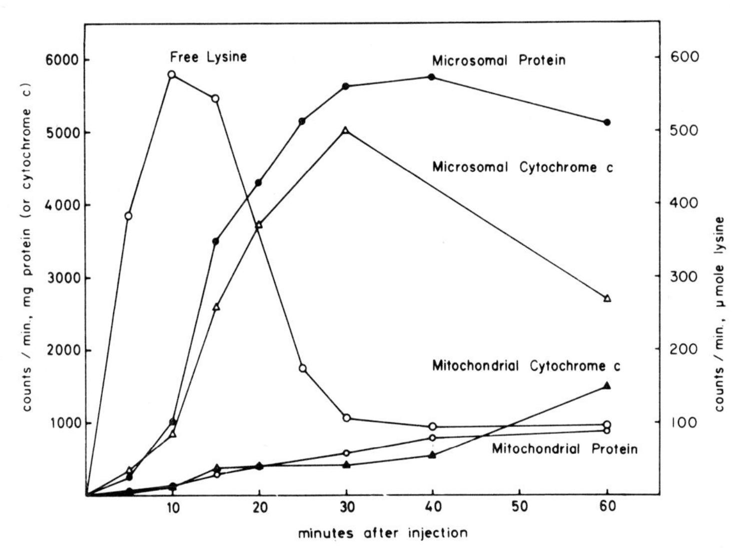

Figure 5. Kinetics of [^{14}C] lysine-activity and of labelling of cytochrome *c* and protein in rat liver after injection of [^{14}C] L-lysine. Starved rats were injected intraperitoneally with 10 μc [^{14}C] L-lysine/100 g rat weight. From the radioactivity of the supernatant of a rat liver homogenate after precipitation of proteins, and from the known lysine-content, the specific activity of [^{14}C] lysine was calculated.

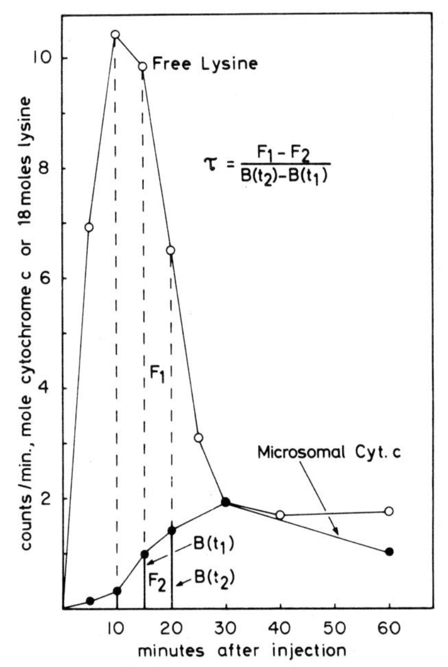

Figure 6. Graphic estimation of turnover time of microsomal cytochrome *c* according to Zilversmit *et al.* [15].

It is of interest that the newly synthesized cytochrome *c* is attached to the endoplasmic reticulum for a relatively long time. As we have shown before, the average time of attachment is more than one hour. Since it is improbable that mitochondrial enzymes reveal their activity outside the organelle, a masking of the enzymes has to be assumed. We therefore postulated that the mitochondrial enzymes at the endoplasmic reticulum are attached to phospholipids which are synthesized at the endoplasmic reticulum as well. The enzyme is released or bound to its functional site and reveals its full enzymatic activity only after the transfer of the phospholipid enzyme complex into the mitochondria. This postulated mechanism is favoured by an experiment shown in Fig. 8. Microsomal

Table 3. Pool-size and turnover time of rat liver cytochrome *c*

	Pool-size				Turnover time
	measured		calculated		
	mμ moles/g protein (mμ moles/g total protein)				
Microsomal cyt. *c*	11·7	(2·7)*	0·31	(0·067)	75 min
Mitochondrial cyt. *c*	217	(63·0)†			40 days

* Assuming 23% microsomal protein from total protein.
† Assuming 29% mitochondrial protein from total protein.

phospholipids were labelled *in vivo* with [^{32}P]phosphate, which was injected into rats 30 min prior to death. The isolated microsomes were incubated with unlabelled mitochondria and subsequently the mitochondria separated by repeated washings. The result is quite similar to that obtained for the transfer of proteins. The phospholipid-transfer reaction is time dependent and requires energy, since it is inhibited in the presence of KCN. Both the inner and outer membrane of mitochondria are labelled energy-dependent, the latter being to a higher degree, similar to the higher labelling of phospholipids of the outer membrane *in vivo*.

We have tried to isolate a microsomal cytochrome *c*–phospholipid complex, however without success. Obviously the medium required to detach cytochrome *c* from the microsomes would rather split the cytochrome *c*–phospholipid linkage than the linkage between the microsomal membrane and the complex.

From our results the following conclusions may be drawn:

1. The major part of the cytochrome *c* found in the microsomal fraction of rat liver is of mitochondrial origin due to redistribution during homogenization of the tissue. Only a very small amount, about 1/1000th of total rat liver cytochrome *c* represents newly synthesized cytochrome *c*.

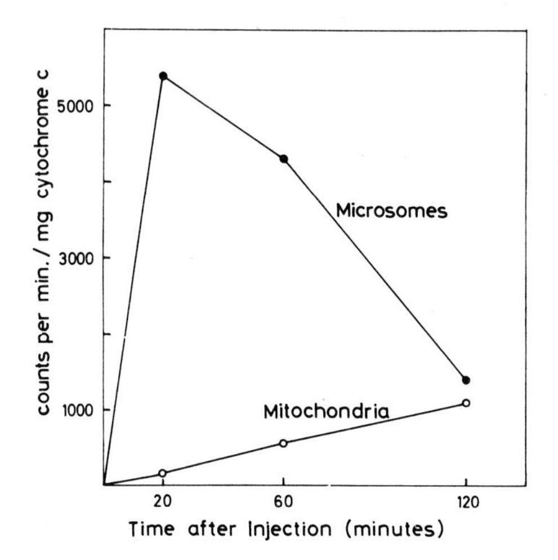

Figure 7. Incorporation of ^{59}Fe into mitochondrial and microsomal cytochrome *c in vivo.* From ref. 4.

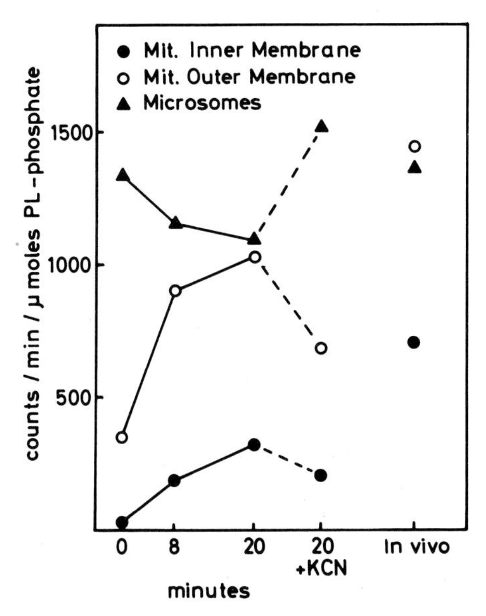

Figure 8. Transfer of [^{32}P]phospholipids from microsomes into mitochondria. Microsomes were isolated from the liver of a rat, injected with [^{32}P]phosphate 30 min prior to death. After incubation with unlabelled mitochondria for the indicated times, the mitochondria were separated by repeated washing and the inner and outer membranes isolated with digitonin.

2. If ^{14}C-labelled L-lysine is injected into rats, a precursor-product relationship is observed between free lysine and microsomal cytochrome *c*, and between cytochrome *c* of microsomes and of mitochondria.
3. The turnover time of microsomal cytochrome *c* is about 75 min. After synthesis, cytochrome *c* is bound to the endoplasmic reticulum until it is transferred into the mitochondria.
4. The transfer of cytochrome *c* as well as other mitochondrial enzymes from microsomes into mitochondria seems to be mediated by a concomitant transfer of phospholipids.

REFERENCES

1. González-Cadavid, N. F. and Campbell, P. N., *Biochem. J.* **105** (1967) 443.
2. Kadenbach, B. and Urban, P. F., *Z. analyt. Chem.* **243** (1968) 542.
3. Kadenbach, B., *Biochim. biophys. Acta* **138** (1967) 651.
4. Kadenbach, B., *in* "Biochemical Aspects of the Biogenesis of Mitochondria" (edited by E. C. Slater, J. M. Tager, S. Papa and E. Quagliariello), Adriatica Editrice, Bari, 1968.
5. Flatmark, T., *Acta chem. scand.* **20** (1966) 1476.
6. Flatmark, T., *Acta chem. scand.* **20** (1966) 1487.
7. Estabrook, R. W. and Holowinski, A., *J. biophys. biochem. Cytol.* **9** (1961) 19.
8. Jacobs, E. E. and Sanadi, D. R., *J. biol. Chem.* **235** (1960) 531.
9. Kadenbach, B., *Biochem. Z.* **344** (1966) 49.
10. Phillips, A. H. and Langdon, R. G., *J. biol. Chem.* **237** (1962) 2652.
11. González-Cadavid, N. F. and Campbell, P. N., *Biochem. J.* **105** (1967) 427.
12. Beinert, H., *J. biol. Chem.* **190** (1951) 287.
13. Gonzalez-Cadavid, N. F., Bravo, M. and Campbell, P. N., *Biochem. J.* **107** (1968) 523.
14. Schimassek, H. and Gerok, W., *Biochem. Z.* **343** (1965) 407.
15. Zilversmit, D. B., Entenman, C. and Fishler, M., *J. gen. Physiol.* **26** (1943) 325.
16. Fletscher, M. J. and Sanadi, D. R., *Biochim. biophys. Acta* **51** (1961) 356.

FEBS Symposium, Volume 17, 1969, pp. 189-198

The Biosynthesis of Mitochondrial Energy Transfer Components in Baker's Yeast

G. SCHATZ* and R. S. CRIDDLE†

Institut für Biochemie, University of Vienna, Vienna, Austria

Recent evidence suggests that the proteins of the mitochondrial energy transfer chain are not merely enzymes mediating the synthesis of ATP, but are also important structural elements of the mitochondrial inner membrane [1, 2]. Thus, one of the most distinguishing components of this membrane is a complex between mitochondrial ATPase (F_1), phospholipid, and an insoluble protein fraction (CF_0) that specifically binds F_1 and renders it cold-stable and sensitive to oligomycin [3, 4].

The present experiments with yeast cells direct themselves to the following two questions:

(a) Is part of the mitochondrial ATPase complex synthesized by the mitochondrial genetic system?

(b) Are anaerobically-growing yeast cells still capable of forming this complex, as well as mitochondrial inner membranes?

Effect of the cytoplasmic "petite" mutation on the mitochondrial ATPase complex

The non-respiring mitochondria of the cytoplasmic "petite" mutant of *Saccharomyces cerevisiae* (strain D 273-10 B-1) are devoid of normal mitochondrial DNA [5] and of ribosomes [6] and thus appear to contain a defective genetic system. It is therefore reasonable to assume that they have lost most, if not all, of the components which are normally synthesized by the mitochondria themselves. In the present study [7] it was found that the mutant mitochondria still contained almost normal amounts of F_1 (Table 1). However, the F_1 was oligomycin-*in*sensitive and cold-*labile* and thus differed from F_1

* Present address: Section of Biochemistry and Molecular Biology, Cornell University, Ithaca, U.S.A.

† Present address: Department of Biochemistry, University of California, Davis, U.S.A.

Table 1. Identification of mitochondrial ATPase (F_1) in the "petite" mutant

Subcellular fraction	ATPase activity µmole ATP/min/mg protein
Wild-type mitochondria	0·83
+ 135 µg of F_1 inhibitor	0·081
+ 270 µg of F_1 inhibitor	0·070
"Petite" mitochondria	0·32
+ 135 µg of F_1 inhibitor	0·067
+ 270 µg of F_1 inhibitor	0·065

The isolation of mitochondria at room temperature, the measurement of ATPase and of protein, and the preparation of the naturally occurring F_1 inhibitor [8] are described in Ref. 7.

associated with the wild-type mitochondria which is oligomycin-sensitive and cold-stable (Table 2). This fact has also been independently reported by Kováč and Weissová [9]. The alteration of the mitochondrial ATPase complex appeared to be rather specific for the "petite" mutation since it could not be induced by growing the wild-type cells in the presence of high concentrations of glucose or in the absence of oxygen (cf. below). Mitochondrial ATPase associated with the mutant mitochondria thus exhibits properties heretofore attributed only to the *soluble* enzyme [10].

In an attempt to identify the lesion in the ATPase complex of the mutant mitochondria, their F_1 was purified and compared with the corresponding enzyme from the wild-type strain. The purified enzymes were found to be indistinguishable with respect to substrate specificity, to inhibition by ADP and Dio-9, as well as to stimulation by 2,4-dinitrophenol. Their sedimentation coefficients in a sucrose gradient were also identical. Finally, an antiserum

Table 2. Effect of oligomycin and incubation temperature on ATPase activity of wild-type and "petite" mutant yeast mitochondria

Subcellular fraction	ATPase activity µmole ATP/min/mg protein
Wild-type mitochondria	0·96
+ 85 µg oligomycin/mg protein	0·056
incubated for 20 h at 0°C	1·73
"Petite" mitochondria	0·29
+ 87 µg oligomycin/mg protein	0·27
incubated for 20 h at 0°C	0·01

The experimental details were as described in Table 1.

against the purified F_1 from the wild-type completely inhibited the enzyme from the mutant (Fig. 1). These findings suggested that the ATPase itself had not been altered by the "petite" mutation.

Additional support for this notion was obtained by the demonstration that the properties of cold-stability and oligomycin-sensitivity, characteristic of F_1 from wild-type mitochondria, could be restored to the "petite" enzyme by binding it to F_1-deficient beef heart particles. Such particles are known to contain functional CF_0 (cf. [12]). As can be seen from Table 3, the ATPase activity of the resultant "hybrid" particles was almost completely the

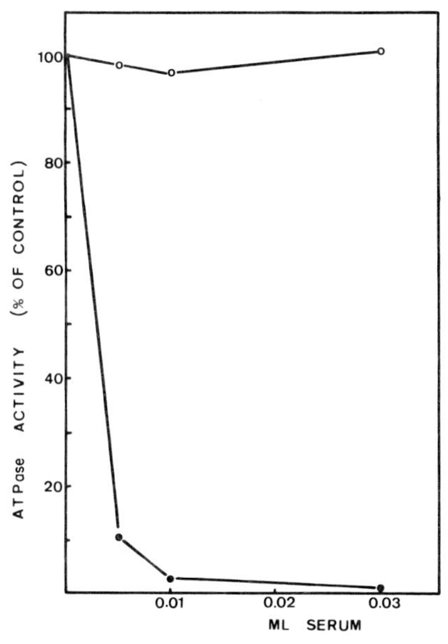

Figure 1. Inhibition of soluble mitochondrial ATPase from the "petite" mutant by an antiserum against purified mitochondrial ATPase from wild-type cells.

The antiserum was obtained as described in Ref. 11 and, like the control serum, contained 85 mg protein per ml. See Table 1 for further details. Open circles: control serum; solid circles: antiserum.

manifestation of bound *yeast* F_1. It is thus clear that the cytoplasmic "petite" mutation does not destroy the inherent ability of F_1 to interact with CF_0 and thereby assume the allotopic properties of membrane-bound F_1.

Since neither F_1 nor the mitochondrial lipids [13] are altered by the "petite" mutation, one is forced to conclude that the mutation inactivates a CF_0-like component of the yeast mitochondria. The isolation of this membrane constituent from normal and mutant mitochondria, coupled with a comparative investigation of their properties, can be expected to provide important information about the linkage between F_1 and the mitochondrial inner

Table 3. Conferral of cold-stability and oligomycin-sensitivity to purified mitochondrial ATPase from the "petite" mutant

Additions to assay medium	ATPase activity μmole ATP/10 min
Purified ATPase (23 μg)	1·26
+ oligomycin (6 μg)	1·30
incubated 15 h at 0°C	0·01
+ yeast F_1-antiserum (2 mg)	0·01
+ control serum (2 mg)	1·28
"Hybrid" particles (58 μg)	1·42
+ oligomycin (6 μg)	0·20
incubated 15 h at 0°C	1·30
+ yeast F_1-antiserum (2 mg)	0·13
+ control serum (2 mg)	1·50

See Ref. 7 for experimental details.

membrane, and about the mechanism of oxidative phosphorylation. The present findings also raise the possibility that CF_0, or one of its subfractions, is a product of the mitochondrial genetic system (cf. [14]). This possibility is emphasized by our observation that a loss of functional CF_0 from the mitochondria can also be induced by chloramphenicol-repression of wild-type yeast cells. On the other hand, the presence of unaltered F_1 in the "petite" mutant, as well as in chloramphenicol-repressed yeast [7, 15], suggests that the mitochondrial ATPase is coded for by the chromosomal system. It is hoped that labelling experiments with isolated yeast mitochondria will soon provide more direct evidence on this important point.

The synthesis of mitochondrial inner membranes by anaerobic yeast cells

The early studies of Ephrussi and Slonimski [16, 17] have shown that anaerobically-grown yeast cells lack the respiratory capacity and the classical cytochrome complement of the aerobically-grown cells, but adaptively regain these characteristics upon aeration. While these findings have been confirmed in numerous laboratories (cf. e.g. [14]), their cytological implications have generated considerable discussion. According to Wallace and Linnane [18], anaerobically-grown yeast cells have completely lost their mitochondria, but reform these organelles *de novo* from non-structured precursors during oxygen adaptation. In contrast, Schatz [19] reported the presence of mitochondria-like particles in the anaerobic cells and suggested that respiratory adaptation in yeast involved the differentiation of incomplete "promitochondria" rather than mitochondrial *de novo* formation.

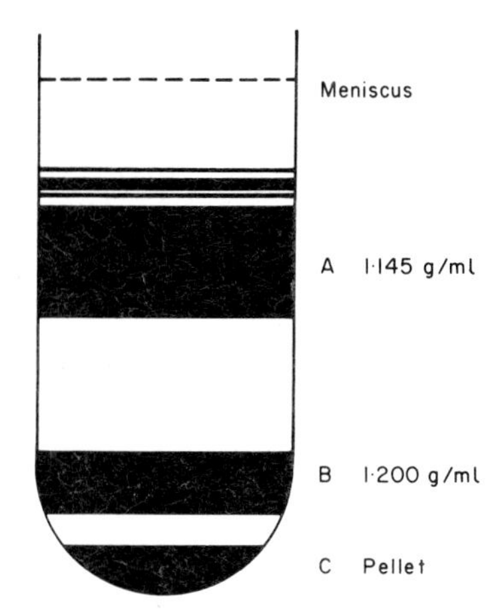

Figure 2. Purification of promitochondria in a sucrose density gradient.

Wild-type yeast cells (strain D 273-10 B) were grown anaerobically in the presence of 10% glucose, 0·26% Tween 80 and 12 p.p.m. ergosterol, harvested and homogenized as outlined in Ref. 20. The subcellular particles were isolated from the cell homogenate by centrifugation for 90 min at 105,000 x g, washed once by re-sedimentation, and separated into various particle classes by centrifugation in a sucrose gradient (26 ml; 20-65% w/v sucrose containing 20 mM Tris SO_4, pH 7·4 and 2 mM EDTA) for 16 h at 90,000 x g (Spinco SW 25 rotor). Band A: promitochondria; band B: nonmitochondrial membranes (cf. ref. 7). See Ref. 20 for further experimental details.

Table 4. Properties of ATPase associated with yeast promitochondria

Addition to promitochondria	ATPase activity μmole ATP/min/mg protein	
	Wild-type promitochondria	"Petite" mutant promitochondria
None	0·73	0·31
Yeast F_1-antiserum (3 mg)	0·06	0·02
Control serum (3 mg)	0·81	0·35
Oligomycin (80 μg/mg protein)	0·09	0·30
None; particles incubated for 16 h at 0°C prior to assay	0·89	0·05

The promitochondria were isolated and purified at room temperature essentially as outlined in Fig. 2. See Ref. 20 for additional details.

The unique properties of the mitochondrial ATPase complex have now allowed us to reach an unequivocal decision between these two opposing views and to consolidate and extend our earlier work. It was found [20, 21] that yeast cells grown under strictly anaerobic conditions still contained subcellular particles whose buoyant density in a sucrose gradient resembled that of respiring yeast mitochondria (Fig. 2, band A). The mitochondria-like particles from the anaerobic wild-type cells carried an oligomycin-sensitive and cold-stable ATPase which was inhibited by the specific F_1-inhibitor of Pullman and Monroy [8] as well as by the antiserum against F_1 from respiring yeast mitochondria (Table 4). The corresponding particles from the anaerobic "petite" mutant contained cold-labile and oligomycin-insensitive F_1 and thus exhibited the characteristic lesion which is also evident with the mitochondria of the aerobically-grown cells (cf. above). These results were obtained regardless of whether the cells had been grown in the presence or in the absence of Tween 80 and ergosterol. In spite of their respiration-deficiency, the anaerobically-grown yeast cells have therefore retained incomplete mitochondrial inner membranes. Since these membranes appear to be functionally analogous to the "proplastids" from dark-grown plant cells, they are termed "promitochondria".

Yeast promitochondria lack the cytochromes aa_3, b, c and c_1 as well as a functional respiratory chain (Fig. 3 and Table 5). On the other hand, they still contain succinic- and NADH-dehydrogenase, ferrochelatase, and a pigment resembling cytochrome b_5. Since they are agglutinated by an antiserum [23] against "structural protein" [24] from aerobic yeast mitochondria, they also seem to be equipped with normal mitochondrial "structural protein" (Table 6). In further support of this conclusion, it was found that erythrocytes coated with promitochondrial "structural protein" were also specifically agglutinated by the antiserum. Preparations of "structural protein" from mitochondria and promitochondria were indistinguishable with respect to their electrophoretic mobility in acrylamide gels and the number of their peptide fragments after tryptic digestion. Regardless of the present uncertainties concerning the homogeneity of "structural protein" [14, 25, 26], these results constitute strong evidence for the presence of mitochondrial membranes in anaerobic yeast.

Perhaps the most compelling argument for this notion is furnished by the observation that the isolated promitochondria contain mitochondrial DNA of the density 1·685 g/cm^3 (Fig. 4). This finding, together with the above-described effect of the "petite" mutation on the promitochondria, suggests the possibility that at least a part of the mitochondrial genetic system is still functional during anaerobic growth.

Taken together, our results justify the conclusion that anaerobic propagation of yeast cells causes a de-differentiation, rather than a complete loss, of the mitochondrial structures. We therefore propose that oxygen-adaptation of the anaerobic cells involves the addition of electron- and perhaps energy-transfer

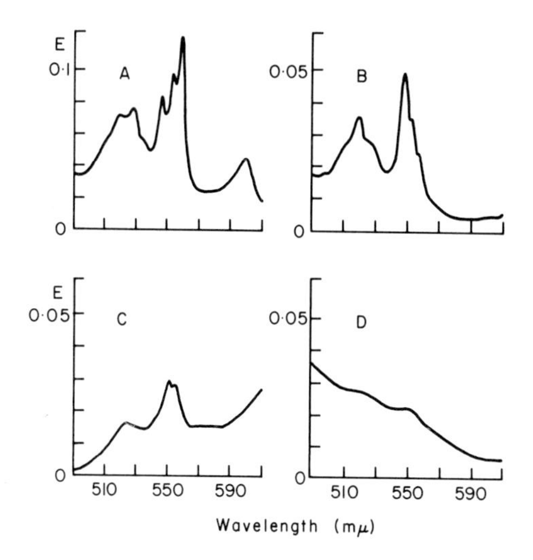

Figure 3. Absolute low-temperature spectra of mitochondria and promitochondria.
A: wild-type mitochondria (2·7 mg/ml)
B: "petite" mitochondria (4·8 mg/ml)
C: wild-type promitochondria (5·5 mg/ml)
D: "petite" promitochondria (7·1 mg/ml)
The absorption spectra of the particles was measured after reduction with sodium dithionite at the temperature of liquid nitrogen essentially as described by Estabrook [22].

Table 5. Enzyme content of mitochondria and promitochondria from yeast

Enzyme or enzyme systems	Specific activity μmole of substrate /mg protein	transformed/min
	mitochondria	promitochondria
F_1–ATPase	3·94	0·95
Succinate dehydrogenase	0·138	0·063
NADH-ferricyanide reductase	0·93	0·293
Succinate oxidase	0·100	0·000
NADH oxidase	0·75	0·002
+ 1 mM KCN	0·003	0·003
Cytochrome *c* oxidase	1·10	0·002

The yeast cells (wild-type) were grown aerobically or anaerobically in the presence of 10% glucose, 0·26% Tween 80 and 12 p.p.m. ergosterol [20]. The isolation of mitochondria and promitochondria and the various enzyme assays are described in Ref. 20.

Table 6. Immunological demonstration of mitochondrial "structural protein" in *Saccharomyces cerevisiae* promitochondria

Antigen used in agglutination experiment	Serum	Agglutination at dilution of serum							
		1	2	4	8	16	32	64	128
Promitochondria	Antiserum	+	+	+	+	+	+	±	–
	Control serum	+	–	–	–	–	–	–	–
"Structural protein" from promitochondria	Antiserum	+	+	+	+	+	+	+	+
	Control serum	–	–	–	–	–	–	–	–

The agglutination experiments were carried out essentially as described by Tuppy *et al.* [23].

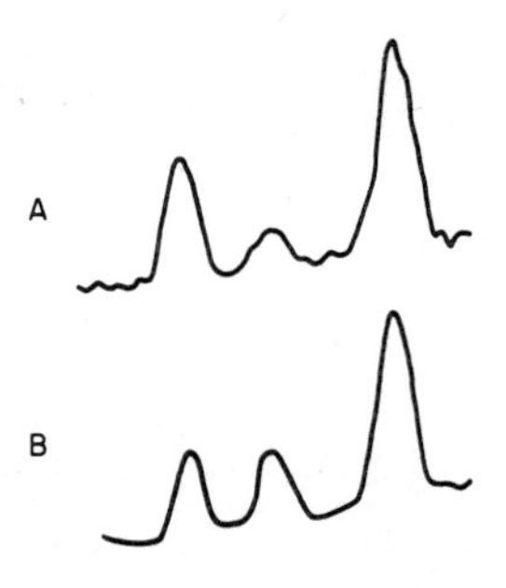

Figure 4. Identification of mitochondrial DNA in yeast promitochondria.

Promitochondria were isolated from wild-type yeast (strain D 273-10 B) grown anaerobically in the presence of 10% glucose, 0·26% Tween 80 and 12 p.p.m. ergosterol [20], and purified by flotation in a "Urografin" gradient [27]. Their DNA was isolated according to the technique of Moustacchi and Williamson [28] and analyzed by cesium chloride gradient centrifugation in a Spinco Model E analytical ultracentrifuge (trace B). The banding pattern of DNA extracted from partially purified aerobic yeast mitochondria is included for the sake of comparison (trace A). *Pseudomonas aeruginosa* DNA (density 1·727 g/cm^3) served as the standard.

Peak 1: *P. aeruginosa* DNA
Peak 2: Nuclear DNA
Peak 3: Mitochondrial DNA

components to the undifferentiated mitochondrial membranes and their transformation into functional, aerobic mitochondria. While our experiments do of course not exclude the possibility of mitochondrial *de novo* formation [18], they clearly invalidate its experimental support.

ACKNOWLEDGEMENTS

The authors are greatly indebted to Dr. R. W. Estabrook for carrying out the spectroscopic measurements, to Drs. H. Tuppy, P. Swetly and I. Wolff for a gift of anti- "structural protein" antiserum, and to Dr. H. Timpl for his help in the immunological experiments.

REFERENCES

1. Pullman, M. E. and Schatz, G., *A. Rev. Biochem.* **36** (1967) 539.
2. Schatz, G., *Angew. Chem.* (Int. Ed. English), **6** (1967) 1035.
3. Kagawa, Y. and Racker, E., *J. biol. Chem.* **241** (1966) 2467.
4. Kagawa, Y. and Racker, E., *J. biol. Chem.* **241** (1966) 2475.
5. Wintersberger, E. and Viehhauser, G., *Nature, Lond.* **220** (1968) 699.
6. Wintersberger, E., *Hoppe-Seyler's Z. physiol. Chem.* **348** (1967) 1701.
7. Schatz, G., *J. biol. Chem.* **243** (1968) 2192.
8. Pullman, M. E. and Monroy, G. C., *J. biol. Chem.* **238** (1963) 3762.
9. Kovač, L. and Weissová, K., *Biochim. biophys. Acta* **153** (1968) 55.
10. Pullman, M. E., Penefsky, H. S., Datta, A. and Racker, E., *J. biol. Chem.* **235** (1960) 3322.
11. Schatz, G., Penefsky, H. S. and Racker, E., *J. biol. Chem.* **242** (1967) 2552.
12. Racker, E., *Biochem. biophys. Res. Commun.* **10** (1963) 435.
13. Kovác, L., Šubik, J., Russ, G. and Kollár, K., *Biochim. biophys. Acta* **144** (1967) 94.
14. Schatz, G., *in* "Structure and Function of Membranes of Mitochondria and Chloroplasts" (edited by E. Racker), Reinhold Publishing Co., New York, in press.
15. Schatz, G., unpublished results.
16. Ephrussi, B. and Slonimski, P. P., *C.r. hebd. Séanc. Acad. Sci., Paris* **230** (1950) 685
17. Slonimski, P. P., "La formation des enzymes respiratoires chez la levure", Masson, Paris, 1953.
18. Wallace, P. G. and Linnane, A. W., *Nature, Lond.* **201** (1964) 1191.
19. Schatz, G., *Biochim. biophys. Acta* **96** (1965) 342.
20. Criddle, R. S. and Schatz, G., *Biochemistry* **8** (1969) 322.
21. Paltauf, F. and Schatz, G., *Biochemistry* **8** (1969) 335.
22. Estabrook, R. W., *in* "Haematin Enzymes" (edited by J. E. Falk, R. Lemberg and R. K. Morton), Pergamon Press, Oxford, 1961, p. 436.
23. Tuppy, H., Swetly, P. and Wolff, I., *Eur. J. Biochem.* **5** (1968) 339.

24. Criddle, R. S., Bock, R. M., Green, D. E. and Tisdale, H., *Biochemistry* **1** (1962) 827.
25. Criddle, R. S., Edwards, D. L. and Petersen, T. G., *Biochemistry* **5** (1966) 578.
26. Haldar, D., Freeman, K. B. and Work, T. S., *Nature, Lond.* **211** (1966) 9.
27. Schatz, G., Haslbrunner, E. and Tuppy H., *Biochem. biophys. Res. Commun.* **15** (1964) 127.
28. Moustacchi, E. and Williamson, D. H., *Biochem. biophys. Res. Commun.* **23** (1966) 56.
29. Plattner, H. and Schatz, G., *Biochemistry* **8** (1969) 339.
30. Schatz, G., Saltzgaber, J. and Rouslin, W., *in* "Round Table Discussion on Electron Transport and Energy Conservation", Fasano (Italy), May 1969.

Note added in proof.

Since the submission of this manuscript the following information was obtained:

(1) Electron micrographs of frozen-etched, anaerobically-grown yeast cells revealed mitochondrial structures regardless of whether the cells had been grown in the presence or the absence of a lipid supplement [29].

(2) Promitochondria from wild-type yeast cells grown anaerobically in a lipid-supplemented medium carry out chloramphenicol-sensitive protein synthesis. Promitochondria from the corresponding "petite" mutant lack this activity [30].

(3) Label transfer experiments with adapting yeast cells have established a physical continuity between promitochondria and the respiring mitochondria of oxygen-adapted cells [30].

FEBS Symposium, Volume 17, 1969, pp. 199-204

Biochemical Mutants as a Tool in the Study of Mitochondrial Function

L. KOVÁČ

Department of Biochemistry, Komensky University, Bratislava, Czechoslovakia

Since mitochondria are organized in space and time from many elementary units, their structure and function can be dissected in a stepwise manner by means of biochemical mutants each carrying the lesion in a single unit. Mutants with modifications in mitochondria can be easily isolated from facultatively-anaerobic yeast. The usefulness of such mutants is exemplified in this communication.

Cytoplasmic respiration-deficient mutants

According to Mounolou *et al.* [1] the hereditary modification in this well-known class of yeast mutants (designed as ρ^- mutants) [2, 3] concerns mitochondrial DNA. The mutants possess no cytochromes $a.a_3$, b and c_1 [3, 4], have either no functionally inactive factor conferring oligomycin sensitivity to mitochondrial ATPase [5, 6], and lack a component in electrophoretic pattern of mitochondrial "structural protein" [7-9]. To account for these biochemical lesions one can assume that the affected mitochondrial components may be coded by the segment of mitochondrial DNA which was exposed to mutation. Some evidence, however, suggests that at least cytochromes may be coded by nuclear genes [10, 11]. Theoretically, it may not be excluded to restore phenotypically their normal function in ρ^- cells under particular conditions.

In fact, ρ^- cells—always claimed to possess only negligible respiration activity—can sometimes respire considerably. As shown in Fig. 1, when grown under rather unusual conditions, anaerobically in the absence of the anaerobic growth factor ergosterol, they showed a marked respiration. We were not able, however, to detect cytochromes $a.a_3$ or b in absolute or difference spectra of these cells measured either at room or at liquid nitrogen temperatures. In difference spectra, ρ^- cells grown under various conditions showed only oxidation of flavin and of the pigment designed previously as cytochrome b_1 [12, 13]. As shown in Fig. 2, respiring mutant cells from anaerobic ergosterol-deficient culture differed from other cells in showing a large oxidation of

$NADH_2$ by air. This indicates the presence of a particular $NADH_2$ oxidase in these respiring cells which may account for their marked respiratory activity. "Classical" respiratory pigments thus seem to be always absent from cytoplasmic mutant cells even if they do respire, and even though the structural information for the pigments may reside unchanged in their nuclear genes.

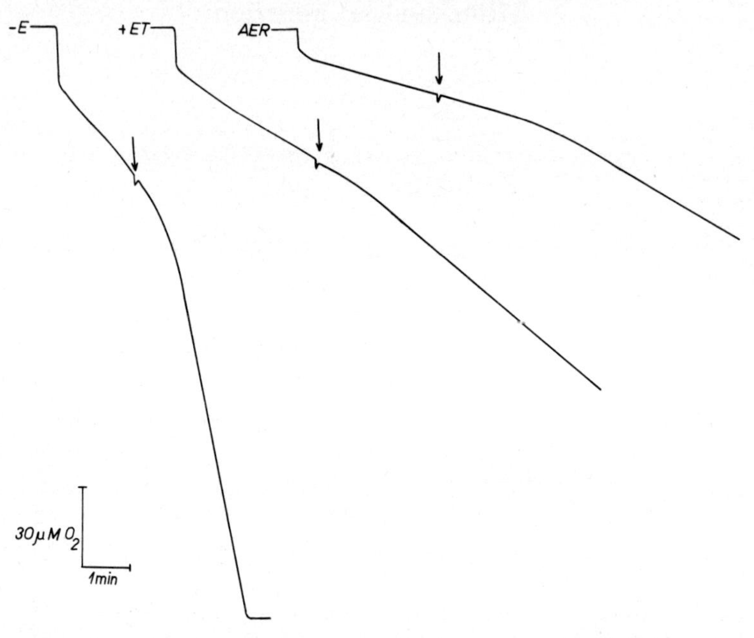

Figure 1. Polarographic recordings of respiration of ρ^- cells. Diploid ρ^- strain was cultured at 30°C for 24 h in a semi-synthetic medium with 5% glucose aerobically (AER), anaerobically in medium supplemented with Tween 80 and ergosterol (+ET), or anaerobically in medium containing Tween 80 but no ergosterol (–E). Respiration was measured in a buffer containing 50 mM K glutarate, 10 mM K phosphate and 100 mM KCl, pH 4·3 with 10 mg (dry weight) cells/ml. The arrows indicate additions of 20 mM glucose.

Oxidative phosphorylation-deficient mutants

Respiration-deficient mutants can be easily differentiated from another class of yeast mutants which posses a normal respiratory chain but are not able to grow on non-fermentable substrates [3, 14]. It could be assumed that some of them may be deficient in oxidative phosphorylation. Mitochondria from one such mutant, with a mutated gene op_1, were studied in more detail [15, 16]. They showed low phosphorylation efficiency ($P/O < 0{\cdot}4$), had preserved the coupling step of oxidative phosphorylation and functioning oligomycin-sensitivity conferring factor, but presumably displayed some modification in subsequent steps of mitochondrial ATP synthesis.

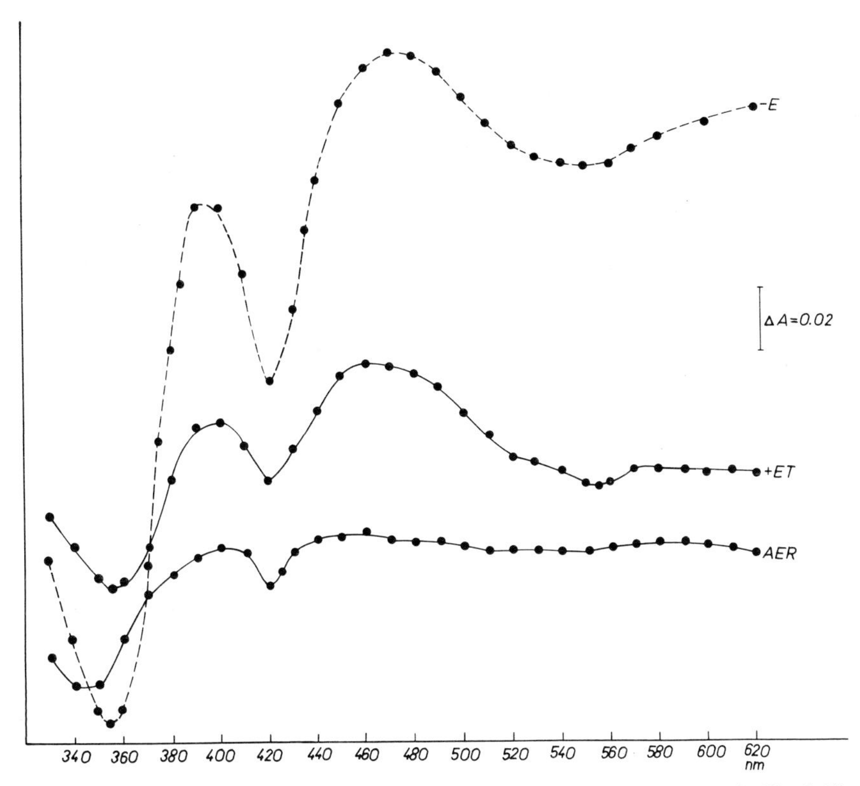

Figure 2. Difference spectra of ρ^- cells. Diploid ρ^- strain was cultured as in Fig. 1. The cells were suspended in a buffer containing 0·05M K phthalate and 0·05M K phosphate, pH 4·3 to contain 40 mg (dry weight) cells/ml. One cuvette of a split-beam spectrophotometer contained anaerobic suspension, the other cuvette suspension bubbled for 10 s with air.

Some doubts may be raised if the low phosphorylation efficiency found with isolated mitochondria is not an artifact due to unusual fragility of the mutant mitochondria. An argument in favour of the view that the operation of the mitochondrial phosphorylation system of the op_1 mutant is inefficient also *in situ,* is furnished by difference spectra of anaerobic yeast suspension in the presence of glucose (Fig. 3). It is known from the work of Chance [17] that glucose addition to an anaerobic yeast suspension is followed by considerable reduction of cytochrome *b*. In our experiments, this reduction was prevented by uncoupling agents and by dicyclohexylcarbodiimide, indicating that it was due to back flow of electrons in the respiratory chain elicited by increased phosphorylation potential. In the suspension of mutant op_1 cells the extent of

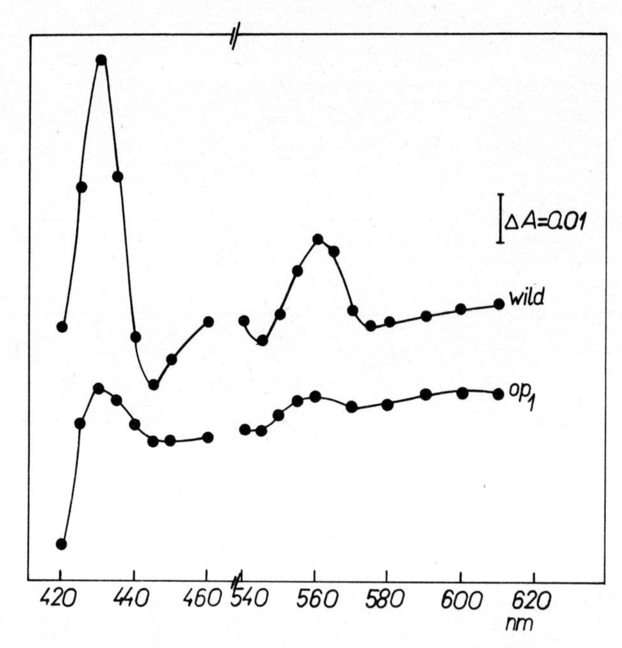

Figure 3. Difference spectra of anaerobic yeast suspensions after addition of glucose. Two cuvettes of a split-beam spectrophotometer contained 30 mg (dry weight) cells/ml of aerobically grown yeast suspended in a buffer containing 0·05M K phthalate and 0·05M K phosphate, pH 4·3. The cells were pre-bubbled with nitrogen and the suspension in the cuvettes covered with a layer of liquid paraffin. 40 mM glucose was added into one cuvette and the difference spectrum measured after 7 min at room temperature.

cytochrome *b* reduction was much lower, showing that the back flow of electrons under the established phosphorylation pressure was largely prevented in the mutant cells.

Combination of the cytoplasmic mutation to respiratory deficiency with nuclear mutation to oxidative phosphorylation deficiency

Since the two mutations mentioned apparently affect different parts of mitochondrial architecture, a superimposition of the cytoplasmic ρ^- mutation over the nuclear op_1 mutation should still more profoundly modify the mitochondria. In order to obtain such a double $op_1\rho^-$ mutant, op_1 cells were grown on glucose medium in the presence of acriflavine which is known to convert yeast cell population completely to ρ^- genotype [18]. Surprisingly, growth of the op_1 cells in the presence of acriflavin stopped after about ten generations and no means were found to resume it (Fig. 4). Genetic evidence indicated that such cultures, unable to grow further, contained double $op_1\rho^-$ mutant cells. The cells

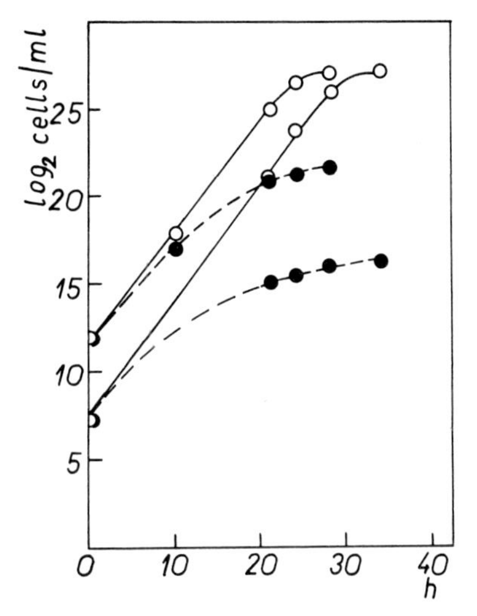

Figure 4. Growth curves of op_1 cells. Diploid op_1 strain was cultured at 30 °C on a shaker in medium containing 2% glucose, 1% peptone, 1·5% yeast autolyzate and 0·004% adenine. ○——○ controls without acriflavin; ●-----● in the presence of 5 μg/ml acriflavin.

were not dead since they did not stain with methylene blue, fermented glucose and assimilated a part of it. Apparently, although viable, the cells had lost growth or multiplication ability.

CONCLUSION

To account for the observations on ρ^- and on double $op_1\rho^-$ mutants we can speculate that there might exist a bidirectional flow of information between the nucleus and mitochondria in a cell. Two kinds of signals may come from mitochondria to the nucleus. One signal may function as a depressor of nuclear genes which code for some cytochromes and perhaps for other mitochondrial components. This signal could not be formed in ρ^- cells and the synthesis of the mitochondrial components in them would be permanently repressed. Another signal between mitochondria and the nucleus would operate to assure a balance between mitochondrial and cellular multiplications. The signal would be preserved in ρ^- cells, but the specific combination of cytoplasmic ρ^- and nuclear op_1 mutations in a cell would modify mitochondrial structures so profoundly that the signal would not function more and the cell would lose multiplication ability.

ACKNOWLEDGEMENT

Collaboration of H. Bednárová, J. Irmlerová and E. Hrušovská is gratefully acknowledged.

REFERENCES

1. Mounolou, J. C., Jakob, H. and Slonimski, P. P., *Biochem. biophys. Res. Commun.* **24** (1966) 218.
2. Nagai, S., Yanagishima, N. and Nagai, H., *Bact. Rev.* **25** (1961) 404.
3. Sherman, F. and Slonimski, P. P., *Biochim. biophys. Acta* **90** (1964) 1.
4. Slonimski, P., "La Formation des Enzymes Réspiratoires chez la Levure", Masson, Paris, 1953.
5. Kováč, L. and Weissová, K., *Biochim. biophys. Acta* **153** (1968) 55.
6. Schatz, G., *J. biol. Chem.* **243** (1968) 2192.
7. Tuppy, H., Swetly, P. and Wolff, I., *Eur. J. Biochem.* **5** (1968) 339.
8. Work, T. S., *in* "Round Table Discussion on Biochemical Aspects of the Biogenesis of Mitochondria" (edited by E. Quagliariello, E. C. Slater, S. Papa and J. M. Tager), Adriatica Editrice, Bari, 1967, in press.
9. Kužela, Š., Šmigaň, P. and Grečná, E., unpublished results.
10. Sherman, F., Taber, H. and Campbell, W., *J. molec. Biol.* **13** (1965) 21.
11. Kadenbach, B., *Biochim. biophys. Acta* **134** (1966) 430.
12. Ephrussi, B. and Slonimski, P. P., *Biochim. biophys. Acta* **6** (1950) 256.
13. Labbe, P. and Chaix, P., *C. r. hebd. Séanc. Acad. Sci., Paris* **258** (1964) 1645.
14. Kováč, L., Lachowicz, T. M. and Slonimski, P. P., *Science, N.Y.* **158** (1967) 1564.
15. Kováč, L. and Hrušovska, E., *Biochim. biophys. Acta* **153** (1968) 43.
16. Beck, J. C., Mattoon, J. R., Hawthorne, D. C. and Sherman, F., *Proc. natn. Acad. Sci. U.S.A.* **60** (1968) 186.
17. Chance, B., *Harvey Lect.* Ser. **49** (1953-54) 145.
18. Ephrussi, B., Hottinguer, H. and Chimenes, A. M., *Annls. Inst. Pasteur, Paris* **76** (1949) 351.

FEBS Symposium, Volume 17, 1969, pp. 205-217

Energy-conservation Mechanisms of Mitochondria

E. C. SLATER

Laboratory of Biochemistry, B.C.P. Jansen Institute,
University of Amsterdam, The Netherlands

The physiological function of mitochondria is to catalyse the oxidation of intermediary metabolites by molecular oxygen in such a way that the energy made available by these oxidation reactions may be conserved in a form that can be utilized by the cell for energy-requiring reactions, such as mechanical work, ion movements, or the formation of carbon-carbon covalent bonds in chemical syntheses. There are two main ways in which this energy may be conserved: by the synthesis of ATP from ADP and phosphate, or by the reduction of nicotinamide nucleotide.

Since mitochondria contain limited (although very considerable) amounts of adenine and nicotinamide nucleotides, the amount of energy that may be conserved in the mitochondrion in the form of ATP or reduced nicotinamide nucleotide is restricted, so that the reactions leading to the conservation of energy soon come to a stop when oxidizable substrate is added to mitochondria in the presence of oxygen. The reactions may be made continuous by adding either the substrates for oxidative phosphorylation (ADP and phosphate) or a hydrogen acceptor for the reduced nicotinamide nucleotide.

Experience in biochemistry has taught us that, when in the cell an enzyme or a sub-cellular structure possesses unusual properties, further investigation will reveal that these unusual properties have physiological significance. That of cytochrome a_3 with its very high affinity for oxygen is obvious. Not yet obvious is why cytochrome *c* oxidase, as isolated, is a polymer or oligomer of a protomer containing two molecules of heme *a*, only one of which reacts with oxygen or respiratory inhibitors, and two molecules of copper, also only one of which reacts readily with respiratory inhibitors. The physiological significance of a respiratory chain with a series of hydrogen or electron carriers is also clear, but it is not obvious why there are so many components in the chain. The characteristic structure of the mitochondrion must also have a special significance, the exact nature of which we do not yet understand.

The State 3 to State 4 transition in mitochondria

As already mentioned, mitochondria can store a limited amount of energy in the form of ATP or reduced nicotinamide nucleotide when substrate and oxygen are added. They are then said to be in the energized state, or in State 4 to use the terminology of Chance and Williams [1]. Lardy and Wellman [2] showed in 1952 that energized mitochondria respire only slowly, and that the addition of ADP (in the presence of phosphate) causes a considerable stimulation of respiration. Discussing these results in 1953, I pointed out that two possible explanations for this respiratory control remained open, a kinetic and a thermodynamic [3]. According to the kinetic (I used the word 'stoichiometric' in 1953) explanation, the steady-state concentration of ADP becomes so low as a result of oxidative phosphorylation as to limit respiration. According to the thermodynamic explanation, respiration comes to a stop because some of the phosphorylation reactions are readily reversible.

Chance and Williams [1] subsequently emphasized the kinetic explanation, whereas Klingenberg and Schollmeyer [4] proposed that both respiratory control and reversal of the respiratory chain [5, 6] are due to a thermodynamic equilibrium between redox and phosphorylation reactions.

Chance and Williams [1] first showed that, in State 4, the degree of reduction of the components of the respiratory chain descends from nicotinamide nucleotide to cytochrome *a*. This was confirmed by Klingenberg and Schollmeyer [4]. The stimulation of respiration associated with the addition of ADP caused, in both Chance and Williams' [1] experiments with rat liver mitochondria and with Klingenberg and Schollmeyer's [4] with locust flight muscle mitochondria, an oxidation of cytochrome *c* and components on the substrate side of cytochrome *c*, and a reduction of cytochrome *a*. More recently, however, Klingenberg and Kröger [7] reported that, with rat heart mitochondria oxidizing glutamate + malate, the crossover is not between cytochrome *c* and cytochrome *a*, but between cytochromes *b* and *c* (or c_1).

The position of the crossover in rat liver mitochondria has been recently systematically studied by Dr. Muraoka in our laboratory. Figure 1 shows that with 10 mM succinate no crossover is obtained, i.e. on the transition from State 3 to State 4, all cytochromes become more reduced. When the oxidase end of the chain is partially inhibited by 100 μM azide, a crossover between cytochromes *c* and *a* becomes apparent.*

Figure 2 shows another experiment, with a lower concentration of succinate (1 mM). In the absence of azide, cytochrome *c* becomes more reduced on the

* The wavelength pairs used are not completely selective. Cytochrome *c* includes a contribution from cytochrome c_1, cytochrome *a* a small contribution from a_3, and cytochromes *a* and a_3 contribute equally to the absorbancy change at 448 mμ *minus* 455 mμ.

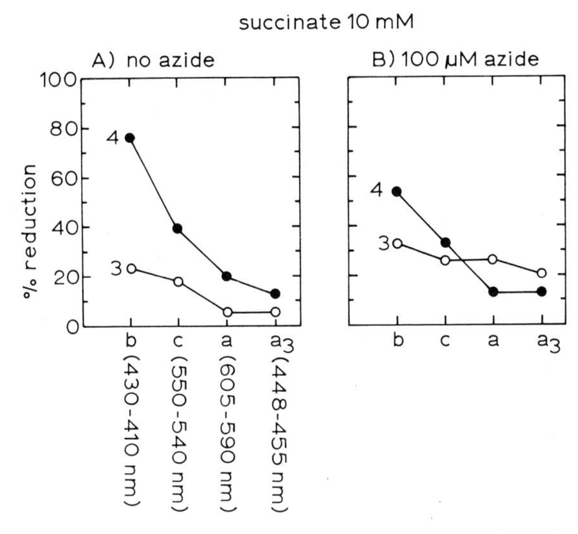

Figure 1. Redox states of cytochromes in State 4 and State 3 rat liver mitochondria oxidizing succinate in the presence and absence of 100 μM azide. 25 mM Tris-HCl buffer, 50 mM sucrose, 5 mM $MgCl_2$, 2 mM EDTA, 15 mM KCl, 10 mM phosphate, 0·3 mM ADP, 10 mM succinate, 1·3 mg/ml mitochondria. Temperature, 23°C. States 3 and 4 measured as in Fig. 2. Degree of reduction measured in Aminco-Chance dual-wavelength spectrophotometer, using the wavelength pairs indicated. 100% = Absorbancy difference between State 2 and anaerobic state measured with $Na_2S_2O_4$. No correction was made for the effect of azide on the absorption spectrum of ferrocytochrome *a* (Ref. 26). Since cytochrome *a* contributes also to the $A_{448-455}$ mμ, the degree of reduction of cytochrome a_3 in the presence of azide is probably overestimated.

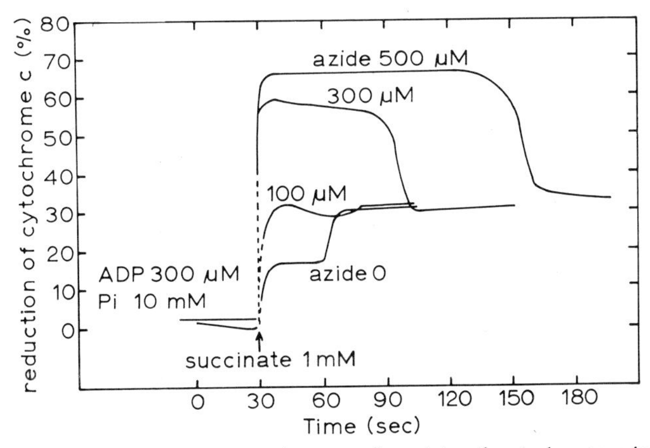

Figure 2. Effect of azide concentration on redox state of cytochrome *c* in State 3 and State 4 mitochondria. Succinate added at State 2 yields first State 3 and then State 4. Other conditions as in Fig. 1, except that 0·1 μg/ml rotenone was present. 1·8 mg/ml rat liver mitochondria.

transition from State 3 to State 4. With 100 μM azide, there is almost no change, and with higher concentrations of azide, cytochrome *c* becomes more oxidized during the transition, i.e. the cross over shifts to the substrate side of cytochrome *c*. It is striking that although the degree of reduction of cytochrome *c* in State 3 increases with increasing azide concentration, as would be expected if the degree of reduction is governed by steady-state or kinetic considerations, the degree of reduction in State 4 is independent of the azide concentration. This suggests rather strongly that in State 4 an equilibrium is reached, governed by the phosphate potential, as proposed by Klingenberg and Schollmeyer [4] for the first two phosphorylation sites. Indeed, it would appear from Fig. 2 that all three sites are in equilibrium.

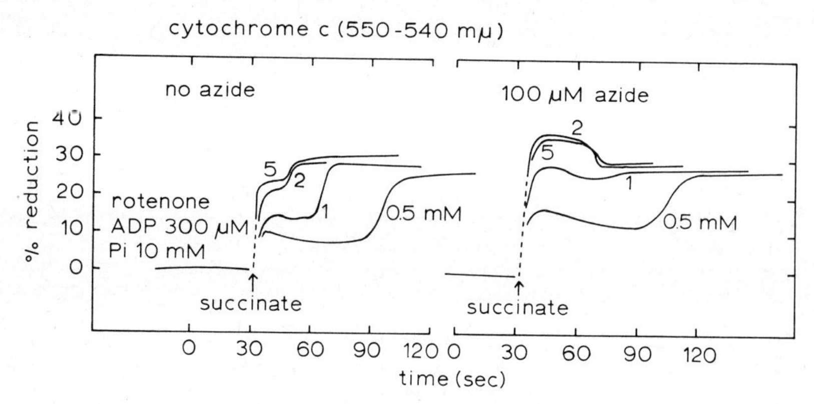

Figure 3. Effect of succinate concentration on redox state of cytochrome *c* in State 3 and State 4 mitochondria. Procedure as in Fig. 2. Concentrations of succinate (mM) are indicated against the trace.

Figure 3 shows that the degree of reduction of cytochrome *c* in State 3, in the presence and absence of azide, is dependent upon the concentration of succinate, being lower with the smaller succinate concentrations. Once again, the degree of reduction in State 4 is largely independent of succinate concentration, increasing only slightly with increasing substrate concentration.

In our view, the crossover phenomenon should be interpreted in terms of a kinetic steady state in State 3 and a thermodynamic equilibrium in State 4. The redox state in State 3 depends upon the relative activities of the dehydrogenase and oxidase ends of the chain. The redox state in State 4, on the other hand, is dependent only on the phosphate potential and is independent of the relative activities of the various parts of the chain. Increasing inhibition of the oxidase end of the chain, brought about by increasing concentrations of azide, causes the chain to become more reduced in State 3, but has no effect on the redox state in State 4. In consequence, the crossover shifts towards substrate with increasing concentration of azide. I remain unconvinced, then, that the position of the crossover need be identical with a phosphorylation site, as proposed by Chance and Williams [1]. However, it is possible to draw the conclusion that the redox state of cytochrome a_3 is under control of the phosphate potential in State 4. In other words, the gap between cytochrome a_3 and oxygen is a phosphorylating site, as was to be expected on thermodynamic grounds. The apparent crossover between cytochrome aa_3 and oxygen, which is illustrated in Fig. 1 for succinate as substrate, was also obtained with pyruvate, β-hydroxybutyrate, and ascorbate + tetramethyl-*p*-phenylenediamine as substrate. Ramirez [8] has previously reported a crossover above cytochrome a_3 on initiation of muscle activity in intact toad, frog or lobster heart muscle.

The phosphate potential

Clearly, it is now important to consider the magnitude of the phosphate potential in oxidative phosphorylation. Let us first consider the simple case of one phosphorylating step in the respiratory chain, both preceded and succeeded by non-phosphorylating electron or hydrogen transfers. This is represented in Fig. 4. Hydrogen or electron transfer from A to B, and therefore through the

$$SH_2 \rightarrow A \rightleftarrows B \rightarrow O_2$$

$$\begin{array}{lll} SH_2 + A & \rightarrow S + AH_2 & \\ AH_2 + B + C & \rightleftharpoons A{\sim}C + BH_2 & (1) \\ A{\sim}C + ADP + P_i & \rightleftharpoons A + C + ATP & (2) \\ BH_2 + 1/2\,O_2 & \rightarrow B + H_2O & \\ \hline \end{array}$$

$$\text{Sum: } SH_2 + 1/2\,O_2 + ADP + P_i \rightarrow S + H_2O + ATP$$

Figure 4. Representation of a phosphorylating step in the respiratory chain.

whole chain, will cease when eqn. 1 is at equilibrium, which is reached when eqn. 2 is also at equilibrium. Thus the phosphate potential that can be maintained by respiration is a measure of the ΔG of the phosphorylating step $AH_2 + B \rightarrow A + BH_2$. Experimentally, we can only determine lower limits since there will be a measurable hydrolysis of ATP by extraneous ATPase, as well as probable losses of A~C.

Surprisingly, only one measurement of the phosphate potential in State 4 has appeared in the literature. Two years ago, Cockrell *et al.* [9] reported that the phosphate potential is as high as 15·6 kcal/mole at pH 7·8. We have confirmed and extended this finding.

As in the experiments of Cockrell *et al.* [9], we measured the concentrations of ATP, ADP and P_i in the fluid in which respiring mitochondria are suspended, after State 4 has been reached. Since the phosphate potential at the end of the experiment is higher than at the beginning, our values represent the minimum value of the potential against which respiring mitochondria are capable of making ATP. In the experiments described here we used succinate as substrate at 25°C, in the presence of rotenone. All experiments were carried out in the absence of added magnesium. We separated the mitochondria by rapid centrifugation into silicone [10] after 10-min respiration and analysed the supernatant layer. The same results were obtained after respiration for 5, 10, 15 or 30 min.

We have expressed the phosphate potential, at a particular pH, $\Delta G'$, according to the formula

$$\Delta G' = \Delta G'_o + 1{\cdot}36 \log \frac{[\mathrm{ATP}]}{[\mathrm{ADP}]\,.\,[\mathrm{P}_i]}$$

$$= \Delta G'_o + \Delta G'_c$$

where $\Delta G'_o$ is the standard free-energy change when the activities of ATP, ADP and P_i are all equal to 1M and $\Delta G'_c$ is the concentration term.

The effect of pH on $\Delta G'$ is shown in Table 1. The effect of pH on $\Delta G'_o$ is taken from a review by George and Rutman [11]. The results show that $\Delta G'$ is maximal at a pH a little greater than 7. Table 2 shows that the potential is constant when we vary the concentration of the components, by choosing different initial phosphate or ADP concentrations. The potential at pH 7·85, like that in the second experiment of Table 1, is 15·8. Indeed values between 15·8 and 16·1 have been obtained in several experiments.

The fact that mitochondria can make ATP against a phosphate potential of at least 15·8 kcal/mole has important implications, not only for the chemiosmotic hypothesis, as I have already discussed elsewhere [12], but for any hypothesis. In the first place, it provides further support for the suggestion, already made on the basis of the experiments on the transition from State 3 to State 4, that the entire phosphorylating respiratory chain between NADH and oxygen is reversible. This follows from the fact that the free energy for the reaction between NADH by oxygen, under the conditions of our experiments (51·9 kcal), is almost balanced by the energy utilized in making 3 molecules of ATP at 15·8 kcal/mole, i.e. 3 x 15·8 = 47·4 kcal.

Indeed, when one considers that the value of 15·8 kcal for the measured phosphate potential is probably an underestimate, because leaks are always present in State 4 mitochondria, the agreement between the two $\Delta G'$ values is quite good.

A phosphate potential of 15·8 kcal/mole delineates quite sharply the three phosphorylating sites, which must lie between NADH and the *b*-Q region, between *b*-Q and the *c*-*a* region, and between the latter and oxygen. This is clearly seen in Fig. 5, in which the data of Klingenberg and Kröger [7] with rat heart and of Muraoka with rat liver are plotted on a $\Delta G'$ scale, relative to oxygen

Table 1. Effect of pH on phosphate potential

$$\Delta G' = \Delta G'_o + 1{\cdot}36 \log \frac{[\text{ATP}]}{[\text{ADP}]\ .\ [\text{P}_i]}$$
$$= \Delta G'_o + \Delta G'_c$$

Expt.	pH	$\Delta G'_o$	$\Delta G'_c$	$\Delta G'$
1	6·45	8·5	6·15	14·65
	7·1	8·85	6·55	15·4
	7·7	9·3	5·83	15·1
	8·25	9·9	4·63	14·5
2	6·85	8·7	6·35	15·05
	7·4	9·1	6·75	15·85
	7·9	9·6	6·2	15·8
	8·45	10·1	5·35	15·45

Table 2. Effect of varying concentrations of phosphate and ADP on phosphate potential

$$\Delta G' = \Delta G'_o + \Delta G'_o$$
$$= 9{\cdot}4 + 6{\cdot}4$$
$$= 15{\cdot}8 \text{ kcal/mole}$$

ADP (mM)	P_i (mM)	ATP (mM)	$\Delta G'_c$ (kcal)
0·13	1·01	5·94	6·35
0·18	0·81	6·48	6·34
0·20	0·58	6·50	6·48
0·37	0·45	6·77	6·29
		Mean	6·40

pH 7.85, 25°C.

at 0·2 atm. Muraoka's calculations for aa_3 are based on measurements of the Soret band and are made for the assumption that the potential for cytochrome a_3 is the same as that of cytochrome *a*. The $\Delta G'$ for the reaction between NADH and oxygen is slightly overestimated in Muraoka's measurements, since both NADP and NAD are included in these measurements.

If, as we have concluded, State 4 is an equilibrium, cytochrome *b* and Q should fall together on the free-energy scale in Fig. 5, and so should *c* and *a*. Indeed they are quite close, although the difference between *c* and *a* is disturbingly great in some cases but not in all. The surplus of hydrogen or

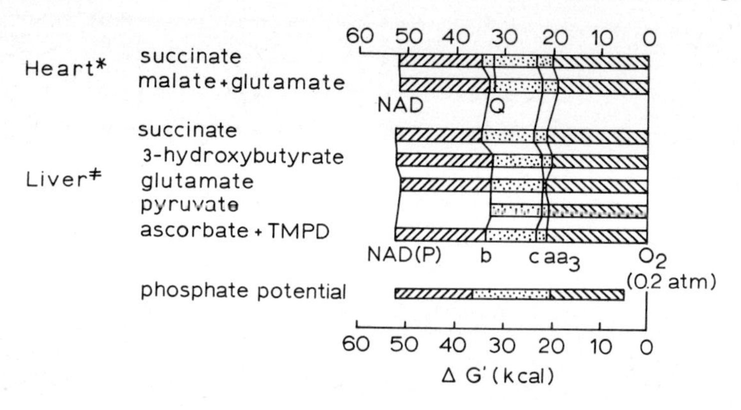

Figure 5. The redox state of components of the respiratory chain in State 4 mitochondria (rat heart and rat liver) plotted on a $\Delta G'$ scale, relative to oxygen at 0·2 atm. Calculations carried out as in Ref. 25, Table IV. (An error appears in the last two columns of the last line of Table IV. This should read −23410 and −12870, respectively.)

electron carriers in the respiratory chain becomes even more embarassing when viewed in this light. The extra components not indicated in Fig. 5 (the two flavoproteins between NADH and Q, the iron-sulphur groupings both in Site I and Site II, cytochrome c_1 and the copper atoms in cytochrome aa_3) must presumably lie in the region of NADH, *b* or *a* on the free-energy scale. Since the transition from State 4 to State 3 on the addition of ADP occurs rapidly and immediately, it seems unlikely that any component of the chain will be overwhelmingly (more than 99% say) in the fully oxidized or fully reduced form. Thus, their standard redox potentials will probably be not very far from one of the key regions, NAD, cytochrome *b* or cytochrome *a*. If this is the case, their redundance for any simple chemical mechanism of oxidative phosphorylation, of the type of $AH_2 + B + C \rightarrow A{\sim}C + BH_2$, is even more striking. Our lack of understanding of the function of these apparently redundant components of the respiratory chain is a measure of our lack of understanding of the primary act of energy conservation. Interestingly enough, a long respiratory chain is not only required for oxidative phosphorylation in mitochondria, but also for hydroxylation reactions catalysed by microsomes or adrenal-cortex mitochondria.

Table 3. Coupling activity of F_1-X on submitochondrial particles

Expt.	Additions	P:O
1	None	0·04
	F_1 (88 μg)	0·10
	F_1-X (120 μg)	0·68
2	None	0·04
	F_1 (18 μg)	0·08
	F_1 (72 μg)	0·09
	F_1-X (500 μg), cold-treated	0·06
	F_1 (72 μg) + F_1-X (500 μg), cold-treated	0·47

See Ref. 14 for details.

Thanks to the work of Racker *et al.* [13], we do know something about the last reaction leading to ATP synthesis. This is probably catalysed by F_1, a coupling factor needed to restore oxidative phosphorylation to certain types of submitochondrial particles. Vallejos [14] in our laboratory has isolated another form of F_1, which we have named F_1-X, because it probably contains another factor complexed to F_1. Like Racker's F_1, F_1-X is cold labile, but the X component present in cold-treated F_1-X is still capable of supplementing Racker's F_1 preparation in stimulating oxidative phosphorylation (Table 3.)

Everyone, even Mitchell [15], agrees with the mechanism of oxidative phosphorylation formulated by the equations

$$\mathrm{NADH} + \mathrm{H}^+ + \frac{1}{2}\mathrm{O}_2 + 3\,X + 3\,I \rightleftharpoons \mathrm{NAD}^+ + \mathrm{H_2O} + 3\,X{\sim}I \qquad (1)$$

$$3\,X{\sim}I + 3\,\mathrm{ADP} + 3\,\mathrm{P}_i \rightleftharpoons 3\,X + 3\,I + 3\,\mathrm{ATP} \qquad (2)$$

$$\mathrm{NADH} + \mathrm{H}^+ + \frac{1}{2}\mathrm{O}_2 + 3\,\mathrm{ADP} + 3\,\mathrm{P}_i \rightleftharpoons \mathrm{NAD}^+ + \mathrm{H_2O} + 3\,\mathrm{ATP}$$

It seems safe to say that Racker's F_1 is an enzyme involved in Reaction 2. However, nothing is known about the nature of $X{\sim}I$ or of the mechanism of Reaction 2.

Reaction 1 is the fundamental and controversial one. It is fundamental because $X{\sim}I$, or energy conserved in the respiratory chain before the synthesis of the high-energy compound *X*~*I*, can be utilized directly for many energy-requiring reactions in the mitochondria without first conversion to ATP. In recent years, much has been learned about these reactions. Although this new knowledge has enriched our understanding of the bioenergetics of the mitochondrion, it has not greatly contributed to our knowledge of the mechanism of the primary energy-conserving reaction. Indeed, I think that we can be easily misled into thinking that we are studying energy conservation when we are in reality studying energy utilization.

There are at present three hypotheses concerning the nature of the primary energy-conserving process (see Fig. 6). I still favour the chemical hypothesis,

EXISTING HYPOTHESES CONCERNING THE NATURE OF THE PRIMARY ENERGY-CONSERVING PROCESS

(1) *CHEMICAL*

$$AH_2 + B + C(I) \rightleftharpoons A{\sim}C(I) + BH_2$$

(2) *CHEMIOSMOTIC*

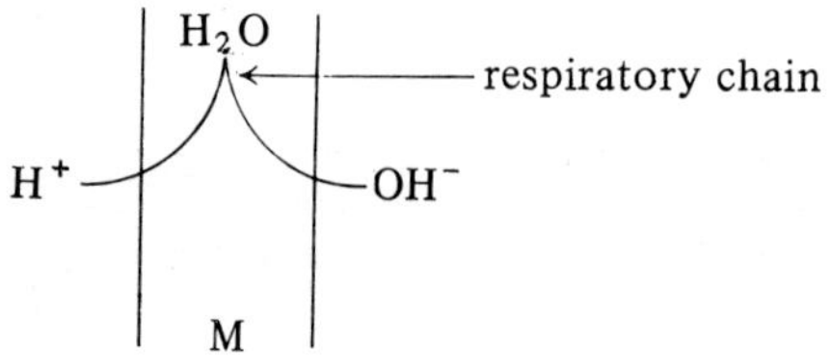

H_2O
respiratory chain
H^+
OH^-
M

(3) *CONFORMATIONAL*

$$AH_2 + B \rightleftharpoons A^* + BH_2$$

Figure 6. Current hypotheses on the mechanism of energy conservation in the respiratory chain.

although as already mentioned it does not give a satisfactory explanation for the multi-component respiratory chain. Mitchell's chemiosmotic hypothesis [15] is in many ways attractive and it has stimulated much discussion. In my opinion, however, the experimental evidence is against the basic postulates of the hypothesis, and the thermodynamic difficulties are even greater than with the chemical hypothesis. The conformational hypothesis, first proposed by Boyer [16] and recently supported by Green *et al.* [17], is formally rather similar to the chemical hypothesis, particularly in the form used by Chance [18], since the energy of the oxidoreduction reaction is conserved in one of the products, not by combining with a ligand to form an energy-rich compound as in the chemical hypothesis, but in a high-energy conformation of the protein. Thus, all the formal equations developed around the chemical hypothesis to explain reversibility of the respiratory chain, exchange reactions and the action of uncouplers and inhibitors can be applied without change.

At present the hypothesis that energy is conserved in a high-energy conformation of a protein is an attractive speculation without much experimental support. Personally, I do not believe that the experimental evidence will come from electron microscopy of the intact mitochondrion which has a diameter of about 10,000 Å. I think that we have to look for interactions at a few Å distance, and these are more likely to be revealed by studying the isolated enzymes or enzyme complexes, rather than the intact mitochondrion or even submitochondrial particles. In this context, I consider the studies of the

$$R \rightleftharpoons T + nI \rightleftharpoons T_I$$

$$R + nI \rightleftharpoons T_I$$

$$\Delta G'_o = -1{\cdot}38 \log L\left(1 + \frac{[I]}{K_i}\right) n$$

$$\text{where } L = \frac{[\mathrm{To}]}{[\mathrm{Ro}]} \text{ at equilibrium in absence of } I$$

Example:

$$L = 10^{-2}, \frac{[I]}{K_i} = 10^3, n = 4$$

$$\Delta G'_o = -13{\cdot}8 \text{ kcal}$$

Figure 7. An allosteric model for energy conservation in the respiratory chain.

mechanism of inhibition of antimycin by Rieske and colleagues [19] in Green's laboratory, and by Bryła and Kaniuga [20] in Warsaw and Amsterdam, most significant. Rieske and co-workers showed that antimycin inhibits the cleavage of cytochromes b and c_1 in Complex III, one of the four lipoprotein complexes that can be isolated from the respiratory chain. Bryła and Kaniuga have provided impressive evidence that antimycin should be considered an allosteric inhibitor of the respiratory chain. I shall not discuss this and I feel it is only sufficient here to point out that this hypothesis provides an explanation for the sigmoid inhibition curve for particulate preparations [20] and the hyperbolic inhibition curve for Complex III (ref. 21).

On the basis of the allosteric model of Monod *et al.* [22], sigmoid inhibition curves are to be expected when the inhibitor combines more firmly with a conformational state of an oligomeric protein that is present in low concentrations, than with a conformational state present in high concentrations.

This is represented schematically in Fig. 7. Thus, the inhibitor shifts the equilibrium in favour of the conformational form that, in the absence of inhibitor, is less favoured, i.e. the form with the higher energy.

It seems possible that the application of Monod's model to the respiratory chain might provide an alternative hypothesis that combines features of both the original ~ and conformational hypotheses. This requires further working out. To mention just one of the possibilities, if R and T represent two conformational forms of the b-c_1 complex and b is the site of action of antimycin, then T_I represents a high-energy state of cytochrome b, indeed the ferrocytochrome $b{\sim}I$ intermediate originally proposed by Chance and Williams [1]. If antimycin can be considered a model for a naturally occurring inhibitor, it is not difficult to visualize in general terms a mechanism in which electron transfer from b to c_1, leading to ferricytochrome b, which does not combine with the inhibitor, might be utilized to make high-energy intermediates, while cytochrome b reverts to the R state. The $\Delta G_0'$ of the reaction $R + I \rightleftharpoons T_I$ is a function of L (the equilibrium constant in the absence of inhibitor), the concentration of inhibitor, the inhibition constant K_I and n, the number of promoters in the oligomeric or polymeric b-c_1 complex. Even though, according to this model, L is small, ΔG_0 can be made high if K_I is low and n is high, as is illustrated by the example given.

I am rather intrigued by the possibilities, as illustrated here, that an oligomeric structure gives of conserving energy. According to the model, the energy is already conserved in the form of T_I before electron transfer, which releases the energy by removal of I and reversion of the protein to its R state.

Antimycin is not the only inhibitor that shows a sigmoid inhibition curve. The same is also true for oligomycin, which also has a special feature that it inhibits F_1 only after the addition of CF_0 (Ref. 23). Perhaps, we should think of the $F_1.CF_0$ complex as analogous to aspartate carbamyltransferase which consists of two polypeptide chains, one of which binds substrate and the other the inhibitor CTP (Ref. 24).

In conclusion, it would appear that the application of the allosteric model of Monod, Wyman and Changeux to the respiratory chain and to oxidative phosphorylation opens up interesting possibilities.

REFERENCES

1. Chance, B. and Williams, G. R., *Adv. Enzymol.* **17** (1956) 65.
2. Lardy, H. A. and Wellman, H., *J. biol. Chem.* **195** (1952) 215.
3. Slater, E. C., *A. Rev. Biochem.* **22** (1953) 17.
4. Klingenberg, M. and Schollmeyer, P., *in* "Proceedings of the 5th International Congress of Biochemistry, Moscow, 1961", Vol. 5, Pergamon Press, Oxford, 1963, p. 46.

5. Chance, B. and Hollunger, G. R., *Fedn Proc. Fedn Am. Socs exp. Biol.* **16** (1957) 163.
6. Bücher, Th. and Klingenberg, M., *Angew. Chem.* **70** (1958) 552.
7. Klingenberg, M. and Kröger, A., *in* "Biochemistry of Mitochondria" (edited by E. C. Slater, Z. Kaniuga and L. Wojtczak), Academic Press and Polish Scientific Publishers, London and Warsaw, 1967, p. 11.
8. Ramirez, J., *J. Physiol., Lond.* **147** (1959) 14.
9. Cockrell, R. S., Harris, E. J. and Pressman, B. C., *Biochemistry* **5** (1966) 2326.
10. Harris, E. J. and van Dam, K., *Biochem. J.* **106** (1968) 759.
11. George, P. and Rutman, R. J., *Prog. Biophys. biophys. Chem.* **10** (1960) 1.
12. Slater, E. C., *Eur. J. Biochem.* **1** (1967) 317.
13. Penefsky, H. S., Pullman, M. E., Datta, A. and Racker, E., *J. biol. Chem.* **235** (1960) 3330.
14. Vallejos, R. H., Van den Bergh, S. G. and Slater, E. C., *Biochim. biophys. Acta* **153** (1968) 509.
15. Mitchell, P., "Chemiosmotic Coupling in Oxidative and Photosynthetic Phosphorylation", Glynn Research, Bodmin, Cornwall, 1966.
16. Boyer, P., *in* "Oxidases and Related Redox Systems" (edited by T. E. King, H. S. Mason and M. Morrison), Vol. 2, Wiley, New York, 1965, p. 994.
17. Harris, R. A., Penniston, J. T., Asai, J. and Green, D. E., *Proc. natn. Acad. Sci. U.S.A.* **59** (1968) 830.
18. Chance, B., Lee, C. P. and Mela, L., *Fedn Proc. Fedn Am. Socs exp. Biol.* **26** (1967) 1341.
19. Baum, H., Rieske, J. S., Silman, H. I. and Lipton, S. H., *Proc. natn. Acad. Sci. U.S.A.* **57** (1966) 798.
20. Bryła, J. and Kaniuga, Z., *Biochim. biophys. Acta* **153** (1968) 910.
21. Rieske, J. S., Baum, H., Stoner, C. D. and Lipton, S. H., *J. biol. Chem.* **242** (1967) 4854.
22. Monod, J., Wyman, J. and Changeux, J. P., *J. molec. Biol.* **12** (1965) 88.
23. Kagawa, Y. and Racker, E., *J. biol. Chem.* **241** (1966) 2461, 2467.
24. Changeux, J. P., Gerhart, J. C. and Schachmann, H. K., *in* "Regulation of Nucleic Acid and Protein Biosynthesis", (edited by V. V. Koningsberger and L. Bosch), BBA Library, Vol. 10, Elsevier, Amsterdam, 1967, p. 344.
25. Slater, E. C., *in* "Comprehensive Biochemistry" (edited by M. Florkin and E. H. Stotz), Vol. 14, Elsevier, Amsterdam, 1966, p. 327.
26. Wilson, D. and Chance, B., *Biochim. biophys. Acta* **131** (1967) 421.

FEBS Symposium, Volume 17, 1969, pp. 219-232

The Chemical and Electrical Components of the Electrochemical Potential of H^+ Ions across the Mitochondrial Cristae Membrane

P. MITCHELL

Glynn Research Laboratories, Bodmin, Cornwall, England

One of the crucial questions concerning the so-called energy conservation mechanisms of mitochondria [1], is whether a current of protons, generated by respiration, drives ATP synthesis osmotically by reversing the chemically separate mitochondrial ATPase reaction, or whether there is direct coupling between the respiratory and ATPase systems at the chemical level of dimensions.

If a proton current is the connecting link between the respiratory chain and the reversible ATPase or hydrodehydration system in the mitochondrial membrane certain requirements must be fulfilled.

The fact that the activity of the respiratory chain system and the activity of the reversible ATPase system of rat liver mitochondria are each coupled to proton translocation of appropriate stoichiometry – so that $(\rightarrow H^+/O)/(\rightarrow H^+/P) = P/O$ – may be most simply interpreted to mean that the oxidoreduction (o/r) chain and the hydrodehydration (h/d) chain are chemically separate proton translocating systems as illustrated in Fig. 1A [2-5]. However, it has been pointed out by Mitchell and Moyle [4] and by Slater [6] that the same stoichiometric results would be obtained if the o/r chain and the h/d chain were coupled chemically through a hypothetical $X{\sim}I$ factor, provided that, as suggested by Chappell and Crofts [7, 8] and by Lardy *et al.* [9], there was an H^+ ion pump driven by the hypothetical $X{\sim}I$ intermediate, as illustrated in Fig. 1B. A fundamental difference between the chemiosmotic coupling scheme A and the chemical coupling scheme B is that the hypothetical H^+ ion pump on the sideline in scheme B is so far required only to conform *ad hoc* with the proton translocation stoichiometries observed experimentally. Considerations of the possible function of the $X{\sim}I$-driven H^+ pump (see Chappell and Crofts [8] and Slater [6] have not, so far, imposed any testable requirements on this hypothetical system in scheme B. On the other hand, according to scheme A, the protons are an intermediary on the direct pathway between respiration and phosphorylation, and it follows that two requirements must be fulfilled in

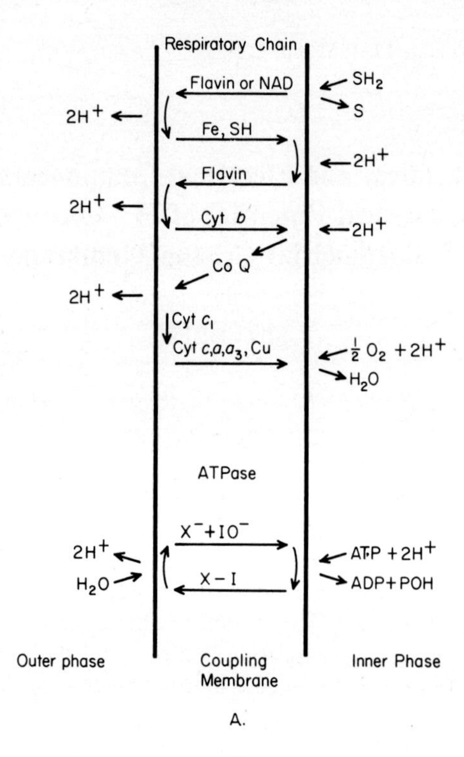

Figure 1. A. Proton translocating respiratory chain and proton translocating ATPase of the chemiosmotic hypothesis. The respiratory chain system is illustrated for an NAD-linked substrate. Succinate taps into the chain at the second loop (see [5, 10, 11]). B. Chemical hypothesis with H^+ ion pump.

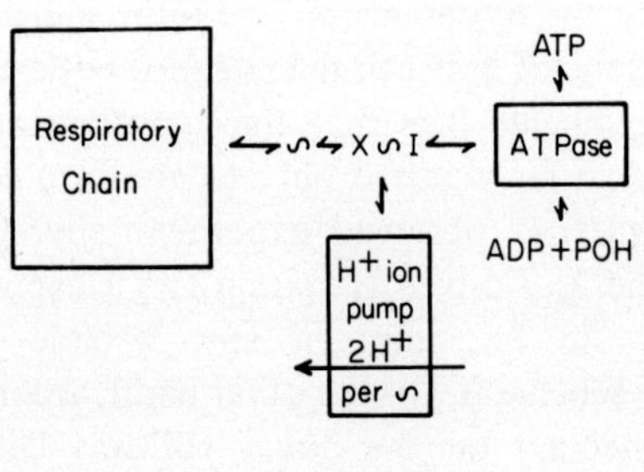

addition to the stoichiometric one already mentioned: the thermodynamic potential difference of the protons or protonmotive force (p.m.f.) across the coupling membrane must be sufficient to drive ATP synthesis at the appropriate ATP/(ADP + Pi) poise; and the rate of flow of the proton current during respiration must be sufficient to drive ATP synthesis at the known rate.

RATE OF FLOW OF PROTON CURRENT

Figure 2 shows that the initial rate of proton translocation, observed when respiration is initiated by a pulse of oxygen in a rat liver mitochondrial suspension in which proton translocation has been freed from electrical back pressure by the presence of a high concentration of valinomycin (see below), corresponds to 5·7 μg ions of H^+/s per g of mitochondrial protein. The State 3 rate of respiration corresponds to 0·59 μg atoms of O/s per g of mitochondrial protein, equivalent to a proton translocation rate of 3·5 μg ions of H^+/s per g of

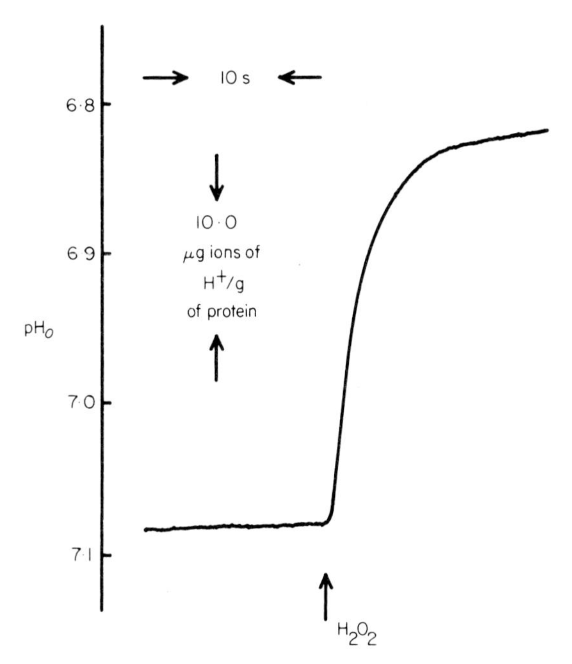

Figure 2. Rate of respiration-driven proton translocation. The figure shows an uncorrected recording from a H^+ ion-sensitive glass electrode in 3·2 ml of mitochondrial suspension (7·0 mg of protein/ml) at 25°C in 150 mM choline chloride, 25 mM sucrose, 3·3 mM glycylglycine, 2 mM β-hydroxybutyrate, 0·2 mM EGTA, 5 μl of catalase (see legend of Fig. 5) and 100 μg of valinomycin/g of protein. The mitochondria were equilibrated anaerobically for 10 min at pH 7·0–7·1 and 2 μl of 0·5M H_2O_2 was injected at the arrow. (Data of Mitchell and Moyle [13].)

mitochondrial protein at the known $\rightarrow H^+/O$ quotient of 6. The rate of ATP synthesis in State 3 (allowing for a State 4 respiration rate of 0·17 μg atoms of O/s per g of protein that does not contribute to phosphorylation) is 1·25 μmoles of ATP/s per g of mitochondrial protein. Thus, in this typical experiment, the observed initial rate of proton translocation driven by respiration corresponds to 1·6 times the total rate of proton translocation required to drive ATP synthesis at the State 3 rate and provide for energy loss at the State 4 rate. It follows that the rate of flow of the proton current generated by β-hydroxybutyrate oxidation is fast enough to permit the protons to mediate between respiration and phosphorylation. Similar results have been obtained when succinate is used as substrate.

THE TOTAL PROTONMOTIVE FORCE

The total p.m.f. Δp is made up of the electric potential $\Delta\psi$ and the pH difference ΔpH across the membrane [10-12] according to equation (1)

$$\Delta p = \Delta\psi - Z\Delta pH \tag{1}$$

The value of Z is 59 at 25°C when Δp and $\Delta\psi$ are expressed in mV and ΔpH is expressed in pH units; and the $\Delta\psi$ values mean the values in the inner phase minus those in the outer phase. Dr. Moyle and I have recently described the experimental details of the methods of measuring $\Delta\psi$ and ΔpH in resting and respiring rat liver mitochondria [13], and I shall therefore concentrate attention here on the rationale of the measurements and on the broader implications of the findings.

The measurements of $\Delta\psi$ depend on the fact that it is possible, under appropriate conditions, to cause the cristae membrane of rat liver mitochondria to become freely permeable to K^+ ions without appreciably changing the permeability to H^+ ions and without interfering with oxidative phosphorylation. Under such conditions, the Donnan equilibrium distribution of the K^+ ions across the membrane can be used to estimate $\Delta\psi$ from the classical Donnan equilibrium relationship, as shown with a preliminary result described previously [5].

There is now much evidence [8, 14-18] that valinomycin forms a lipid-soluble K^+ ion complex and thus makes lipid membranes permeable to K^+ ions, but it has been reported that valinomycin increases not only the K^+ ion permeability but also the H^+ ion permeability of artificial membranes [17]. Further, Pressman and collaborators [19, 20] have consistently taken the view that the effects of valinomycin on K^+ ion movements in mitochondria are due, not to passive K^+ ion permeability, but to activation of a metabolically-driven K^+ ion pump, perhaps by forming a K^+ ion-conducting pathway via a K^+ ion pump that is otherwise cryptified in the membrane. How, then, can one justify the statement that valinomycin can be used to equilibrate K^+ ions but not H^+ ions across the cristae membrane, and thus, to determine $\Delta\psi$?

It was observed by Mitchell and Moyle by pulsed acid-base titrations [21] and by measurements of the decay of respiration-driven proton pulses [2] that 10 μg of valinomycin/g of mitochondrial protein did not significantly affect the proton conductance of the M phase although one could infer that it induced a large increase in the K^+ ion conductance. The fact that this result differs from

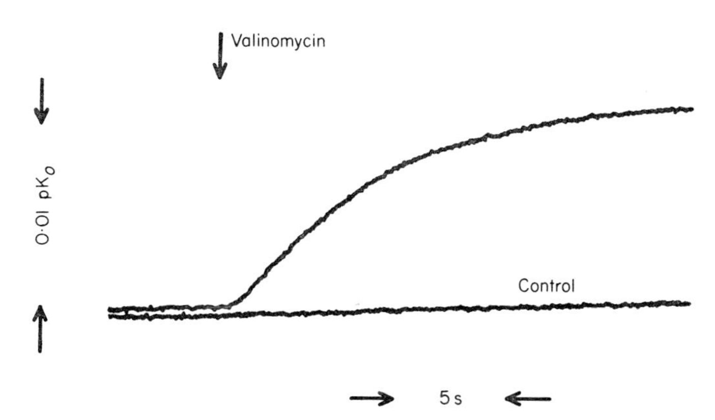

Figure 3. Passive permeability to K^+ ion induced by valinomycin. The figure shows uncorrected recordings from a K^+ ion-sensitive electrode. The measurements were done on 3·2 ml of anaerobic mitochondrial suspension (13·6 mg of protein/ml) at 25°C in a 250 mM sucrose medium containing 3·3 mM glycylglycine, 9 mM choline chloride, 1 mM KCl, 0·4 mM EGTA and 1 μM FCCP. After equilibration for 30 min at pH 7·08, a pulse of 4·0 μl of 0·5M choline hydroxide was added and pH_o shifted rapidly to 7·59. At arrow (30 s after the alkali pulse), 4·3 μl of oxygen-free ethanolic valinomycin (500 μg/ml) was injected, giving a final valinomycin concentration of 50 μg/g of mitochondrial protein. Control, no valinomycin added. (Data of Mitchell and Moyle [13].)

the observations of Andreoli *et al.* [17] in which valinomycin was shown to increase the H^+ ion conductance of artificial lipid membranes may be explained by the circumstance that the mitochondrial experiments were done near pH 7 whereas the artificial membrane experiments were done at more than a thousand-fold higher H^+ ion concentration (between pH 3 and 4).

Figure 3 shows an experiment with anaerobic rat liver mitochondria treated with 1 μM FCCP* in which the K^+ ion equilibration induced by 50 μg of valinomycin/g of mitochondrial protein was observed with a K^+ ion-sensitive electrode after the equilibrium poise was shifted by an alkali pulse. The time for half equilibrium of the K^+ ions on adding the valinomycin was only about 5 s.

* *Non-Standard Abbreviations.* Carbonylcyanide *p*-trifluoromethoxy phenylhydrazone, FCCP; ethylene glycol-bis(aminoethyl)-tetraacetic acid, EGTA; time for half equilibration, $t_{\frac{1}{2}}$

Other related experiments (see [13]) in media containing about 1 mg ion or 10 mg ion of K^+/litre confirmed that 100 μg of valinomycin/g of mitochondrial protein did not appreciably change the H^+ ion conductance of the M phase, but permitted equilibration of K^+ ions across the M phase with a $t\frac{1}{2}$ of only about 5 s. The possibility still remained, however, that in addition to the passive K^+ ion permeability induced by valinomycin, a metabolism-driven K^+ ion pump might be activated, as contended by Pressman and collaborators [19, 20]. If this were the case, the simultaneous activation of the hypothetical K^+ ion pump and

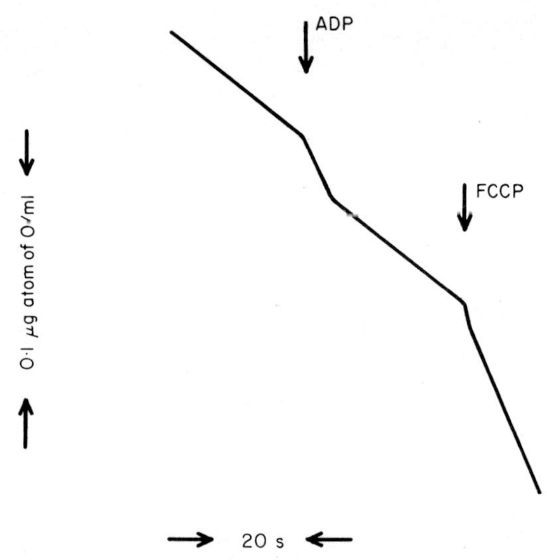

Figure 4. Normal respiratory control and uncoupling in valinomycin-treated mitochondria. Oxygen utilization was followed with an oxygen electrode in 3·2 ml of mitochondrial suspension (7·3 mg of protein/ml) at 25°C in 250 mM sucrose, 3·3 mM glycylglycine, 10 mM choline chloride, 5 mM β-hydroxybutyrate, 10 mM choline phosphate, 5 mM $MgCl_2$, 5 μl of catalase (see legend of Fig. 5) and 100 μg of valinomycin/g of protein. The mitochondria were equilibrated anaerobically for 10 min at pH 7·0–7·1 and respiration was induced by injecting 2 μl of 0·5 M H_2O_2. At the first arrow, ADP (9·6 μmoles/g of mitochondrial protein) was added. At the second arrow, FCCP (to give a final concentration of 0·2 μM) was added. (Data of Mitchell and Moyle [13]).

induction of high passive K^+ ion permeability in respiring mitochondria treated with valinomycin should cause a consumption of the hypothetical energy-rich intermediates produced by respiration, and activation of respiration should result. Figure 4 shows that normal respiratory transitions can be induced in rat liver mitochondria treated with 100 μg of valinomycin/g of protein by adding ADP, and that normal uncoupling can be observed with FCCP, in a medium containing 250 mM sucrose, 10 mM choline chloride, 5 mM β-hydroxybutyrate, 10 mM choline phosphate, 5 mM $MgCl_2$ and about 0·6 mg ion K^+/litre. The P/O quotient estimated by the method of Chance and Williams [22] was 2·88 in this experiment [13]. It follows that valinomycin does not activate a K^+ ion pump of significant capacity under these conditions.

We can now proceed to the next stage of the argument – the estimation of $\Delta\psi$ from the Donnan distribution of K^+ ions.

The Donnan equilibrium relationship for K^+ ion distribution can be written simply as

$$\Delta\psi = Z\Delta\text{pK} \tag{2}$$

where

$$\Delta\text{pK} = \log_{10}\frac{[K^+]_I}{[K^+]_O} \tag{3}$$

The suffixes I and O refer to the inner and outer phases and we assume that the activity coefficients in the inner and outer phases are the same. When mitochondrial suspensions are equilibrated anaerobically in State 5, the pH values of the inner and outer phases (pH_I and pH_O) are not equal and $\Delta\psi$ is not zero because the unequal distributions of certain ionic polymers and small molecular weight ions maintain a Donnan potential such that

$$\Delta p^o = 0$$

and

$$\Delta\psi^o = Z\Delta\text{pH}^o \tag{4}$$

where the superscript o refers to the state in which H^+ ions are equilibrated across the M phase. Similarly, in valinomycin-treated mitochondria, the K^+ ions are also equilibrated, and so in State 5

$$\Delta\psi^o = Z\Delta\text{pK}^o \tag{6}$$

When we induce a transition from State 5 to a respiring state (State 3, 4 or 6) by injecting oxygen into the mitochondrial suspension, the decrease in the value of $[K^+]$ in the outer phase ($-\Delta[K^+]_O$), measured as ΔpK_O with a suitably calibrated K^+ ion-sensitive glass electrode can be expressed (see [13]) in terms of the membrane potential, as follows

$$\Delta\psi = Z\log_{10}\frac{[K^+]_I^o - (V/v)\Delta[K^+]_O}{[K^+]_O^\omega} \tag{7}$$

In this equation, superscript ω denotes the value in the new state, V and v are the quantities of solvent water/g of mitochondrial protein in the whole suspension and in the inner phase respectively, and we have assumed that the activity coefficients are the same in the inner and outer phases as before.

Thus, we may estimate $\Delta\psi$ provided that we know the value of $[K^+]_I^o$ and v. Estimates of the quantity of total water in the inner (sucrose-inaccessible) phase of rat liver mitochondria suspended in media of osmolality corresponding to 250 mM sucrose vary between 1 and 0·25 g/g of mitochondrial protein [23-27]. Allowing for 0·26 g of non-solvent water per g of mitochondrial protein, as estimated by Bentzel and Solomon [27], and using their data and the data of Harris and van Dam [26], we obtain a value of 0·4 g of solvent water/g of mitochondrial protein for ν. It should be noted that the estimated value of $\Delta\psi$ is not very sensitive to error in the value of ν. An error of a factor of 2 in ν would cause an error of only 18 mV in the value of $\Delta\psi$ estimated from equation (7).

The value of $[K^+]_I^o$ can most easily be estimated from ΔpK^o and $[K^+]_O^o$, using the fact that $\Delta pK^o = \Delta pH^o$ in State 5 [see equations (5) and (6)]. When mitochondrial suspensions equilibrated in State 5 are lysed with the non-ionic detergent Triton X-100, the inner and outer phases titrate each other as they mix, and, in the absence of complicating factors (see [13] and below), the pH of the outer phase would change by an amount (ΔpH_O) dependent on ΔpH^o and the buffering powers B_I and B_O of the inner and outer phases according to the relationship

$$\Delta pH^o = -\Delta pH_O\left(1 + \frac{B_O}{B_I}\right) \tag{8}$$

This method of estimating ΔpH^o and ΔpK^o may be subject to errors, the magnitude of which it is difficult to estimate (see [13]), but at all events the error would not be likely to exceed 0·3 pH or pK unit or about 20 mV.

The value of ΔpH in states other than State 5 was obtained by measuring ΔpH_O in the transition from State 5 to the new state (superscript ω), and using the equation

$$\Delta pH = \Delta pH^o - \int_{pH_O^o}^{pH_O^\omega}\left(1 + \frac{B_O}{B_I}\right) dpH_O \tag{9}$$

The integral is evaluated graphically using plots of B_I and B_O against pH [13]. This concludes the outline of the rationales used for estimating $\Delta\psi$ and ΔpH.

Figure 5 shows plots of $\Delta\psi$, $-Z\Delta pH$ and their sum Δp, estimated from pH and pK recordings obtained with glass electrodes in a rat liver mitochondrial suspension (6 mg protein/ml) in a medium containing 250 mM sucrose, 10 mM choline chloride, 5 mM β-hydroxybutyrate, 3·3 mM glycylglycine and 1 mg of oligomycin/g of mitochondrial protein at 25°C near pH 7. The equilibrium K^+ ion concentration in the medium in State 5 was 0·5 mg ion/litre, and $\Delta\psi^o$ and $-Z\Delta pH^o$ were +33 mV and −33 mV respectively. The mitochondria were brought into State 4 by injecting hydrogen peroxide, catalase being present in the suspension medium to prevent cytochrome oxidase inhibition. The total protonmotive force in State 4 was 223 mV, made up of a membrane potential of

144 mV and a pH difference equivalent to 79 mV. When the oxygen was used up the system returned to the initial State 5 condition.

Figure 6 shows the results of a similar experiment, except that KCl was added to the medium to bring the initial equilibrium State 5 K^+ ion concentration in the medium to 10·5 mg ion/litre. In this case the State 4 value of Δp was as before, but the presence of the larger quantity of permeating cation resulted in

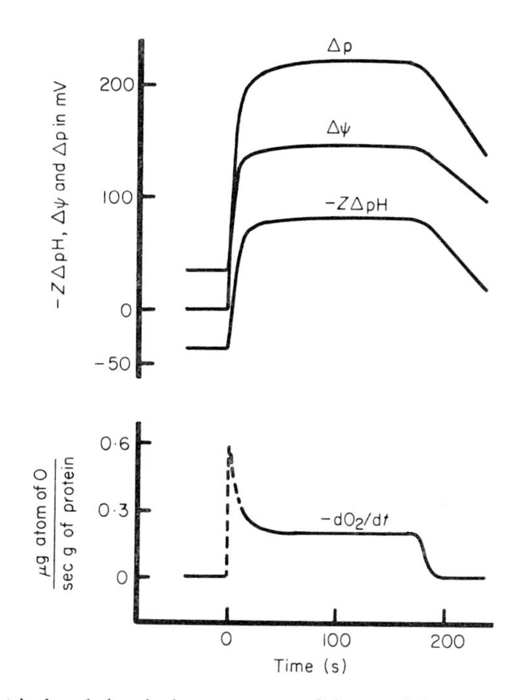

Figure 5. Electrical and chemical components of the p.m.f. in resting and respiring (State 4) mitochondria. $[K^+]^o_o$ = 0·5 mg ion/litre. The measurements were done on 3·2 ml of mitochondrial suspension (6·7 mg of protein/ml) at 25°C in 250 mM sucrose, 3·3 mM glycylglycine, 10 mM choline chloride, 5 mM β-hydroxybutyrate, 1 mg of oligomycin/g of protein, 100 μg of valinomycin/g of protein, and 5 μl of catalase (0·1 ml of C-100 catalase suspension of Sigma diluted to 3 ml with water). The mitochondria were equilibrated for 10 min anaerobically at pH 7·0–7·1 and at zero time 2 μl of 0·5M H_2O_2 was injected. (Data of Mitchell and Moyle [13].)

143 mV of the total protonmotive force being in the form of the pH difference, and only 87 mV being in the form of the membrane potential.

Figure 7 summarizes results of similar experiments which were arranged so as to permit different extents of cation shift across the M phase during the State 5 to State 4 transition. In A and B, the K^+ ion concentration was minimized by pretreating the mitochondria with an ion exchange resin, and in A, 0·2 mM

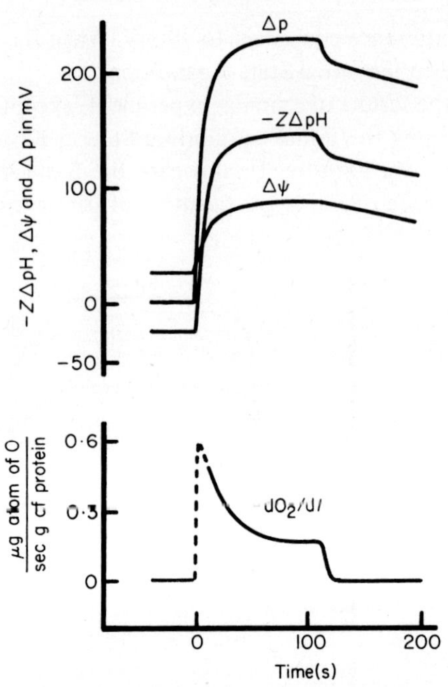

Figure 6. Electrical and chemical components of the p.m.f. in resting and respiring (State 4) mitochondria. $[K^+]_o$ = 10·5 mg ion/litre. The conditions were as for Fig. 5 except that 10 mM KCl replaced 10 mM choline chloride in the suspension medium. (Data of Mitchell and Moyle [13].)

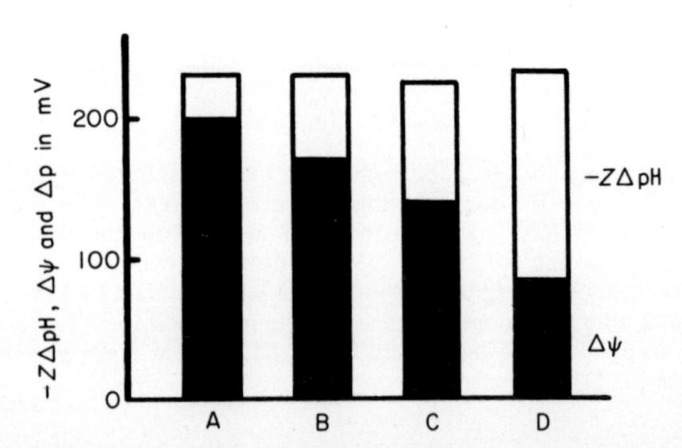

Figure 7. Electrical and chemical components of the p.m.f. in mitochondria respiring in State 4 or State 6. The conditions of the measurements were as in Fig. 5 with the following variations. A. The mitochondria were K^+-depleted and 0·2 mM EGTA was present in the medium. $[K^+]_o$ = *c.* 0·1 mg ion/litre. B. The mitochondria were K^+-depleted, but no EGTA was added. $[K^+]_o$ = *c.* 0·1 mg ion/litre. C. Normal mitochondria were used, and conditions were as in Fig. 5. $[K^+]_o$ = *c.* 0·6 mg ion/litre. D. As in C, but 10 mM KCl in place of 10 mM choline chloride (see Fig. 6). $[K^+]_o$ = *c.* 10 mg ion/litre. (Data taken from Table 1 of Mitchell and Moyle [13].)

EGTA was present to combine endogenous Ca^{2+}. In C and D, normal mitochondria were used and the conditions were as for Figs. 5 and 6 respectively.

In each case the value of Δp was the same within experimental error. It thus appears that the state of respiratory inhibition due to lack of phosphate acceptor is characterized by a particular value of Δp. It does not appear to matter what proportion of the total electrochemical potential of H^+ ions across the M phase

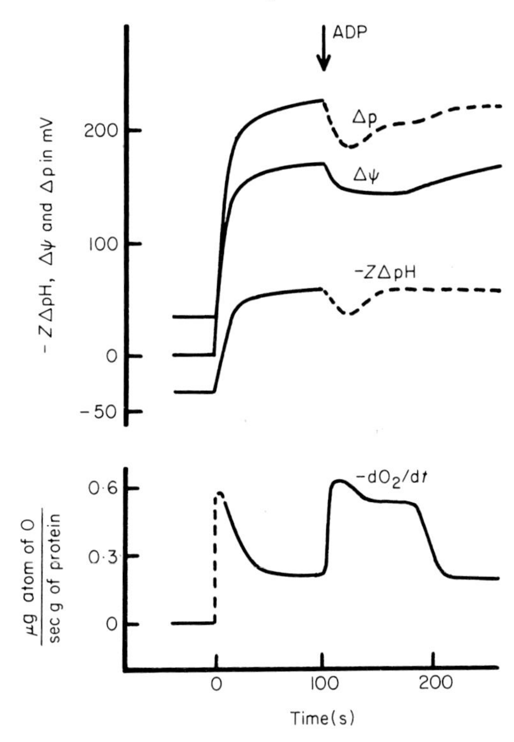

Figure 8. Electrical and chemical components of the p.m.f. in mitochondria respiring in States 4 and 3. The conditions were as for Fig. 4. At the arrow, 96 μmoles of ADP/g of protein were added. During State 3 respiration, further additions of a total of 5 μl of H_2O_2 were made to provide sufficient oxygen to complete the phosphorylation of the ADP added, indicated by restoration of State 4 respiration. (Data of Mitchell and Moyle [13].)

is due to the electrical and what proportion is due to the chemical component over the range so far studied. The physiological state of the mitochondria in case A in Fig. 7 probably corresponds, more or less, to what is usually known as State 4 [28], whereas the physiological state of the mitochondria in case D resembles the inhibited state associated with cation uptake, called State 6 [29].

The effect of bringing the mitochondria into the activated state of respiration (State 3) by adding a pulse of ADP under conditions corresponding to those of Fig. 4 is shown in Fig. 8. The addition of the ADP resulted in a fall of Δp by a maximum of about 30 mV, at the beginning of the burst of State 3 respiration, and as Δp returned to the initial value, respiration returned to the State 4 rate. The values of $-Z\Delta pH$ and Δp are shown by broken lines in this case because it was necessary to allow for alkali production accompanying conversion of ADP + Pi to ATP in computing these values from the measurements of ΔpH_o and ΔpK_o given by Mitchell and Moyle [13], and they are therefore less accurate than the other estimations of proton potentials given in this paper.

CONCLUSIONS

These studies lead to the conclusion that, as required by the chemiosmotic hypothesis, there is a p.m.f. of some 200 mV across the cristae membrane of respiring rat liver mitochondria, and that this p.m.f. is about 30 mV higher in States 4 or 6 than it is in State 3. Our estimate of 230 mV for Δp in State 4 is in reasonably good agreement with the value of 270 mV based on the generally accepted free energy of hydrolysis of ATP (see [10, 11]).

It is instructive to relate the p.m.f. to the o/r potential available across the region of the respiratory chain usually called Site 2, between cytochrome *b* and cytochromes *c* or c_1 (corresponding to a hybrid o/r loop consisting of the hydrogen-carrying arm of Loop 3 and the electron-carrying arm of Loop 2 [10, 11]). The standard o/r potential of cytochrome *b* at pH 7 is about +80 mV [30], and the poise of cytochrome *b* is approximately central in State 4 [31]. The standard o/r potential of cytochrome *c* is about +250 mV (more positive than cytochrome c_1 [32]), and cytochromes *c* and c_1 in State 4 are about 10% reduced [31], giving a corresponding o/r potential of 310 mV. Thus, in State 4 the o/r potential or p.m.f. available in the Site 2 region is about 310 – 80 = 230 mV, in good agreement with our value for Δp.

Cockrell *et al.* [19] and Slater [1, 6] have suggested that, under conditions of low magnesium and inorganic phosphate concentration, and at alkaline pH, the value of the free energy of hydrolysis of ATP may correspond to 15·6 kcal/mole; and it has been proposed that the mitochondrial respiratory system can phosphorylate ADP against this high potential–equivalent to a p.m.f. of some 340 mV. The thermodynamic implications of this proposal are difficult to deal with in terms of any hypothesis of the coupling mechanism. Our recent finding [3] that the entry of ATP through the ATP/ADP antiporter system is probably associated with the translocation of one net negative charge per ATP passing through the cristae membrane may require a reappraisal of the mechanism by which the gross stoichiometry of proton translocation corresponds to an $\rightarrow H^+/P$ quotient of 2. It is conceivable, for example, that the ATPase may be of type I, translocating one H^+ ion outwards per ATP hydrolysed [10, 11], and that the second proton is effectively translocated outwards by the net entry of one electron in the ATP/ADP exchange reaction through the antiporter system, and the exit of one H atom, carried by inorganic phosphate (H_2PO_4), which may effectively be translocated as phosphoric acid [7, 8, 12]. It is important to recognize, however, that although such a mechanism would result in differences of poise of the components of the ATP/(ADP + Pi) couple in the inner and outer mitochondrial phases, the poise of this couple in the outer phase depends only on the gross value of the $\rightarrow H^+/P$ quotient at a given value of the p.m.f. Thus, it would seem that an ATP/(ADP + Pi) poise corresponding to some 340 mV could be achieved only if our estimate of 230 mV for the p.m.f. in State 4 were much too low, or if the gross $\rightarrow H^+/P$ quotient were considerably greater than 2 under the conditions corresponding to a free energy of hydrolysis of ATP of 15·6 kcal/mole.

One may well ask the question, in the context of the chemical hypothesis of the coupling mechanism: why should respiration be associated with a proton translocation system of such high kinetic activity, and why should this system be capable of exerting such a large protonmotive force across the mitochondrial cristae membrane? The proton translocating characteristics of the respiratory chain system are entirely in accord with the requirements of the chemiosmotic hypothesis, but I do not think that any of the current versions of the chemical or conformational coupling hypotheses provide a satisfactory explanation for these characteristics.

ACKNOWLEDGEMENTS

I would like to thank my research colleague, Dr. Jennifer Moyle, for help during the preparation of this paper and Glynn Research Ltd. for general financial support.

REFERENCES

1. Slater, E. C., this volume, p. 205.
2. Mitchell, P. and Moyle, J., *Biochem. J.* **105** (1967) 1147.
3. Mitchell, P. and Moyle, J., *Eur. J. Biochem.* **4** (1968) 530.
4. Mitchell, P. and Moyle, J., *in* "Biochemistry of Mitochondria" (edited by E. C. Slater, Z. Kaniuga and L. Wojtczak), Academic Press and Polish Scientific Publishers, London and Warsaw, 1967, p. 53.
5. Mitchell, P., *Fedn Proc. Fedn Am. Socs exp. Biol.* **26** (1967) 1370.
6. Slater, E. C., *Eur. J. Biochem.* **1** (1967) 317.
7. Chappell, J. B. and Crofts, A. R., *Biochem. J.* **95** (1965) 393.

8. Chappell, J. B. and Crofts, A. R., *in* "Regulation of Metabolic Processes in Mitochondria" (edited by J. M. Tager, S. Papa, E. Quagliariello and E. C. Slater), Elsevier, Amsterdam, 1966, p. 293.
9. Lardy, H. A., Connelly, J. L. and Johnson, D., *Biochemistry* **3** (1964) 1961.
10. Mitchell, P., *Biol. Rev.* **41** (1966) 445.
11. Mitchell, P., "Chemiosmotic Coupling in Oxidative and Photosynthetic Phosphorylation", Glynn Research, Bodmin, Cornwall, 1966.
12. Mitchell, P., "Chemiosmotic Coupling and Energy Transduction", Glynn Research, Bodmin, Cornwall, 1968.
13. Mitchell, P. and Moyle, J., *Eur. J. Biochem.* **7** (1969) 471.
14. Bangham, A. D., Standish, M. M. and Watkins, J. C., *J. molec. Biol.* **13** (1965) 238.
15. Lev, A. A. and Buzhinsky, E. P., *Cytology, U.S.S.R.* **9** (1967) 102.
16. Mueller, P. and Rudin, D. O., *Biochem. biophys. Res. Commun.* **26** (1967) 398.
17. Andreoli, T. E., Tieffenberg, M. and Tosteson, D. C., *J. gen. Physiol.* **50** (1967) 2527.
18. Kilbourn, B. T., Dunitz, J. D., Pioda, L. A. R. and Simon, W., *J. molec. Biol.* **30** (1967) 559.
19. Cockrell, R. S., Harris, E. J. and Pressman, B. C., *Biochemistry* **5** (1966) 2326.
20. Pressman, B. C., Harris, E. J., Jagger, W. S. and Johnson, J. H., *Proc. natn. Acad. Sci. U.S.A.* **58** (1967) 1949.
21. Mitchell, P. and Moyle, J., *Biochem. J.* **104** (1967) 588.
22. Chance, B. and Williams, G. R., *Nature, Lond.* **175** (1955) 1120.
23. Werkheiser, W. C. and Bartley, W., *Biochem. J.* **66** (1957) 79.
24. Bartley, W., *Biochem. J.* **80** (1961) 46.
25. Klingenberg, M. and Pfaff, E., *in* "Regulation of Metabolic Processes in Mitochondria" (edited by J. M. Tager, S. Papa, E. Quagliariello, and E. C. Slater), Elsevier, Amsterdam, 1966, p. 180.
26. Harris, E. J. and van Dam, K., *Biochem. J.* **106** (1968) 759.
27. Bentzel, C. J. and Solomon, A. K., *J. gen. Physiol.* **50** (1967) 1547.
28. Chance, B. and Williams, G. R., *Adv. Enzymol.* **17** (1956) 65.
29. Chance, B., *Fedn Proc. Fedn Am. Socs exp. Biol.* **23** (1964) 265.
30. Holton, F. A. and Colpa-Boonstra, J., *Biochem. J.* **76** (1960) 179.
31. Kröger, A. and Klingenberg, M., *Biochem. Z.* **344** (1966) 317.
32. Green, D. E., Järnefelt, J. and Tisdale, H., *Biochim. biophys. Acta* **38** (1960) 160.

FEBS Symposium, Volume 17, 1969, pp. 233-273

The Nature of the Respiratory Chain: Location of Energy Conservation Sites, the High Energy Store, Electron transfer-linked Conformation Changes, and the "Closedness" of Submitochondrial Vesicles*

B. CHANCE, A. AZZI, I. Y. LEE,
C. P. LEE and L. MELA

*Johnson Research Foundation, School of Medicine,
University of Pennsylvania, Philadelphia, U.S.A.*

As it is a great pleasure to return to a "FEBS" symposium on mitochondrial functions after two years absence, it seems useful to point to questions that remained open after the symposium at Warsaw two years ago on the Biochemistry of Mitochondria [1].

One noted particularly the lack of information on the kinetics of quinone oxidation in isolated mitochondria, which was necessary to resolve the function of this hydrogen carrier in the region of Sites I and II. It was apparent that the methods employed by Klingenberg and Kröger [2] lacked the time resolution required for an accurate tracing of quinone kinetics from the very early stages of the reaction, where it might participate at the level of the cytochromes or flavoproteins. It was obvious that the application of a flow apparatus to this problem was needed, which led, in turn, to the question of whether there was any hydrogen carrier at all involved in phosphorylation at Sites II and III.

There were soon to emerge questions with respect to the role of flavoproteins in the Site I phosphorylation: was this Site between NADH and flavoprotein, or, more likely, between the low and high potential flavoproteins, leading to the possibility that electron transfer might play an important role in Site I as well as in Sites II and III?

The question of the high energy state of the mitochondria has been reopened by doubts about the validity of the ATP-jump phenomenon, as studied some years ago in this laboratory by Schachinger *et al.* [3]. The size of the pool of high energy intermediates was about to be drastically reduced, by a factor of ten. Obviously, newer, more sensitive, and more rapid techniques for estimating energy storage were needed.

*This research has been supported by grants and fellowships from the National Institutes of Health (Gm-12202, GM-38,822, FO-5-TW-01291, GM-00957) and the Jane Coffin Child Memorial Fund for Medical Research.

The "conformation cult" was rapidly joining ranks from two quite divergent viewpoints: that of the gross structure of the mitochondrial membrane, as visualized by electron microscopy [4], and that of the very elementary conformation changes which were about to be discovered in the spin state changes of well-known hemoproteins such as myoglobin [5].

Perhaps most important of all, we were about to learn that the cleft in which the heme of cytochrome *c* is hidden is, as a matter of fact, sufficiently open [6, 7] that cytochrome *c* can serve effectively in contact types of electron transfer reactions which, in the respiratory chain, are strongly suggestive of the need for rotational and vibrational freedom of the molecule about its points of attachment to the membrane [8].

I. THE FLAVOPROTEIN CHAIN

Figure 1 shows, by way of introduction, the spectroscopically—or fluorometrically detectable components of the respiratory chain with which we will deal here. Generally, the chain is in two portions: that from cytochrome c_1 to oxygen, which has been termed the cytochrome chain, and the segment from

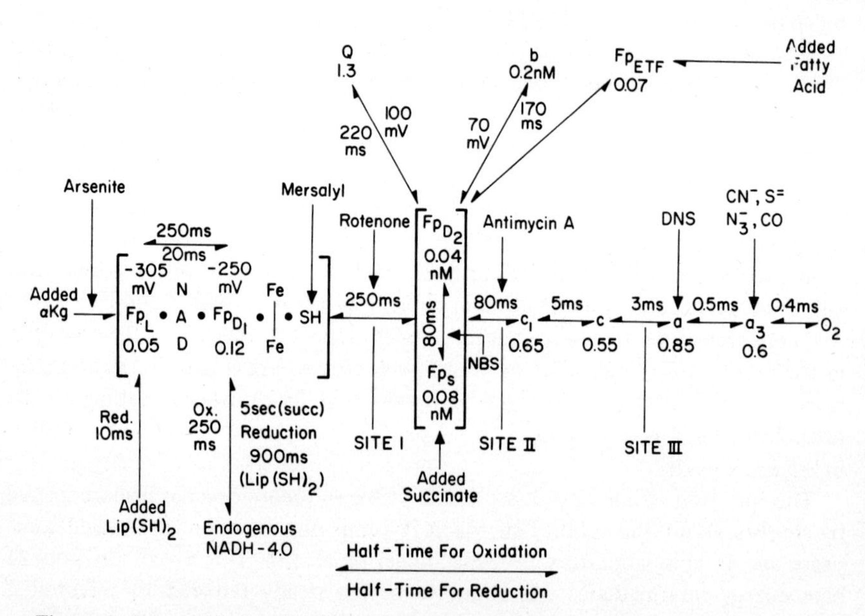

Figure 1. Schematic diagram of the components of the respiratory chain of rat liver mitochondria, showing half-times for oxidation taken from oxygen pulse studies; half-times for reduction of the flavoproteins taken from dihydrolipoamide studies; potential measurements taken from acetoacetate/β-hydroxybutyrate titrations; rapidly reacting amounts of the components in nmoles per mg protein, taken from oxygen pulse studies. Other units are indicated on the figure.

NADH to the antimycin A-sensitive point which we now term the flavoprotein chain. The cytochrome chain is in the sequence previously determined by our rapid kinetic studies [9, 10] and the components are found in the amounts indicated (in nmoles per mg protein) under each symbol with their half-times for oxidation above each arrow, and for reduction by dihydrolipoamide as indicated. However, since only the amounts that react rapidly are shown here, only a portion of the cytochrome *c* is included.

In the flavoprotein chain, two low potential components interact in a tightly coupled, rapidly equilibrated complex consisting of the lipoamide dehydrogenase flavoprotein, Fp_L, and the NADH dehydrogenase flavoprotein, Fp_{D1}. The two high potential flavoproteins, Fp_S and Fp_{D2}, are similarly rapidly equilibrated. These four flavoproteins can be distinguished not only by their fluorescence and absorption properties, particularly in submitochondrial particles, but also by the use of selective inhibitors [11]. Quinone, cytochrome *b*, and Fp_{ETF} are, for the present indicated to be in redox equilibrium with the high potential flavoproteins, until a further delineation of their function can be achieved. The amounts of quinone that are rapidly oxidized are only a fraction of the total quinone pool. The detailed experimental justification for this reaction sequence has been put forward in a number of papers [10, 12-15].

Function of flavoprotein, quinone, and cytochrome *b* in electron transport at Sites I and II

A resolution of many of the conjectures concerning chemical and chemi-osmotic mechanisms of oxidative phosphorylation depends critically upon the sequence of electron and hydrogen atom transfers in the respiratory chain. Recent developments in highly efficient stopped flow apparatuses which permit not only dual wavelength recording of cytochrome kinetics, but also precise measurements of quinone oxidation and fluorescence measurement of flavoproteins and reduced pyridine nucleotides, have been achieved over the past year [10]. Such apparatuses can cover a time range from 150 μs to 2 ms for the cytochromes, and from 2 ms onwards for quinones, flavoproteins, and pyridine nucleotides.

An experimental result in which anaerobic pigeon heart mitochondria are rapidly mixed with 15 μM oxygen is shown in Fig. 2, giving the time course of oxidation of these critical components. On the left, in the rotenone-blocked system, only the components of the succinate oxidation chain are involved; on the right, in the absence of rotenone, the complete system is oxidized. The experimental conditions are so arranged that the 15 μM oxygen pulse (60 electron equivalents) is considerably greater than that of the 10-12 electron equivalents of the mitochondria at the concentration employed in this experiment. Furthermore, electron flow through the system is minimized by the

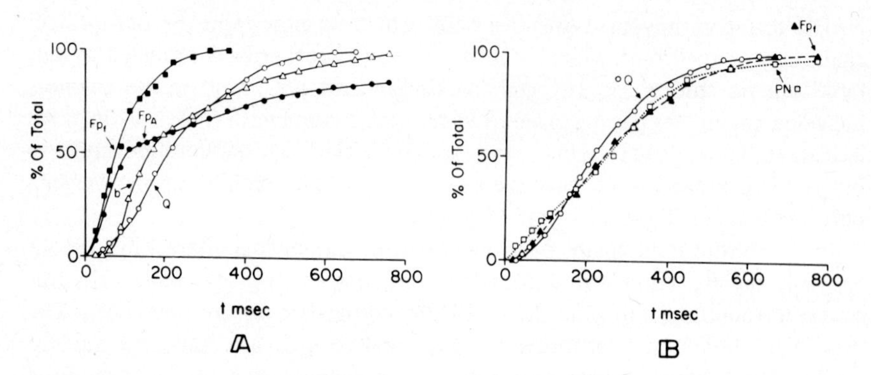

Figure 2. Time course of oxidation, taken from flow apparatus studies of cytochrome *b*, flavoprotein, and ubiquinone in response to a pulse of 15 μM oxygen in the presence of 3·3 μM rotenone (A) and in the absence of rotenone (B). Pigeon heart mitochondria, 3·8 mg protein per ml, in 0·3M mannitol-sucrose, 20 mM Tris-Cl, pH 7·4, supplemented with 1·3 mM glutamate, 1·3 μM FCCP to ensure a maximal rate of respiration, and 6·7 mM malonate to block the oxidation of endogenous substrate. Dual wavelength spectrophotometric measurements of cytochrome *b* at 561-575 mμ, absorbing flavoprotein (FpA) at 475-495 mμ, ubiquinone (Q) at 275-290 mμ, and pyridine nucleotide (PN) at 340-374 mμ; fluorometric measurements of flavoprotein (Fp_F), with excitation at 436 mμ and emission at 570 mμ.

addition of sufficient malonate to suppress succinate oxidation. The cytochromes, other than *b*, are not shown on these diagrams; their profiles would be compressed against the left ordinate, since their half-times are in the range of 1 to 5 ms.

The earliest event in the Site I-Site II region is the appearance of fluorescence due to oxidized flavoprotein, Fp_F. It is noteworthy that this fluorescence is associated with the fully oxidized form, and not with a free radical form of flavoprotein, and that the fluorescence is attributed jointly to succinate dehydrogenase and to a high potential flavoprotein of the NADH dehydrogenase sequence, which lies on the oxygen side of the rotenone block. None of the low potential flavoproteins is seen to be oxidized in this time scale in the presence of rotenone. The absorbing flavoprotein, Fp_A, follows a time course similar to that of the fluorescing flavoprotein, Fp_F, for the first 100 ms, suggesting that no detectable amount of absorbing component, such as non-heme iron, is oxidized before flavoprotein, identified conclusively by its fluorescence emission. Components which do not appear to have significant fluorescence (which may, indeed, include the non-heme iron but which are more likely to be the non-fluorescent flavoproteins of fatty acid oxidation) are more slowly oxidized from 100 ms onwards. Thus, the first components to be oxidized following cytochrome c_1 are flavoproteins, identified both by their characteristic fluorescence emission and by their absorbancy change.

Initially, ubiquinone and cytochrome *b* are oxidized almost simultaneously, but proceed along slightly different time courses, cytochrome *b* being in advance of quinone in the middle portion of the kinetics, and the two coming together again as both reactions reach the steady state oxidized level. Even at the sensitivity of 0·05 nmoles of quinone per mg protein employed here, no measurable quinone oxidation is observed in the first 80 ms; somewhat later, about 5 times more flavoprotein than quinone is oxidized, on a mole basis. Thereafter, the quinone trace rises more rapidly and blends with that of flavoprotein towards the end of the chart.

In Fig. 2B, in the absence of rotenone, the pyridine nucleotide kinetics are recorded by absorption, and the recording for fluorescent flavoprotein is repeated. For comparison, the quinone kinetics from the recording in the presence of rotenone are included. The chief component of the fluorescent flavoprotein is now the highly fluorescent, low potential flavoprotein of the NADH and lipoamide dehydrogenases; its oxidation proceeds roughly parallel with that of reduced pyridine nucleotide. It seems that the respiratory chain is more effective in oxidizing pyridine nucleotide in the absence of rotenone than ubiquinone in the presence of rotenone. Since pyridine nucleotide absorption changes do not interfere seriously with the recording of quinone kinetics, we can state that ubiquinone oxidation is not accelerated in the absence of rotenone.

Remarkable changes in the electron flow pattern occur on rupture of the mitochondrial membranes, as indicated by the kinetics of submitochondrial particles prepared by the method of Lee and Ernster [16], which exhibit a high degree of respiratory control and some phosphorylative activity as well. In these particles, the entire sequence of electron transfer reactions has been speeded up several-fold [17, 18]; fluorescent flavoprotein is 50% oxidized in 4 ms, while the quinone pool is half-oxidized in 180 ms; there is also a corresponding acceleration of the cytochrome *b* kinetics. Although further disruption of the structure of the submitochondrial particles leads to further acceleration of the quinone oxidation, the cytochrome *b* kinetics become slower; half-times of between 10 and 20 ms are observed [15, 19].

The important conclusion to be derived from these studies is that flavoprotein oxidation is always found to precede that of quinone, even when measurements are carried down to the ms time range. Since Sites II and III are shown by crossover data [20] to be on the oxygen side of flavoprotein (see also Fig. 3), any function which quinone might have in either electron transport or energy conservation would be restricted to Site I. We now know that the oxidation of hydrogen carriers in intact mitochondria and submitochondrial particles is confined to the time range between 80 and 500 ms; in this interval, the content of flavoprotein, quinone and reduced pyridine nucleotide will be brought to its full steady state of oxidation. Thus, the comparison of the rate of oxidation of hydrogen carriers in an oxygen pulse experiment with direct measurements of hydrogen ion ejection rates by external or internal pH indicators [21] is validated by these experiments. There remain large discrepancies between measured and expected hydrogen ion movements.

The fact that these kinetics clearly confine quinone oxidation to the flavoprotein region of the respiratory chain, together with the absence of any known hydrogen carrier at Sites II and III, leaves chemiosmotic mechanisms without the essential vectorial hydrogen atom and electron movement at these two sites. It is appropriate, in addition, to consider new experimental results on the nature of electron flow through Site I.

Location of Site I

The application of the "crossover theorem" to oxidation-reduction changes of the respiratory components of rat liver mitochondria located Site I between NADH and an absorbing flavoprotein [20]. The discovery of several species of flavoprotein by simultaneous measurements of the ratio of fluorescence to absorbancy changes has suggested a more precise location for Site I between the pair of low potential flavoproteins, Fp_{D1} and Fp_L, and the pair of high potential flavoproteins, Fp_S and Fp_{D2}, that bracket the rotenone-sensitive site [12, 22-24]. The flavoprotein reaction sequence determined on the basis of kinetics of oxidation has been verified by pulsed reduction studies in which dihydrolipoamide provides an extremely reactive electron donor to reduce the lipoamide and NADH dehydrogenase flavoproteins in a time of approximately 10 ms, much more rapidly than either is reduced by NADH. The flow of electrons from the low potential flavoproteins into the high potential flavoproteins can be effectively traced by the pulse of reduction provided by dihydrolipoamide, giving the reaction times indicated in Fig. 1.

Furthermore, the degree of oxidation and reduction of the flavoproteins can be determined by an acetoacetate/β-hydroxybutyrate titration, which gives values of -305 and -250 mV for the two low potential, highly fluorescent flavoproteins, Fp_L and Fp_{D1}, respectively. The values for the high potential flavoproteins, Fp_{D2} and Fp_S, are not known with certainty, but the effective equilibration of these two components in the rotenone-blocked system indicates their similarity.

The identification of an energy coupling site in the flavoprotein chain depends upon the response of the system to State 4 → 3 and 3 → 4 transitions. In Fig. 3, the ATP-induced oxidation of absorbing flavoprotein and reduction of fluorescent flavoprotein is indicated under two conditions: with succinate as electron donor (A), and with palmitoyl-L-carnitine as electron donor (B). In both cases, the reduction of fluorescent flavoprotein occurs rapidly upon adding ATP, and is accompanied by a cycle of oxidation and reduction of the absorbing flavoprotein. Evidence for the absence of cytochrome *b* interference, which

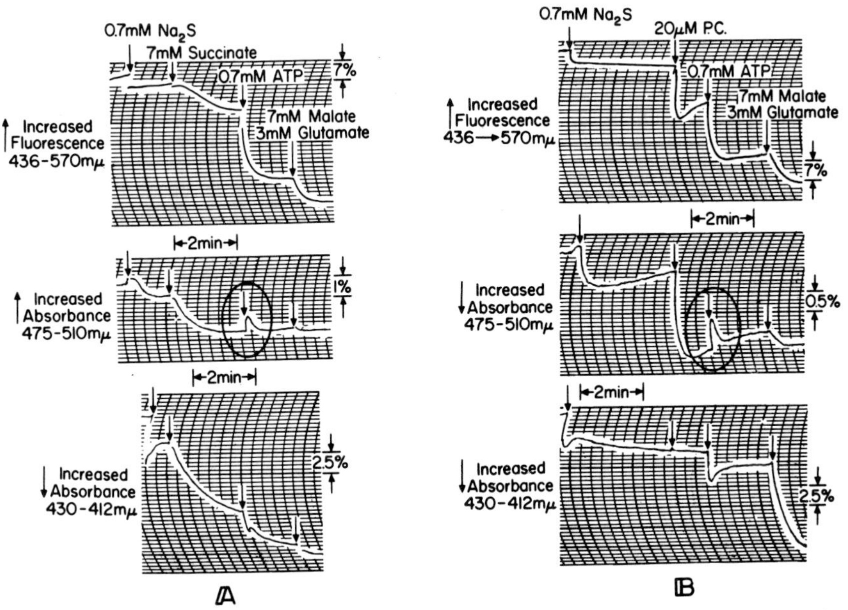

Figure 3. Reduction of fluorescent flavoprotein (top traces) and oxidation of absorbing flavoprotein (middle traces) in response to additions of 0·7 mM ATP (third arrow, each trace), with 7 mM succinate as electron donor in A, and 20 μM palmitoyl-L-carnitine as electron donor in B. The bottom traces indicate the reduction of cytochrome *b* by reversed electron transfer in the sulfide-blocked systems. Pigeon heart mitochondria, 2·8 mg protein per ml, in 0·3M mannitol-sucrose, 20 mM TRA-MES, 10 mM phosphate, pH 7·4. The circles call special attention to the crossover point in the State 4 → 3 transition.

might have been responsible for some absorbancy changes in the flavoprotein region, is indicated by the fact that cytochrome *b* is reduced in the reversed electron transfer in the presence of sulfide and is, therefore, under the influence of Site II rather than of Site I. Here, the electron donor is, presumably, the cytochrome chain.

Other controls indicate that the crossover point cannot be exhibited between the two low potential flavoproteins. This fact, together with the similarity of their redox potentials, makes this an unlikely location for an energy-coupling site.

The question of whether the mersalyl-sensitive site, and the non-heme iron which gives a characteristic "g = 1·94" signal, are involved in either electron transfer or energy coupling at Site I is not yet resolved. However, the absence of the g = 1·94 signal and the mersalyl site in the Site I-deficient *Saccharomyces cerevisiae*, and the presence of electron transfer through the cytochrome chain in this yeast, suggest that these components may be functional in energy coupling [25-27] rather than in electron transfer. On the other hand, Imai's finding of a g = 1·94-deficient bacterial system exhibiting Site I phosphorylation [28] suggests that the component giving the g = 1·94 signal may not be generally required for energy coupling. Nevertheless, we have included the non-heme iron and the mersalyl-sensitive site as a part of the electron transfer and energy coupling system shown in Fig. 1.

The location and function of the large content of non-heme iron which accompanies isolated preparations of the dehydrogenases is not known; no effective methods for dealing with the function of these components in the intact system are available.

II. THE NATURE OF THE HIGH ENERGY STATE

Energy storage in calcium uptake

Two new approaches to the problem of energy storage in mitochondrial membranes are described here. Both, while in principle following the "ATP-jump" technique of Schachinger *et al.* [3], employ calcium instead of ATP to indicate the size of the high energy store, and record sensitively the time course of calcium accumulation by mitochondria. In theory, a pre-steady state burst of calcium uptake in a State 4 → 3 transition could indicate by its extent the amount of energy stored in State 4.

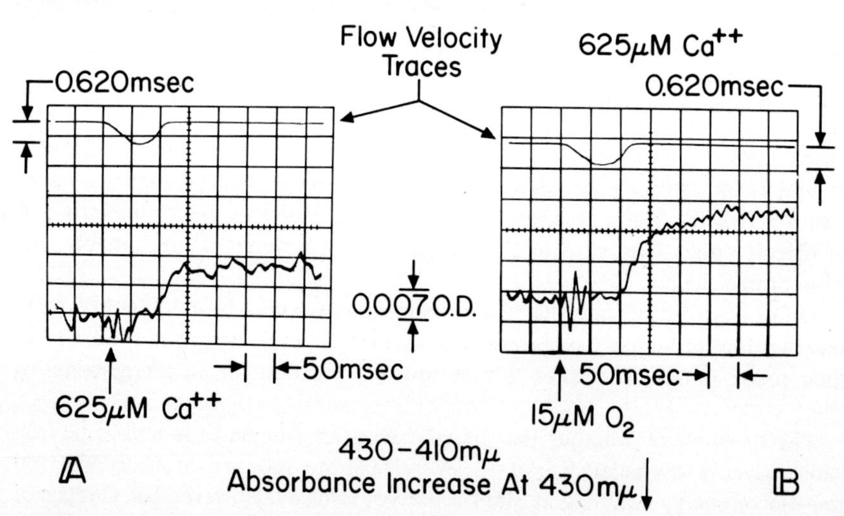

Figure 4. Kinetics of oxidation of cytochrome *b* on adding 625 μM calcium (A) or 15 μM oxygen (B) in the presence of 625 μM calcium remaining in the medium from the experiment in A. Rat liver mitochondria, 5·9 mg protein per ml, in 0·3M mannitol-sucrose, 20 mM Tris-Cl, pH 7·4, in the presence of 5 mM succinate. Top, flow velocity traces; bottom, cytochrome *b* absorbancy changes measured by the dual wavelength spectrophotometer at 430-410 mμ.

In order to indicate the speed with which calcium reacts with the cytochromes of rat liver mitochondria, Fig. 4A shows that the State 4 → 3 transition on adding calcium can be complete within 40 ms, with a half-time of approximately 20 ms. Fig. 4B shows that this is as rapid as the State 5 → 3 transition, the rise time on adding oxygen being identical to that on adding calcium. Thus, the initial reaction of calcium with cytochrome *b* is as fast as the oxidation of cytochrome *b* by the cytochrome chain.

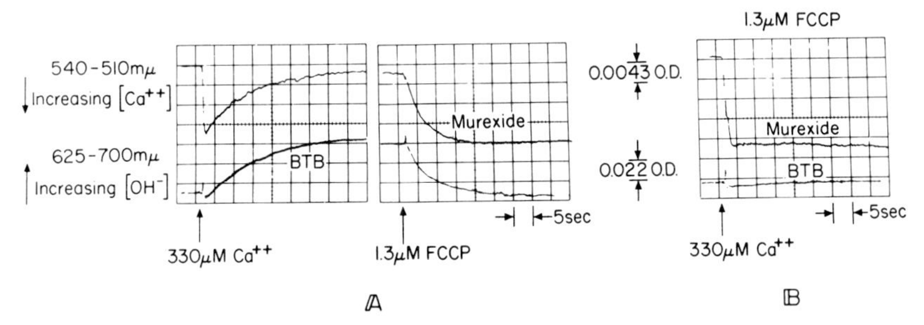

Figure 5. A. Responses of 13 μM murexide (top trace) and 1·7 μM BTB (bottom trace) to the addition of 330 μM calcium, followed by 1·3 μM FCCP; B. Control in the presence of 1·3 μM FCCP, indicating the lack of calcium uptake and BTB response under these conditions. Rat liver mitochondria, 2·1 mg protein per ml, in 0·3M mannitol sucrose, 20 mM Tris-Cl buffer, supplemented with 6·7 mM succinate. Dual wavelength recording at the wavelength pairs, sensitivities and time scales indicated on the figure.

We have found that the calcium electrode is of inadequate speed for the readout of calcium concentrations in rapid reactions [29]. Ohnishi and Ebashi [30] and Jöbsis and O'Connor [31] have employed the colorimetric indicator of external calcium concentrations, murexide. Geier [32] has studied the physical-chemical properties of murexide, and finds that the relaxation times, as obtained in the temperature-jump apparatus, are of the order of microseconds; thus, the indicator will respond much more rapidly than the mitochondria can move calcium. One of us (Mela) has further noted that there is no significant binding of murexide to the mitochondrial membranes in either the presence or the absence of added calcium, and has succeeded in improving the sensitivity of murexide indication by using the dual wavelength technique in its most sensitive operating mode to measure murexide changes at 540 mμ (a decrease of absorbancy representing an increase of calcium external to the mitochondria). We have studied the initial phases of the reaction of calcium with the mitochondrial membranes by flow apparatus techniques. Since the value of $\Delta\xi_{540\text{-}510}$ is 1·3 $mM^{-1}cm^{-1}$, approximately 1 to 5 μM calcium can be detected in a few ms [33] (see Fig. 8).

Figure 5A shows that an abrupt decrease of murexide absorbancy at 540 mμ on adding 330 μM calcium is followed by the rise of the trace with a half-time of

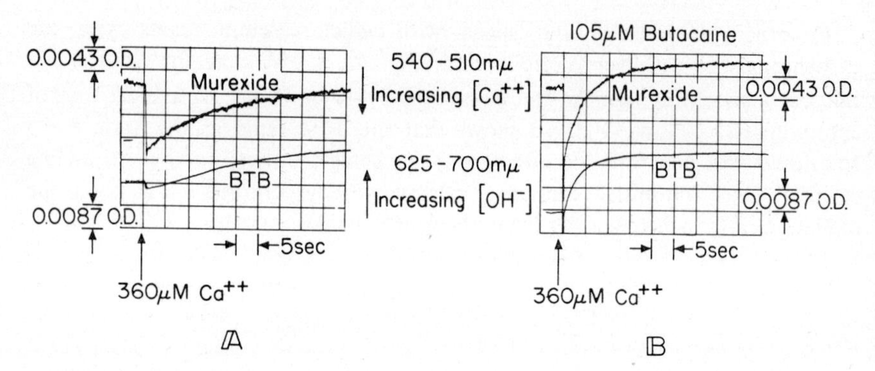

Figure 6. The acceleration of the murexide and BTB responses in the presence of butacaine. A. The response of 18 μM murexide and 1·4 μM BTB to the addition of 360 μM calcium in the absence of butacaine; B. The same, in the presence of 105 μM butacaine.

Rat liver mitochondria, 3·1 mg protein per ml in 0·3M mannitol-sucrose, 20 mM Tris-Cl, pH 7·4, supplemented with 3·6 mM succinate. Dual wavelength recording at the wavelength pairs, sensitivities and time scale noted in the figure.

approximately 10 s, due to the uptake of calcium from the reaction medium. If, now, the mitochondria are uncoupled by the addition of FCCP*, calcium is released with a half-time of 3 s, as indicated by the downward deflection of the trace. It is seen that the amount of calcium that is released is slightly larger than that which was added (by 10 nmoles per mg protein), since calcium already bound to the mitochondria is released as well on adding FCCP. In Fig. 5B, a control, no calcium uptake nor change in murexide absorbancy occurs with FCCP present in the medium.

This figure also shows that the kinetics of calcium uptake are closely correlated with those of the internal alkalinization of the mitochondria, as indicated by the intensification of the blue colour of the dye, bromthymol blue (BTB) [34, 35]. This correlation extends as well to the release of calcium from the mitochondria, with attendant internal acidification, on addition of the uncoupling agent.

The murexide and BTB traces are parallel over their entire course, with a 1 s delay between the two which may represent the membrane transit time for calcium, or the time necessary to saturate the membrane buffer capacity. The phenomenon of calcium translocation, first to the bromthymol blue space and then to the pyridine nucleotide space [36] of the mitochondrial membrane is discussed at this meeting by Azzi and Chance [37].

Several previously unsuspected activation and inhibition phenomena involved in calcium uptake are revealed by the murexide technique. For example, Fig. 6

*Abbreviations used: BTB, bromthymol blue, 3,3′-dibromothymolsulfonphthalein; FCCP, carbonyl cyanide *p*-trifluoromethoxyphenylhydrazone; TTFB, 4,5,6,7-tetrachloro-2-trifluoromethylbenzimidazole; DNS, dimethylaminonapthalene sulfonyl chloride.

shows the considerable acceleration of calcium accumulation induced by 105 μM butacaine. This local anaesthetic, which has already been shown to be effective in altering the BTB response [38, 39], is now seen directly to alter the calcium uptake rate. Butacaine binds phospholipid groups in the membrane that are responsible for energy-independent calcium binding [40] and which, in the free form, would diminish the net rate of calcium accumulation. The two records of Fig. 6 show how much more rapidly calcium is accumulated under these conditions. When small pH changes in the membrane are to be measured, we suggest that butacaine should be added, and thus the experiments of Mitchell *et al.* reported at this Symposium [41] are worth repeating under these conditions. We believe that judgment of their conclusions should be reserved until this is done.

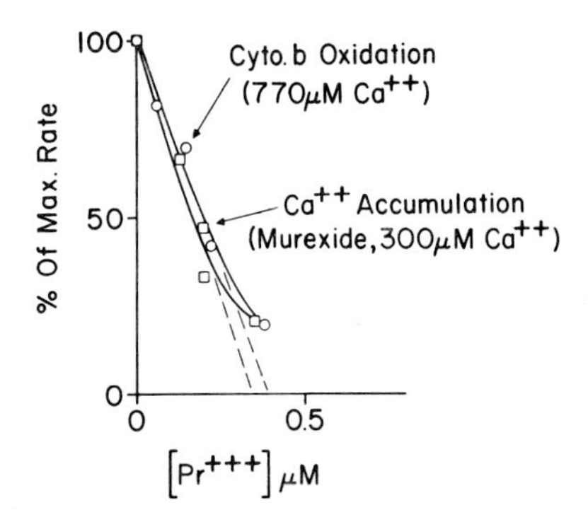

Figure 7. Titration of calcium accumulation as indicated by murexide (squares) and cytochrome *b* oxidation (circles) in the presence of praeseodymium. Rat liver mitochondria, 3 mg protein per ml, in 0·3M mannitol sucrose, 20 mM Tris-Cl buffer, pH 7·4, supplemented with 7·7 mM succinate, and 10 μM murexide when indicated.

A second phenomenon clarified by the use of murexide is the several-fold inhibition of the rate of calcium uptake by the lanthanides [42, 43]. The endpoint of the titration, shown in Fig. 7, of calcium accumulation with murexide and cytochrome *b* response gives a praeseodymium concentration of 0·4 μM per 3 mg protein. In this way, the number of calcium carriers per mitochondrion is determined to be approximately 0·1 nmoles per mg protein, a value about half that of the cytochrome *c* concentration.

Rapid kinetics

Since the time scale of Fig. 4 is too slow to indicate whether the initial portion of the kinetics contains a jump in calcium uptake as rapid as the reaction of calcium with cytochrome *b*, this study has been repeated in the rapid flow

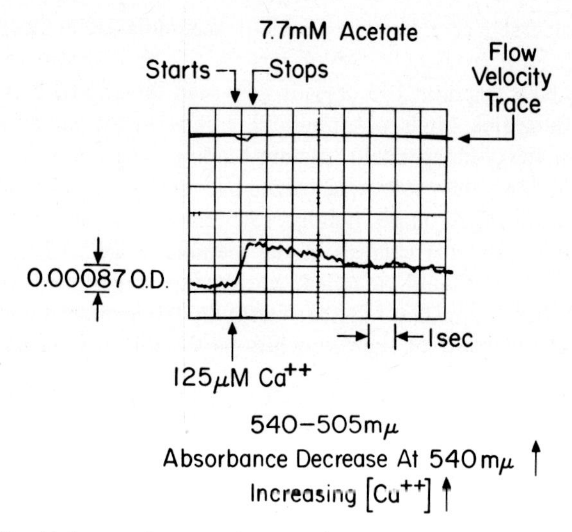

Figure 8. Rapid flow study of the kinetics of murexide response to addition of 125 μM calcium. Rat liver mitochondria, 4·8 mg protein per ml, in 0·3 M mannitol-sucrose, 20 mM Tris-Cl, pH 7·4, supplemented with 3·9 mM succinate and 7·7 mM acetate, 15 μM murexide is present in the medium.

apparatus on a five-fold faster time scale. In Fig. 8, 125 μM calcium is injected into rat liver mitochondria in the presence of a permeant anion. The top trace indicates the discharge velocity of the syringe plungers causing the injection of calcium into the mitochondrial suspension. When the fluid is moving through the observation tube, the time after mixing is approximately 20 ms. The upward deflection of the absorbancy trace indicates the decreased absorbancy of the murexide at 540 mμ. The absorbancy is just as large 20 ms after the addition of calcium as it is when the flow has come to a stop; the times after mixing can be read off directly from the time scale (1 s/division). The absence of a significant decrease of absorbancy at 20 ms and later, suggests the absence of any detectable rapid phase of calcium uptake. Judging from the extrapolation of the calcium accumulation rate and the level during the flow, no more than 5 μM calcium could have disappeared at 20 ms, an amount corresponding to less than 1 nmole of calcium per mg protein. The time for half-completion of calcium uptake is 4 s.

This experiment suggests that there is no rapidly-acting energy storage in excess of the 1 nmole capacity of calcium equivalents per mg protein that can be detected by murexide. Since 1·8 calcium is equivalent to 1 ADP or ATP [44], the limit of energy storage available between 20 ms and 1 s is less than 2 ATP-equivalents per mg protein. While this value is greater than the amount of cytochrome *b* which changes its steady state in a calcium jump experiment, it is considerably less than the amount of the flavoprotein, ubiquinone, and reduced pyridine nucleotide which might be available for a jump response in calcium uptake. It is apparent that these pools do not function as measurable energy stores for the fast phase of calcium uptake; instead, the transient phase of calcium uptake is no faster than the steady state phase. This result is in general agreement with van Dam's interpretation [45] of the earlier results in which the kinetics of ATP production were measured in the first few seconds after adding ADP; there again, no appreciable amount of energy storage was observed that could not be attributed to the steady state turnover of the system. It is apparent that a more sensitive method than that provided by murexide or the ATP-jump phenomenon is necessary for the determination of energy storage in mitochondria in State 4.

Aequorin studies

We have, therefore, employed as a second approach, the extremely sensitive indicator of external calcium, the jellyfish protein aequorin, kindly donated by Dr. F. H. Johnson [46]. This indicator luminesces brilliantly at free calcium levels of 10^{-7} to 10^{-6}M and is, in itself, inactivated. Thus, no calcium need be added to the mitochondria; the release of endogenous calcium (approximately 5 nmoles per mg protein) suffices, in our experiments, to give readily measurable flashes with a few tenths of a microgram of the indicator. The indicator will respond very rapidly to the release of less than 10^{-6}M calcium from the membrane.

The aequorin luminescence is green-blue and thus, with appropriate filters, not only the indicator response but also the oxidation-reduction states of the carriers can be measured simultaneously. The indicator responds rapidly, coming to full luminescence within a few ms after the addition of calcium as determined in control experiments.

In order to conserve the aequorin, it has been employed at such low concentrations that it is expended in 10 to 15 s in the presence of appropriately high calcium concentrations. Therefore, periodic injections of the indicator have been made through a micro-syringe, puncturing the plastic tape sealing the cuvette, in order to avoid disturbing the photomultiplier reading out the intensity of luminescence.

Kinetics of calcium release in rat liver mitochondria

An example of the use of aequorin to indicate the lifetime of the energized state in the State 4 → 5 transition is shown in Fig. 9. Initially, calcium is so tightly bound that no luminescence is observed. On the addition of 2 mM cyanide, the upward deflection of the top trace indicates the reduction of cytochrome *a* in

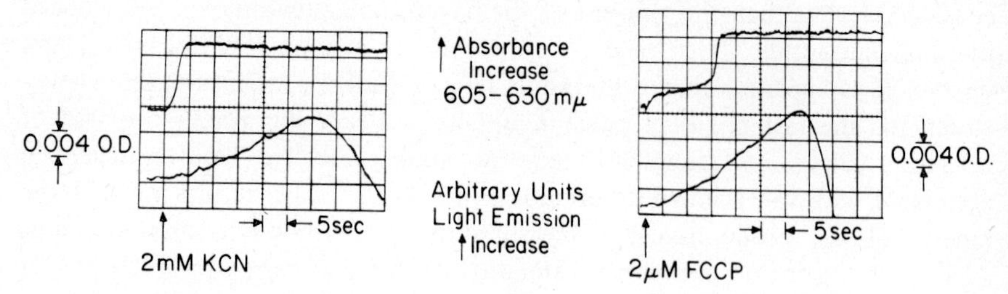

Figure 9. Responses of cytochrome *a* (top trace) and aequorin (bottom trace) to the addition of 2 mM cyanide (left) and 2 μM FCCP (right). Rat liver mitochondria, 7·4 mg protein per ml, in 150 mM KCl, 20 mM Tris-succinate, supplemented with 200 μM $MgCl_2$, 10 μg oligomycin, and 0·2 μg aequorin. Dual wavelength recording of cytochrome *a* reduction at 605-630 mμ. Aequorin bioluminescence measured by an EMI 9524B phototube connected to an electrometer, with a Corning filter (CS 559) interposed to avoid interference from the 605-630 mμ light.

approximately 3 s. The level of aequorin luminescence rises steadily during the reduction of the cytochrome and continues at a steady rate thereafter, indicating an increased rate of calcium release. After about 30 s, the aequorin is expended and the intensity of the luminescence falls. In the right-hand portion of the diagram, the State 4 → 3 transition caused by the addition of the uncoupler, FCCP, is shown, again causing a reduction of cytochrome *a* and a release of calcium that is characteristic of this transition. Fifteen seconds after adding FCCP, anaerobiosis occurs, during which a slightly more rapid rate of calcium release continues unabated until all the aequorin is expended.

Aequorin calibrations

When calcium and aequorin are rapidly mixed together, the logarithm of the intensity of the light emission from the aequorin is directly proportional to the logarithm of the free calcium concentration in the range from 10^{-6} to 10^{-5}M. The light decay process, on the other hand, is inversely proportional to the free calcium concentration, the half-time for the decay being longer at low calcium concentrations and shorter at high calcium concentrations. Thus a prolonged decay is observed at very low calcium concentrations, so that it is possible to make several additions of calcium before the aequorin is exhausted (see Fig. 10). The integral of the light emission is proportional to the total aequorin concentration in the system.

Because of the rather complex characteristics of the reaction of calcium with aequorin, it appeared easier to interpret the results of the calcium efflux from mitochondria on the basis of an empirical calibration. Thus, in Fig. 10, the release of calcium from the mitochondria has been simulated by the continuous injection of a solution of 0·1 mM calcium by a 100 μl micro-syringe into the

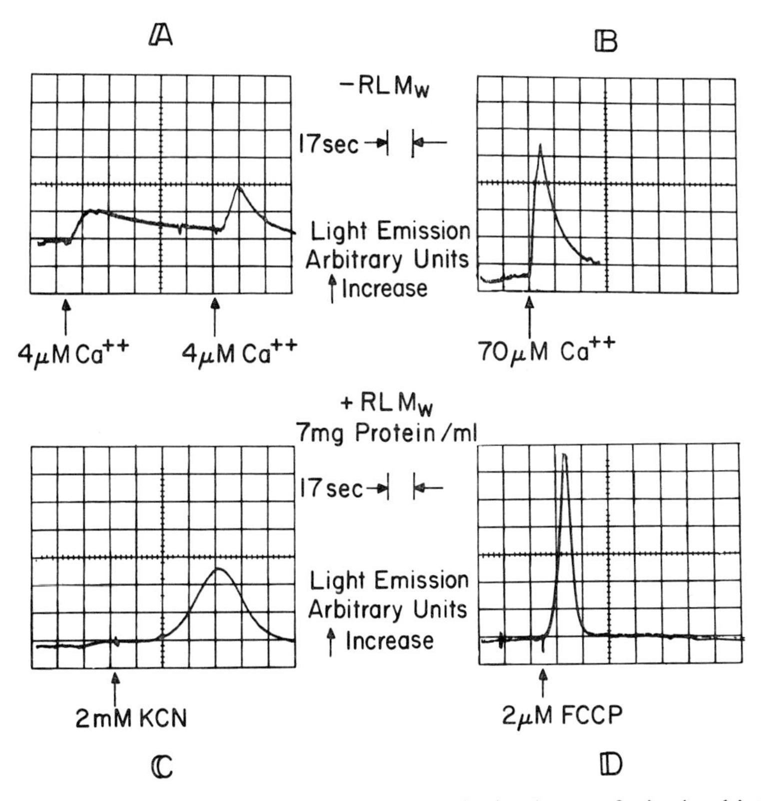

Figure 10. Calibration of the aequorin response in the absence of mitochondria (A, B), by simulation of the aequorin response in the presence of mitochondria to additions of cyanide (C) and FCCP (D). In A and B, calcium is injected by a micro-syringe into a solution containing 0·2 μg aequorin in 1·7 ml distilled water: in A, two injections of calcium over 10-s intervals simulate the response in the presence of mitochondria to the addition of 2 mM KCN (C); in B, the injection of 70 μM calcium simulates the response to 2 μM FCCP. In C and D, the 1·7 ml cuvette contained rat liver mitochondria at 7 mg protein per ml in 0·25M sucrose, 10 mM Tris-succinate, pH 7·4, supplemented with 200 μM Mg^{2+}, and 0·2 μg aequorin.

cuvette containing 0·4 μg aequorin in distilled water, and the results compared with the response of the indicator to calcium release from the mitochondria caused by adding cyanide (bottom left) or FCCP (bottom right). In both cases, the delay in release is somewhat longer than in Fig. 9, since oligomycin has been omitted in Fig. 10 (this phenomenon is discussed below).

On the two top traces of Fig. 10, the manual injection of 4 μM calcium over a 10-s interval gives a deflection of two-thirds that obtained on addition of cyanide; thus, the latter would correspond to approximately 6 μM calcium released per 10 s, or 0·6 μM per second. In order to simulate the release of calcium by the uncoupler, 70 μM calcium is injected over an interval of 8 s (top right); the peak intensity is approximately half that obtained experimentally with FCCP, and thus the FCCP-stimulated rate of calcium release would be approximately 15 μM calcium per second, a value of the same order as that of the maximal rate of calcium uptake in mitochondria at the same temperature (23°C, ref. 44). These controls further show a negligible delay in indicating calcium release under our experimental conditions.

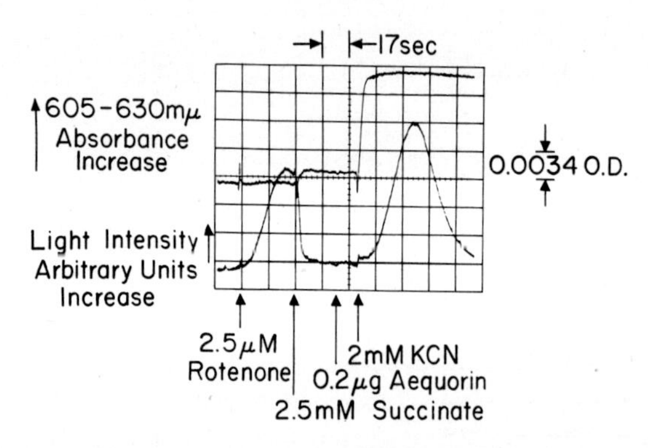

Figure 11. The response of aequorin to a series of metabolic transitions in rat liver mitochondria, 6·2 mg protein per ml in 150 mM KCl, 10 mM Tris-Cl, pH 7·4, supplemented with 200 μM $MgCl_2$, 10 μg oligomycin, and 0·2 μg aequorin. Top trace, dual wavelength recording of cytochrome *a* at 605-630 mμ; bottom, bioluminescence of aequorin measured as in Fig. 10. The various additions are indicated on the figure.

Figure 11 illustrates the response of aequorin (bottom trace) to a series of metabolic transitions in rat liver mitochondria supplemented with oligomycin, so that the energy storage of the ATP pool is not available for calcium retention. When supplemented with rotenone to block NADH oxidation, these State 4 mitochondria show immediate calcium release. However, when 2·5 mM succinate is added, the indicator intensity drops with a half-time of 3 s, showing the immediate uptake of the released calcium. In order to ensure that no calcium was left outside after the indicator had flashed, a second addition of aequorin is made, and gives no response whatsoever. Lastly, 2 mM cyanide causes the reduction of cytochrome *a* within one second (top trace) and a detectable rate of calcium release one second later. Taking the State 4 rate to be 20 ATP-equivalents per mg protein per minute, we find this to indicate that the high energy store is no greater than 0·3 nmoles ATP-equivalents per mg protein. While this value is consistent with a chemical intermediate equal to the cytochrome concentration, it is much less than the 10 to 20 ATP-equivalents expected to be stored in a hydrogen ion gradient or a 100-mV membrane potential. The data further show that whenever electron transport is interrupted and reactivated, a rapid calcium release and uptake occurs, as it must also have done in the determinations of the proton/electron stoichiometry presented in the previous paper [41]. Thus, the available values for proton/electron ratios are seriously affected by such cation movements [47].

On the basis of these data, we consider calcium release from the mitochondrial membranes to provide the most sensitive indication of the depletion of the energy reserves of the membrane in various metabolic transitions.

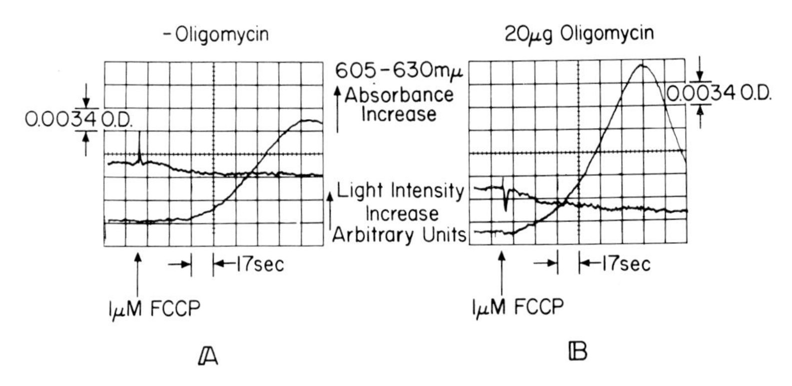

Figure 12. Responses of cytochrome *a* (top trace) and aequorin (bottom trace) to 1 μM FCCP in the absence (A) and presence (B) of oligomycin. Rat liver mitochondria, 3·8 mg protein per ml, in 150 mM KCl, 10 mM Tris-Cl, pH 7·4, supplemented with 10 mM succinate, 200 μM $MgCl_2$, 0·2 μg aequorin, and in B, 20 μg oligomycin at 10°C.

In order to determine more accurately the size of the high-energy pool in relation to the ATP pool, in Fig. 12 the kinetics have been "spread out" over a longer interval by dropping the temperature to 10°C. On the left, in the absence of oligomycin, the first appearance of detectable calcium occurs 40 s after adding FCCP; on the right, in the presence of 20 μg oligomycin, it occurs 3 s after adding FCCP. Since the traces indicate the rate rather than the extent of calcium release, the acceleration in the presence of oligomycin in Fig. 12B gives a higher peak than that found in the absence of oligomycin (Fig. 12A). Assuming that FCCP does not selectively destroy the intermediates, a simple calculation gives the oligomycin-independent energy store. Rat liver mitochondria contain 4 nmoles of ATP per mg protein, and this ATP is readily expended upon addition of inhibitors of electron transport [48]. Taking the ratio of the times for the first appearance of detectable calcium release due to depletion of the available energy store in the presence and absence of oligomycin to be 3/40, the internal pool of high energy intermediates would be 3/40 that of the ATP content, or 0·3 nmoles per mg protein.

Responses of other types of mitochondria

The calcium response can be exhibited in heart as well as in rat liver mitochondria, as shown in Fig. 13, which includes the typical responses of pigeon heart mitochondria to the terminal inhibitor, cyanide, the uncoupler, TTFB, the Site I inhibitor, rotenone, and the mid-chain, Site II inhibitor, antimycin A. Only in the latter case is the calcium release phenomenon delayed, presumably because of the longer interval required for antimycin A to block the chain. It is noteworthy that the energy store in the heart mitochondria would appear to be minimal; the uncoupling agent causes calcium release more rapidly than do the electron-transport blocking agents, cyanide and rotenone. Furthermore, it is of interest that rotenone is as effective in dropping the energy state of heart mitochondria as of liver mitochondria. Experiments with beef heart mitochondria give similar results.

Three important conclusions may be drawn from these studies of energy storage and calcium retention in mitochondria. The first is that the measurement of calcium release from the mitochondria provides the most sensitive indication, both in terms of extent and of kinetics, of the state of the energy reserves of the membrane, of any technique presently available.

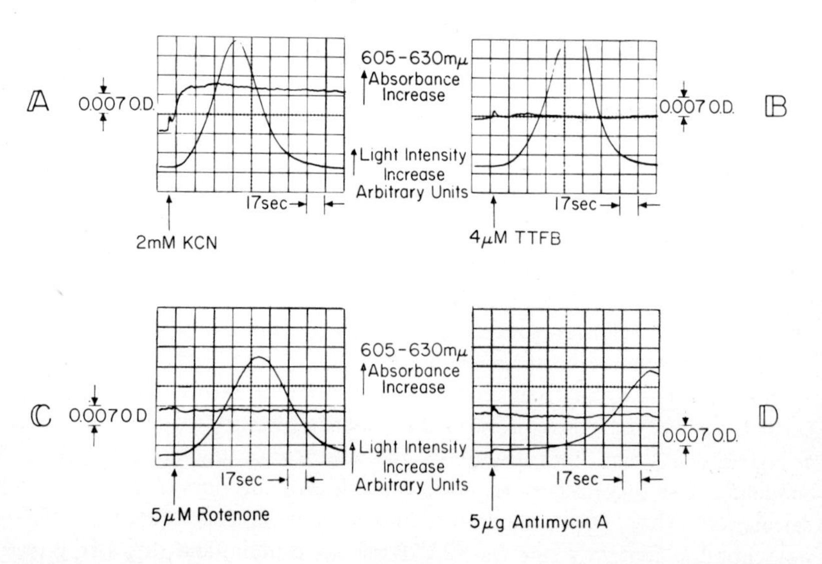

Figure 13. Responses of pigeon heart mitochondria containing endogenous substrate to the additions indicated; top traces, cytochrome *a*; bottom traces, aequorin. Pigeon heart mitochondria, 5·0 mg protein per ml, in 150 mM Kcl, 10 mM Tris-Cl, pH 7·4, supplemented with 200 μM $MgCl_2$, 20 μg oligomycin, and 0·2 μg aequorin.

The second is that the bound calcium of State 4 rat liver and pigeon and beef heart mitochondria is highly labile, and is released rapidly a few seconds after the cessation of energy conservation on adding inhibitors of electron transfer or uncouplers, such as FCCP. In the latter case, the rate of calcium release approaches the maximal rate of calcium accumulation in coupled mitochondria.

A third conclusion, drawn from the comparison of the speed of calcium release by FCCP in the presence and absence of oligomycin, is that the energy reserve available for calcium retention in the coupled mitochondria is approximately 0·3 nmoles of ATP-equivalents per mg protein.

These conclusions have important bearings upon both the chemical and the chemi-osmotic theories of energy coupling. First, the energy store is found to be much smaller than that which would have been expected from vectorial movement of hydrogen ions and electron transport mechanisms involving the quinone, flavin, and pyridine nucleotide pools of rat liver mitochondria. Calculated on this basis, the energy store would amount to between 10 and 20 nmoles of ATP-equivalents per mg protein. An energy store of this magnitude, retained in a hydrogen ion gradient or an electrostatic potential, should cause the delay in calcium release upon adding an uncoupler to persist in the presence of oligomycin. Instead, oligomycin reduces the delay in calcium release by more than ten-fold, and leaves an energy store of only 0·3 nmoles of ATP-equivalents per mg protein.

The paucity of energy storage in the mitochondrial membrane also makes the detection of chemical intermediates by analytical means more difficult. The most recent value from the ATP-jump technique is about 0·6 nmoles of ATP per mg protein, but the jump is insensitive to dinitrophenol [49].

The observation that calcium release occurs almost step by step with the reduction of cytochrome *a* suggests that energy storage is negligible, and that energy conservation occurs when there is an energy debit, without the accumulation of an appreciable energy store, except for that represented by the phosphorylated ADP and the establishment of concentration gradients in the matrix space of the mitochondria.

Secondly, the ready release of the tightly bound calcium of the mitochondria upon interruption of electron transport by inhibitors also occurs in anoxia, and the released calcium is rapidly taken up when oxygen is added. Thus, these calcium movements would occur under the conditions of Mitchell and Moyle's measurements of $H^+/0$ stoichiometries, and are apparently responsible, either fully or in part, for the hydrogen ion movements they observed, as suggested by our previous studies [50]. Thus, direct evidence of hydrogen ion movements in the absence of concomitant calcium movements, essential to the support of the chemi-osmotic mechanism, seems to be lacking.

III. MEMBRANE CONFORMATION CHANGES

About ten years ago, we observed a light-scattering change that closely followed oxidation-reduction changes of cytochrome *b* in rat heart mitochondria [51, 52]. At that time, the half-time for this light-scattering change was found to be 10 s in the State 4 → 3 transition. This may be contrasted with more recent data indicating that the ADP response of cytochrome *b* requires only 70 ms in rat liver mitochondria [44]. Furthermore, the light-scattering change lags behind the oxidation-reduction changes in the State 3 → 4 transition.

The possible relevance of the conformation changes, which may be revealed by this altered light-scattering, to energy conservation in states of the membrane has been reviewed elsewhere [7, 53], and it has been confirmed that the light-scattering changes are secondary consequences of energy conservation as the primary reaction. In fact, Azzone and Azzi [54] suggest that the movements of ionic constituents of the oxidative phosphorylation medium cause volume changes in the mitochondria which are responsible for the light-scattering changes; as further evidence for this conclusion, they report that swelling phenomena are not observed in a medium in which metallic cations are replaced by "Tris".

Ultrastructural transformations which occur in mitochondria during changes of respiratory state have been explored by Hackenbrock [4, 55] and by Green and his colleagues [56, 57]. As Fig. 14 shows, a variety of membrane conformations can be observed electron microscopically, depending upon the experimental conditions and the preparative methods. In both studies, it is postulated that these conformations represent states of the membrane functional in energy conservation. As yet, there is no proof that this is true.

Table 1 compares the electron micrographic and other methods of determining conformation changes. Hackenbrock has identified the "condensed" (C) and "orthodox" (O) types of mitochondrial structure, and finds a C → O transformation in the States 2P → 4P and 5P → 4P transitions. He has made electron microscopic observations of the O → C transformation in the State 4 → 3 transition within 35 s after the addition of ADP [4], and reports a C → O, State 4P → 5P transformation to occur within 30 s in antimycin-condensed mitochondria converted to the orthodox configuration by the addition of TMPD (C. R. Hackenbrock, personal communication). Other transformations appear to be slower; for example, the establishment of the State 4 orthodox configuration appears to require between 10 and 20 min.

Phosphate is present in all of Hackenbrock's studies, but was absent in some cases and present in others of a series of experiments carried out by Green and his co-workers with "heavy" beef heart mitochondria. However, Green proposes a different classification of configurational changes confined within the "aggregated" mode, and describes the various modes as follows: "In the aggregated mode, the headpieces are in the lumen of the cristae. In the orthodox

Table 1. Conformation changes in mitochondria. Transition between various states of mitochondria.

Technique	Transition	Type	Time Min	Time Sec	Medium	Source	Site
EMG	2 → 4	C → O	20		S-P-Mg	RL	I, II, III
	5 → 4	C → O	10		S-P-Mg		II, III
	4 → 3	O → C	2		S-P-Mg		I, II, III
	4 → 5	O → C	8		S-P-Mg		I, II, III
EMG	4 → 3U		2		S-Tris	HBH	I, II, III
	4P → 3U		1		S-Tris P	HBH	I, II, III
	4P → 3U			6 (7°C)	S-Tris P	HBH	I, II, III
DNS	5 → 3 (1)			<0·002	S-Tris	E-SMP	I → III
	3 → 5			10	S-Tris	E-SMP	
Aequorin	4 → 5 (2)			~3	S-Tris	RL	I → III
	2 → 4			~3	S-Tris	RL	I → III
	4 → 3			>20(7°C)	S-Tris	RL	I → III
O_2 Pulse	5 → 3 (3)			0·001	S-Tris	PHM	I → III
	3 → 5			0·003	S-Tris	PHM	I → III
Ca^{2+} Reaction	4 → 3 (4)			0·03	S-Tris	RL	II

C = Condensed Configuration
O = Orthodox Configuration
S = Sucrose
P = Phosphate
U = Uncoupled

(1) Increased Hydrophobicity of Dye Binding Site.
(2) Membrane Energization.
(3) Altered Reactivity of Electron Transport
(4) Cytochrome *b* Oxidation Data.

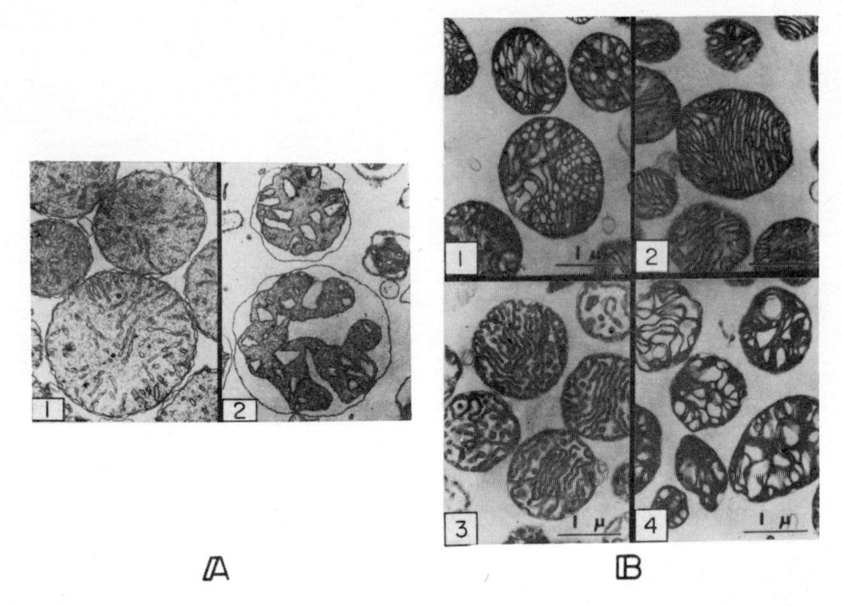

Figure 14. (A) Various membrane conformations of rat liver M_{W} observed by electron microscopy: 1, the "orthodox" conformation. The spatial folding of the electron transport membrane is organized into cristae. The volume of the matrix increases by 100% during the "condensed-to-orthodox" mechanochemical ultrastructural transformation (× 20,500); 2, the "condensed" conformation. The spatial folding of the electron-transport membrane is random. The volume of the matrix is approximately 50% of the total mitochondrial volume (× 20,500) (courtesy of Charles R. Hackenbrock).

(B) The energized configurations of beef heart mitochondria: 1, non-energized configuration obtaining in the presence of rotenone (2 μg/mg protein); 2, induction of the energized configuration by the addition of succinate (5mM); 3, induction of the energized-twisted configuration by the addition of P_i(10 mM, pH 7·4); and 4, discharge of the energized-twisted configuration to the non-energized configuration by the addition of mCl–CCP (10^{-6}M). The incubation was carried out at 30°C with 1 mg of mitochondrial protein per ml of incubation medium. The incubation medium was 0·25M in sucrose and 5 mM in Tris-Cl, pH 7·4. The mitochondria were fixed for electron microscopy 15-30 s after the addition of the reagent (e.g. succinate, P_i, or mCl–CCP) which was responsible for the transition in energy state (courtesy of David E. Green).

mode, the headpieces are outside the lumen of the cristae . . . In a rough way, Hackenbrock's condensed mode corresponds to our aggregated mode, but we are dealing with configurational transitions within that mode . . ." (D. E. Green, personal communication).

In view of the difficult semantic problems, it is probably better to compare these states of mitochondria observed electron microscopically in terms of the usual State 3 → 4 notation. Thus, the micrographs of Fig. 14B which represent Green's sequence NE → E → ET → NE (see legend) can be compared with Hackenbrock's data in terms of the state sequence 2 → 4 → 4P → 3U. On this basis it is possible to focus our attention upon the key question raised by the electron microscopic studies: do the transitions in conformational state occur rapidly enough to be effective in the energy conservation process of mitochondria? In Green's studies, the State 4P → 3U transition has been observed regularly to occur in 1 min and, in one case, in 6 s. In the absence of phosphate, a State 4 → 3U transition has been observed to occur in 2 min. However, the usual State 4P → 3, O → C transformation caused by ADP addition, observed by Hackenbrock to occur completely within 35 s, in Green's hands gives only mixed conformational states. Green has also found that activation of these changes by the separate energy-coupling sites is possible. Hackenbrock finds Site III to be sufficient to activate the State 5 → 4, C → O transformation although Sites II and III have been effective as well; operation at Site I is apparently not necessary.

While both Green and Hackenbrock report reasonable speed in the State 4 → 3 transformation, the re-establishment of the State 4 conformation (not reported by Green) seems to require such an unduly long time as to leave a credibility gap concerning the functionality of the structural changes. One surprising feature of Green's work is that he finds a time of 6 s for the State 4 → 3U transition in the absence of oligomycin; one would have expected the endogenous energy reserves to prolong the State 4 → 3 transition, as indicated by the analytical data of Chappell and Crofts [48]. For these reasons, we have sought other indicators which would reveal a better correlation between the time course of metabolic changes in mitochondria and the transitions between various states of the membrane.

Conformational changes of the mitochondrial membrane measured by a fluorescence indicator, DNS

The possibility of employing fluorescent dyes which can bind the side chains of the protein has been suggested by Weber's work on soluble proteins [58]. This indicator reports conformation changes of a few Å, as required by Slater [59]. The quantum yield of fluorescence emission of such dyes depends upon the degree of hydrophobicity of the environment in which the dye finds itself. DNS is scarcely soluble and shows little fluorescence in aqueous solutions; it will, however, bind proteins, presumably in their hydrophobic spaces, and under these conditions its fluorescence is greatly enhanced. If, now, the protein changes its conformation so that the site to which the dye it binds becomes more hydrophobic, the quantum efficiency of fluorescence emission will be increased still further. One of us (Azzi) has found that when DNS binds to submitochondrial particles, its fluorescence becomes larger and, more important, increases as the state of reduction of the respiratory carriers increases—an electron transport-linked structural change. While the amount of dye which is

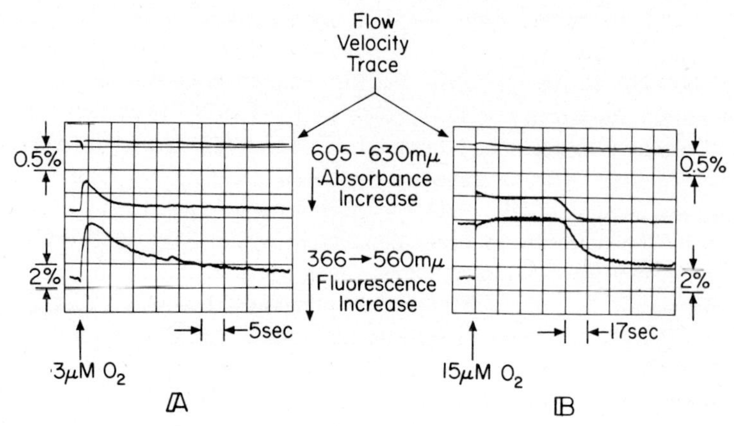

Figure 15. Cycles of oxidation and reduction of cytochrome *a* (upper traces) and DNS (lower traces) in response to pulses of 3 μM oxygen on a 5 s/division time scale (A) and 15 μM oxygen on a 17 s/division time scale (B). Submitochondrial particles (E-SMP) at 6·0 mg protein per ml in 0·3M mannitol-sucrose, 20 mM Tris-Cl, pH 7·4, supplemented with 20 mM succinate and 6 mM malonate. 2 μg of 10% DNS-Cl absorbed on celite was added to 30 mg E-SMP; after 30 min at 0°C, the celite was centrifuged down in a table centrifuge.

bound is not known exactly, the indicator affords a novel approach in indicating the time course of the conformation change in relation to the oxidation-reduction kinetics of the cytochrome components of submitochondrial particles.

The general phenomenon is indicated in Fig. 15 which shows the kinetics of oxidation-reduction of cytochrome *a*, measured at 605-630 mμ, and fluorescence changes of DNS excited at 366 mμ by a water-cooled 1000 W arc and measured at 560 mμ. The submitochondrial particles have already been rendered anaerobic with succinate and supplemented with malonate so that the full oxidation-reduction change of the respiratory carriers can be obtained in the anaerobic-aerobic State 5 → 3 transition following the injection of low concentrations of oxygen. Both the fluorescence and absorbancy traces are recorded on a storage oscilloscope. Oxygen pulses are delivered in the regenerative flow apparatus in which 20 ml of the suspension of anaerobic submitochondrial particles are mixed with a small volume of air- or oxygen-saturated buffer to give the two oxygen concentrations employed.

In the left-hand portion (A), the moment of injection of 3 μM oxygen is indicated on the top trace. As the flow stops, the absorbance of cytochrome *a* diminishes in a fraction of a ms, and then increases with a half-time of 5 s. Simultaneously, the conformation change, represented by the change in DNS fluorescence, shifts toward a smaller quantum efficiency, reaches a minimum, and then increases with a half-time of 12 s.

In (B), a 5-fold larger oxygen pulse is delivered and the cytochrome cycle has a half-time of 20 s, while that of DNS is 22 s. Here, the simultaneity of the rise times of the two traces is apparent, as is the discrepancy of the fall times. It is also apparent, at least in this record, that the change reported by DNS occurs simultaneously with the cytochrome *a* oxidation, but has a relaxation time considerably longer than that for the reduction of cytochrome *a*, and more consonant with the reduction time of the slower components of the chain, such as cytochrome *b*. Generally, this experiment suggests that whenever all the respiratory carriers are in the reduced state, DNS is in a more hydrophobic environment than when any one of the cytochromes is oxidized.

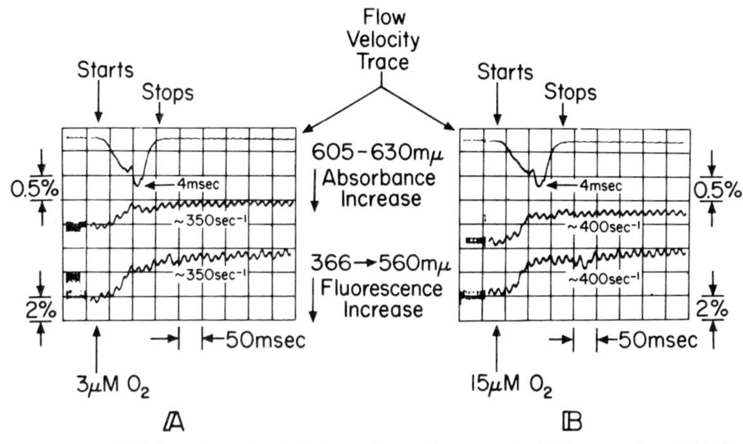

Figure 16. Initial kinetics of oxidation of cytochrome *a* (middle traces) and DNS (lower traces) in response to pulses of 3 μM oxygen (A) and 15 μM oxygen (B). Experimental conditions the same as those of Fig. 20, except that 20 μg oligomycin is present in the medium. Rapid flow studies on a 50 ms/division time scale.

Since the half-time for cytochrome *a* oxidation is a few ms under these experimental conditions [10], it is important to note whether the DNS response accurately follows the cytochrome oxidation time. Therefore, the experiment is repeated on a faster time scale in Fig. 16 under the same experimental conditions, except that 20 μg oligomycin is present in this case. Cytochrome *a* and DNS are monitored in the lower portion of the observation chamber in which the flow rate is approximately 4 m/s, and observations can be started at a time after mixing of 4 ms. At this time, the oxidation of cytochrome *a* is largely complete, and the first order velocity constant for the reaction with 3 μM oxygen (A) is 350 s^{-1}. In the lower trace, the extent of reaction of the DNS fluorescence change is similar and the first order velocity constant is indistinguishable from that for cytochrome *a*. A similar situation obtains at 15 μM oxygen (B), where the pseudo first order velocity constant for both curves is 400 s^{-1}. It is apparent that an extremely rapid change in the DNS environment occurs when electron transport is initiated by oxygen pulses.

To eliminate the possibility that a pH change rather than a change of membrane conformation is involved, we note that the pK for DNS is so low (4·5) as to be far outside the pH range (7·4) over which our experiments are carried out; the DNS anion is present in such large excess that the pH change in the membranes of the submitochondrial particles, measured independently by bromthymol blue or bromcresol purple, would be too small to affect the DNS concentration appreciably. Furthermore, the pH change caused by initiation of electron transport in submitochondrial particles has a half-time of approximately 20 s rather than a few ms [21]. In an experiment on this point in which the energy-dependent BTB response in the submitochondrial particles was specifically inhibited by the uncoupling agent, FCCP, the DNS response was, nevertheless, observed in oxygen-nitrogen cycles. A further distinction between the response of the two indicators is that the addition of alkali, which causes no change of DNS fluorescence, readily evokes a response of BTB equal to that observed in the aerobic-anaerobic transition of the coupled submitochondrial particles.

These data suggest that a structural change of the respiratory chain occurs whenever the cytochromes turn over from the reduced to the oxidized state, and vice versa. This conformation change appears to occur whenever the most rapid members of the chain are oxidized, but does not relax until the slower components are reduced. We propose that the DNS site (or sites) are responsive to a membrane conformation which changes from one state to another as soon as a member of the chain is oxidized, and remains in that conformation until all the members of the chain are reduced. The change appears to be independent of energy coupling reactions, at least in submitochondrial particles, and thus differs from the changes observed electron microscopically.

Responses of aurovertin

In previous experiments with Dr. Henry Lardy, we found another substance, aurovertin [60], whose fluorescence emission was greatly enhanced when bound to the mitochondrial membranes. Furthermore, the binding site for aurovertin has been identified in experiments with the isolated ADP-phosphorylating enzyme (coupling factor F_1) which indicate a high affinity of aurovertin for this component. In addition, titrations of the mitochondrial membranes with aurovertin show the amount bound to be consistent with the amount of F_1 to be expected in the membrane (approximately 5% of the total protein). These observations suggest that the binding site of aurovertin in the mitochondrial membrane is at or near the ADP-phosphorylating enzyme, identified with the projecting subunits of the phosphotungstic acid-stained membranes of surface-spread mitochondria [61]. Thus, we are in the fortunate position of being able to indicate where, in mitochondrial structure, the aurovertin is bound, namely, to the mitochondrial cristae.

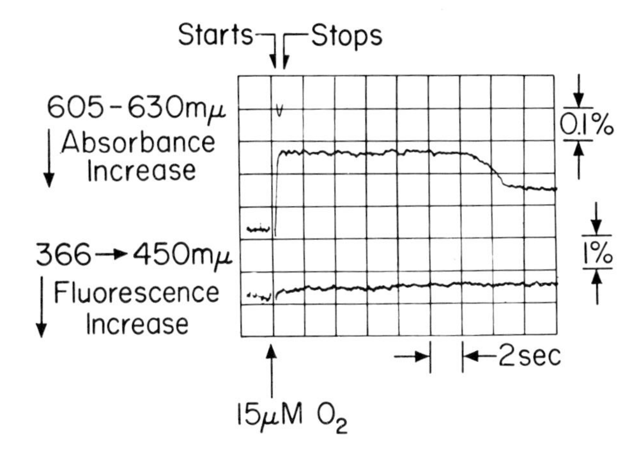

Figure 17. Responses of cytochrome *a* (top trace) and aurovertin (bottom trace) to changes of the oxidation-reduction state induced by a pulse of 15 μM oxygen in submitochondrial particles. E-SMP, 0·75 mg protein per ml, in 0·3M mannitol-sucrose, 20 mM Tris-Cl, pH 7·4, supplemented with 5 mM succinate, 75 μg oligomycin, 5 μM FCCP, and 1·7 μM aurovertin.

It is of great interest, therefore, to determine whether or not the degree of fluorescence of aurovertin, like that of DNS, depends upon the oxidation-reduction state of the respiratory carriers. An experiment to this point is shown in Fig. 17, where the submitochondrial particles have been rendered anaerobic with succinate and supplemented with aurovertin and oligomycin, and, in this portion of the experiment uncoupled with FCCP. At the moment of injection of 15 μM oxygen in the pulsed flow apparatus, cytochrome *a* is oxidized, as in the DNS experiments, and 14 s later, becomes spontaneously reduced upon exhaustion of the added oxygen. But here, there is no evidence for a synchronized change of aurovertin fluorescence, even though a change of less than 2% of the total would be detected; we conclude that there is no change in the aurovertin environment at the F_1 site in the cristae.

In order to demonstrate that the reaction of aurovertin with the submitochondrial membranes gives a highly significant fluorescence response, we show in Fig. 18, the reaction of aurovertin with the submitochondrial particles in the oligomycin and FCCP-uncoupled states. Apparently, no change of conformation in the region of the inner membrane subunits is involved in uncoupling the submitochondrial particles, since the rate and extent of the fluorescence change seem to be similar in both cases.

The downward deflection of the curves in Fig. 18A, on a 200 ms/division time scale, shows a rapid phase of binding; and in Fig. 18B, on a scale of 5 s/division, a slower phase. Apparently there is some heterogeneity in binding sites. Alternatively, the possibility exists that a portion of the inner membrane subunits are exposed on the outside of the submitochondrial vesicles, leading to rapid binding: and others are less accessible, resulting in the slower binding step.

In order to test whether the heterogeneity of sites is due to a distorted structure of the submitochondrial particles, the reaction of intact mitochondrial membranes in the uncoupled state is shown in Fig. 19. Here, the reaction, while somewhat slower (as indicated on the 200 ms trace of Fig. 19A) is, nevertheless, definitely biphasic (as shown on the slower scale of Fig. 19B). Thus, a heterogeneity of binding sites is characteristic of the natural membrane.

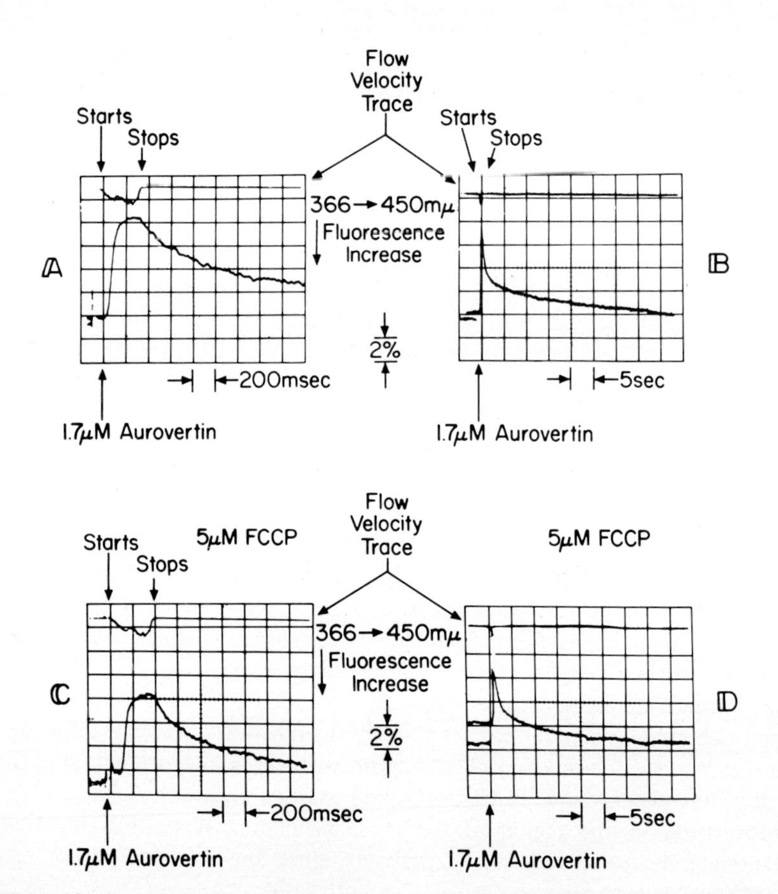

Figure 18. Responses of submitochondrial particles to the addition of 1·7 μM aurovertin in the oligomycin-supplemented (upper traces) and FCCP-uncoupled (lower traces) states; on the left, on a time scale of 200 ms/division; on the right, at 5 s/division. E-SMP, 0·75 mg protein per ml in 0·3M mannitol-sucrose, 20 mM Tris-Cl, pH 7·4, supplemented with 5 mM succinate and 75 μg oligomycin in all traces. In the two lower traces, 5 μM FCCP is present as well. Final volume, 32 ml.

The lack of evidence for conformation changes at the variety of sites bound by aurovertin may well be contrasted with those reported by DNS and by the electron micrographic technique, as shown in Table 1. At the binding sites for aurovertin, which include the inner membrane subunits, no conformational changes are observed in terms of alterations of the fluorescence enhancement of the antibiotic, in either coupled-uncoupled or oxidized-reduced transitions; this portion of the membrane would then appear to maintain its structure during these two important transitions.

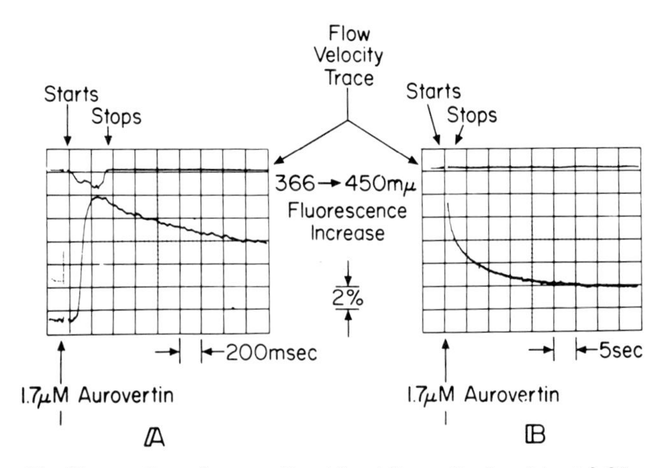

Figure 19. The reaction of aurovertin with rat liver mitochondria at 0·95 mg protein per ml, in 0·3M mannitol-sucrose, 20 mM Tris-Cl, pH 7·4, in the presence of 5 mM succinate, 5 μM rotenone, and 5 μM FCCP, on a 200 ms/division time scale (A) and a 5 s/division scale (B).

A second discrepancy between the fluorescence indicator data and the electron microscopic data is that neither DNS nor aurovertin show any significant conformational changes at the structure levels of the cytochrome chain or the coupling factor, F_1, due to reaction of the membranes with uncoupling agents. This remarkable observation assists greatly in the interpretation of the electron micrographic studies, where the State 4 → 3 transition, particularly when caused by uncouplers, gives generally an orthodox-to-condensed (O – C) transition of the mitochondrial structure. Whatever the O – C change may be, it does not appear to affect greatly the macromolecular structure of the electron transfer components or of the coupling factor, F_1. Thus, it is possible that the O – C rearrangement involves only a large-scale phenomenon and does not affect the molecular structure of the cytochrome chain.

The fact that the membrane can undergo the gross structural changes indicated by electron microscopy without altering the environment of either DNS or aurovertin is, indeed, remarkable. However, it should also be noted that these membrane conformation changes occur as well without any concomitant changes in the fluorescence efficiency of the mitochondrial pyridine nucleotides; the most appropriate conditions for altering the fluorescence yield of NADH appear to be found in State 6 [62], in which uptake of calcium in the absence of a permeant anion leads to an inhibited state of respiration. The effect of brief sonication upon the NADH fluorescence is described below.

DNS, on the other hand, reports a structural change of the electron transfer components associated only with their oxidation-reduction state, and independent of their energy coupling state. The very rapid changes of membrane conformation reported by DNS in the nitrogen-oxygen transition is of special interest with respect to the control of reactivity of the electron transfer reactions associated with the oxidation state of the cytochromes, as required by the nature of the kinetic responses of these components in oxygen pulse experiments [15, 63]. In those experiments, the computer representation requires the conversion from the unreactive to the reactive form of the mitochondrial membrane in 1 ms. This activity change gives the rapid electron transfer reactions in the presence of oxygen, and slow reactions when oxygen has been exhausted. These two results are, therefore consistent and, furthermore, indicate that when DNS is in a less hydrophobic environment, the reactivity of the cytochromes is high.

An analysis of the results in the oxygen-nitrogen transition of the oxygen pulse experiments suggests, however, that the conversion from the active to the inactive state of the membrane requires about 1 ms, a much shorter time than the several seconds required for the increase of DNS fluorescence under the same conditions. Obviously, further exploration of the computer fit to determine the effect of the longer relaxation time is desirable. On the other hand, it is possible that the DNS is responding to conformation changes of the respiratory components which do not control cytochrome reactivity as well as those that do.

Protein conformation changes

In order to illustrate structural effects at the level that may be detected by changes in the fluorescence efficiency of bound dyes, we will describe the recently discovered low-level conformation changes of the well-known hemoprotein, myoglobin. When myoglobin is bound to appropriate ligands which alter the spin state of the iron (e.g. cyanide or hydroxide) small differences of electron density appear in Fourier difference syntheses between the liganded and unliganded species [64]. At first, the nature and significance of these changes was somewhat in doubt because some of the electron density differences were of such small magnitude that they were at the "noise" level, as indicated in the left-hand portion of Fig. 20. If, however, the Fourier difference

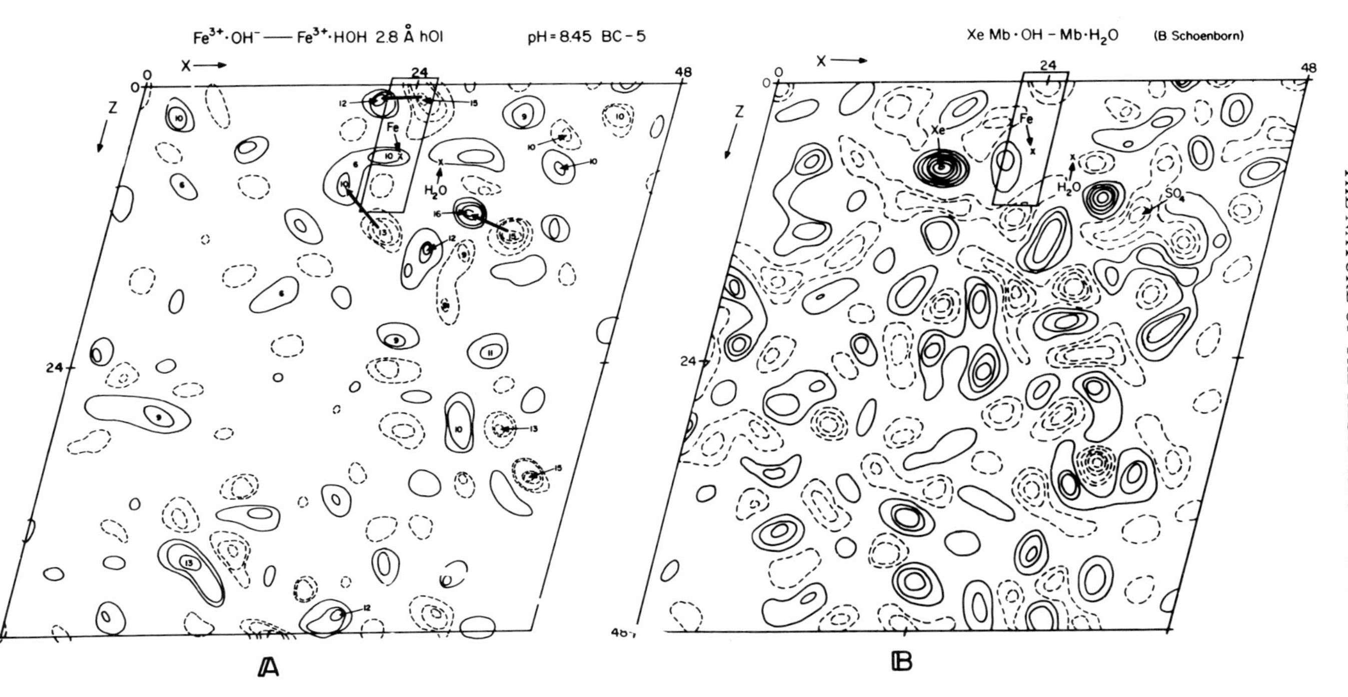

Figure 20. Small-scale electron density changes revealed by Fourier difference synthesis against acid metmyoglobin, pH 7. A, Alkaline metmyoglobin, pH 8·45; B, xenon alkaline metmyoglobin, pH 9·2. Both diagrams represent h01 projections at 2·8 Å resolution. The electron density differences are contoured at arbitrary intervals which are identical in both diagrams.

synthesis is carried out on the xenon-saturated hydroxide compound and the acid metmyoglobin, the electron density differences become so distinct that the conclusion of a definite structural change associated with spin state conversion is unmistakable. More recently, Schoenborn has determined the three-dimensional structure as well, and a number of rearrangements involving polar groups are noted as well as a "kink" in the E-helix and small changes in the G-H corner [65, 66]. This sort of conformation change can be extremely rapid in solution, although rather slow in the crystal structure [67, 68].

The relationship of small-scale structural changes to the larger-scale rotations and translations of the subunits of tetrameric hemoglobin, described by Perutz in his opening lecture [69], may well be analogous to the relationship between the small-scale changes reported by the DNS technique and the larger-scale changes observed electron micrographically in the mitochondrial membranes. As in the case of hemoglobin, where the small-scale secondary and tertiary structural changes may well precede the larger rearrangements of quaternary structure [70], the small and rapid DNS-reported changes appear to precede the gross phenomena observed electron micrographically. An example of slow quaternary structure changes is provided by glyceraldehyde phosphate dehydrogenase [71, 72].

The functionality of the smaller and larger changes becomes a matter of interesting conjecture in both cases. One possibility is that the large changes represent a relaxation of the structure to a minimum entropy form dictated by the small-scale conformation changes and the transmission of their effects throughout the macromolecular system. On this basis, the large-scale changes may, indeed, be useful in emphasizing or amplifying the effects of the small changes but would not, of course, represent primary events in the structural control of membrane reactivity.

Ion gradients during membrane fragmentation: effect of sonication

The chemi-osmotic hypothesis depends upon energy conservation in a closed membrane, while chemical mechanisms do not. In an attempt to study the properties of mitochondrial fragments in which the likelihood of a closed membrane structure is minimized, we have carried out a series of experiments in which ion gradients in mitochondrial and submitochondrial membranes were directly observed during intense sonication.

In order to ensure the opening of not only mitochondrial membranes but also those of submitochondrial vesicles, a large portion of the power of the sonic oscillator usually employed in a 20 cm^3 volume has been concentrated into a 1 cm^3 volume. Under these conditions, the rate of temperature rise during sonication, indicative of the power transfer from the sonic tip into cavitation vacuoles in the liquid, is increased two-fold over that normally employed to prepare submitochondrial particles from beef heart mitochondria (0·3°C per sec). With such intense sound powers, the interval of exposure is necessarily limited to avoid overheating of the preparation. Thus, the temperature rise was set at 0·6°C per sec for rat liver mitochondria and for beef heart submitochondrial particles.

Splashing and aerosol formation are avoided by projecting the 4 mm tip of the Branson sonifier 3 to 4 mm beneath the surface of the mitochondrial suspension contained in a 10 x 10 x 10 mm cuvette. Currents of between 1 and 3 A are employed, and the power transfer into the liquid is monitored by a small thermocouple.

Note should be taken that the sonication process depends upon energy dissipation during the collapse of the cavitation vacuoles; mitochondria or submitochondrial particles at the interface of such vacuoles are subjected to an intense shearing force, thereby causing them to leak their ionic constituents and rapidly neutralize any ionic gradients that they might contain. As a rapid and sensitive indicator of ionic gradients in mitochondrial membranes, we have employed the absorbancy change of bromthymol blue; in mitochondria, an increased absorbance of the dye indicates that calcium has been accumulated in the absence of a permeant anion; in submitochondrial particles, a decrease of BTB absorbance indicates the activation of electron transport and a corresponding ion movement. Since these processes are attributed by the chemi-osmotic hypothesis to ionic gradients across intact mitochondrial or submitochondrial membranes, the corresponding absorbancy changes should be abolished the moment the closed membrane is opened to the external medium and equilibration of H^+ ions occurs.

Because the light-scattering changes are relatively larger in the sonication experiments, two additional precautions are observed. First, the dynode voltage to the photomultiplier is regulated so as to maintain constant sensitivity at the reference wavelengths. Secondly, a triple beam apparatus is used, with the measuring wavelength at 615 mμ, near the peak of the BTB absorption, and two reference wavelengths, one shorter and the other longer. These differ in each experiment, but generally their relative sensitivities are so arranged that the sum of the non-specific absorbancy changes at the reference wavelengths very nearly equals the non-specific absorbancy change at 615 mμ. Control experiments on phosphate-induced swelling of rat liver mitochondria showed negligible disturbance in the differential output (615 mμ vs. the average of the reference wavelengths) but a large change of the total absorbance at the reference wavelengths. Because of the particular electronic circuitry of this apparatus, it is most convenient to measure non-specific changes at all three wavelengths; thus these changes have a component of the BTB absorbancy impressed upon them, but this is small compared to the very large and non-specific changes caused by rupture of the membranes.

Effect of sonication upon the BTB response in rat liver mitochondria

In order to demonstrate the effectiveness of the sonication method for rapidly equalizing concentration gradients across the mitochondrial membrane, Fig. 21 illustrates in four traces reading from top to bottom, alterations in light-scattering measured at the average absorbancy of the three wavelengths indicated (thus including a small but negligible contribution from bromthymol blue, as mentioned above); NADH fluorescence; the temperature of the suspension; and changes in bromthymol blue absorbancy measured at 617 mμ

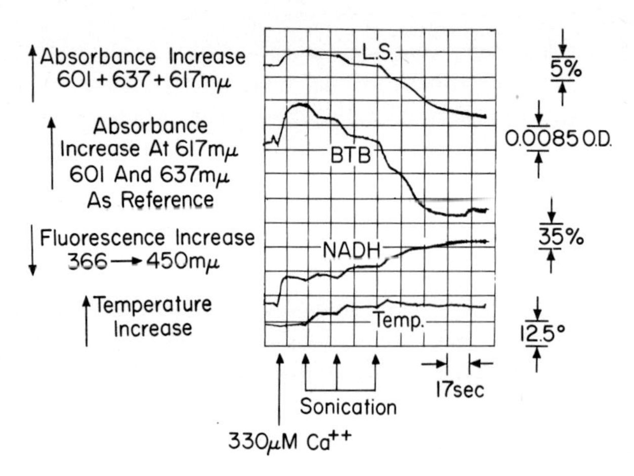

Figure 21. Opening of membranes of intact mitochondria by three intervals of sonication, following the addition of 330 μM calcium. Rat liver mitochondria, 3·7 mg protein per ml, in 0·3M mannitol-sucrose 20 mM Tris-Cl, pH 7·4, in the presence of 6·7 mM succinate, 6·7 μM butacaine and 6·7 μM BTB. Experiment carried out in a 1·7 ml volume.

with reference to the average absorbancies at 601 and 637 mμ. The mitochondria, initially in State 4 in a Tris-Cl medium free of permeant anions and supplemented with butacaine to block non-specific calcium binding to phospholipid, respond to the addition of 330 μM calcium by an alteration in light scattering, by a decreased fluorescence indicating oxidation of NADH, and by an increasing alkalinity as indicated by the increased absorbance of BTB.

Sonication is initiated at a power input giving a rate of temperature increase of 0·6°C per second and there is a rise of approximately 7°C in the 1·7 ml volume of the cuvette, during the first sonication interval. Both the fluorescence intensity and bromthymol blue absorbance decrease; however, there is only a small change in light-scattering, suggesting that no large reduction of particle size has occurred.

On the second interval of sonication, however, the BTB gradient established by the addition of calcium is almost completely depleted; the pyridine nucleotide fluorescence decreases further, and the light-scattering trace shows a marked change. A third sonication initiates light-scattering changes which now continue in an irreversible fashion, and in addition, a further and large loss of BTB absorption is observed. The NADH fluorescence slowly drifts to a plateau. Control experiments in the absence of BTB show that the light-scattering changes follow a similar course, and furthermore, that only a small part of the response of the BTB trace is due to light-scattering. Thus, these records indicate a progressive alteration of the state of bromthymol blue, pyridine nucleotide, and light-scattering during the sonication process.

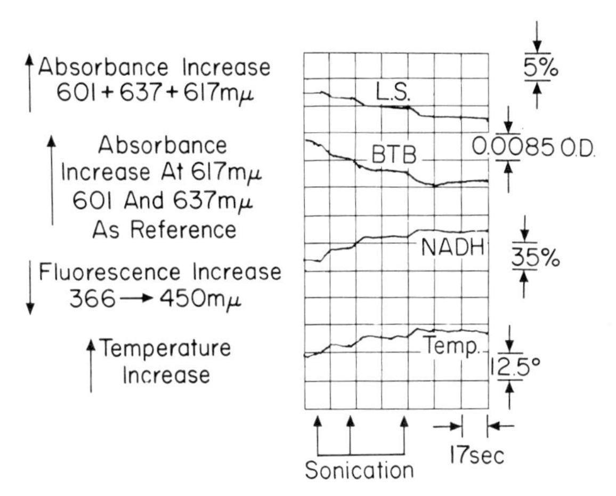

Figure 22. Opening of membranes of intact mitochondria in State 4; endogenous calcium only. Other experimental conditions as in Fig. 21.

In summary, these results show that the bromthymol blue absorbancy and the pyridine nucleotide fluorescence are most responsive to sonication; the gross morphological changes indicated by the light-scattering trace occur when the calcium-induced BTB gradient has been discharged, and when a considerable portion of the pyridine nucleotide fluorescence has disappeared. The further changes of bromthymol blue during the irreversible swelling of the mitochondria suggest the equilibration of the entire internal contents of the mitochondria with the external phase; presumably, calcium already present in the mitochondria prior to the addition of external calcium was responsible for the initial alkalinization of the BTB space which is reversed in the final stages of disruption of the mitochondria.

This technique has also been used to verify the location of BTB inside the membranes of rat liver mitochondria in State 4, again supplemented with butacaine but without added calcium. In Fig. 22, three intervals of sonication raise the temperature by 12°C at a rate of 0·6°C per second. During the intervals of sonication, three distinctive "opening" phenomena are observed: on the top trace, a 5% decrease in light-scattering or absorbancy; on the second trace, a 3% loss of BTB absorbancy; and on the third trace, a total decrease of 40% of the pyridine nucleotide fluorescence. From these traces, we conclude that these sonic power levels immediately open up the spaces within the mitochondrial membranes where BTB and NADH are bound and admit external hydrogen ions to acidify the BTB, which had been maintained at an alkaline pH due to the pH gradient associated with the accumulation of endogenous calcium in State 4.

The effect of intense sonication upon the BTB response of submitochondrial particles

It is of particular interest to learn whether the sonication intensities which cause rapid equilibration of bromthymol blue with the external medium in the case of rat liver mitochondria cause similar effects in submitochondrial particles. Thus, the experiment of Fig. 21 has been repeated using submitochondrial particles [E-SMP, prepared according to the method of Lee and Ernster [16] and supplemented with oligomycin]; the experimental results are shown in Fig. 23. The top trace indicates generalized light-scattering responses, plus, as before, some small contribution due to the dye itself since the reference wavelengths, 617-646 mμ, were somewhat affected by the BTB as well as by non-specific changes. In this figure, the second trace is the differential recording of BTB absorbancy changes at 626 mμ with reference to the sum of the absorbancy changes at 617 and 646 mμ. These measuring and reference wavelengths were chosen specifically to minimize responses to cytochrome *a*, even though this entailed some sacrifice in the sensitivity of the BTB response. The third trace is a fluorometric recording of NADH oxidation, and the bottom trace records the temperature, as measured by the thermocouple located directly in the cuvette.

The addition of 400 μM NADH activates pH changes in the submitochondrial particles, causing an abrupt increase of fluorescence and a decrease of BTB absorbancy, which reaches a plateau in 10 s. Some of the BTB change is reflected in the light-scattering trace. As a steady state of NADH oxidation and BTB change is reached, the sonicator is turned on with sufficient power (1·7 A) that the temperature in the 1·7 ml cuvette rises at the rate of 0·5°C per second for 10 s. The effect of the temperature rise is clearly reflected in the changing slope of the NADH trace. The light-scattering trace indicates a disturbance in absorbancy due to the formation of the cavitation vacuoles; and after the cavitation has stopped, a net decrease of absorbancy, presumably due to some further disruption of the submitochondrial particles.

The differential trace is the most interesting of the four. First, the non-specific absorbancy changes are largely cancelled out, and except for a small disturbance at the moment of starting sonication (the contents "slosh" when the

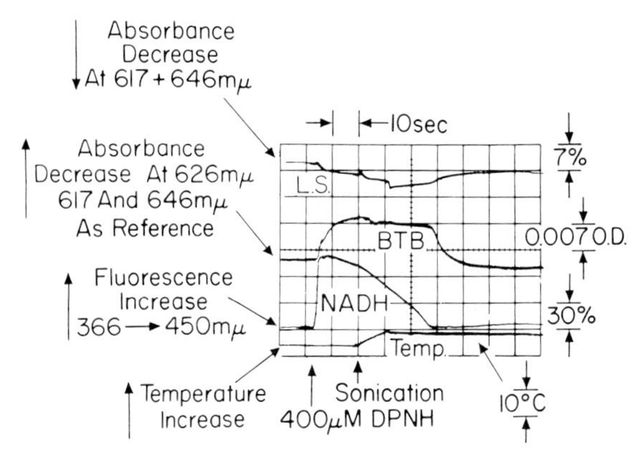

Figure 23. Response of submitochondrial particles to sonication. E-SMP, 1·7 mg protein per ml, in 0·3M mannitol-sucrose, 20 mM Tris-Cl, pH 7·4, in the presence of 3·3 μg oligomycin and 1·3 μM BTB.

sonifier is started), there is very little change of the BTB gradient—less than 5%. It is also noted that there is no observable disturbance whatsoever when sonication ceases, and, in contradistinction to the data on intact mitochondria, there is no time variation in the BTB gradient during sonication. In short, the experimental result indicates that the BTB gradient in these submitochondrial particles is stable against sonication from start to finish. Following the cessation of NADH oxidation, the traces return to their original levels.

Further experiments indicate that a second addition of NADH re-establishes the BTB gradient. Controls show that negligible absorbancy changes are seen in the presence of FCCP.

The facts of these experiments are that the BTB gradient did not appear to be altered, either during or immediately after sonication in the submitochondrial particles, a result quite inconsistent with the chemi-osmotic hypothesis, in which rupture of the vesicles would rapidly equilibrate the inside and outside hydrogen ion concentrations, causing the BTB absorbancy to drop immediately to the baseline. Following cessation of cavitation, resealing of the vesicles would be expected, with a consequent re-establishment of the proton gradient as electron transport is reactivated (as in the first portion of the curve following NADH addition in Fig. 23). Since neither a drop nor a recovery of the BTB trace is observed during the sonication, it seems unlikely that the BTB absorbancy change in intact submitochondrial vesicles, and the attendant observation of hydrogen ions external to the vesicles, actually corresponds to a gross separation of hydrogen and hydroxyl ions across a continuous lipid dielectric. In short, these experiments suggest a vesicle with a closed limiting membrane across which macroscopic hydrogen ion gradients are maintained is not required, but instead, that both the hydrogen ion binding and the BTB absorbancy changes are characteristic of the microscopic membrane structure, which may be sheared off to make smaller segments by sonication, but which nevertheless maintains these energy-dependent responses within the membrane structure itself. The actual structure involves the proteins and lipids of the membrane, but, according to these experiments, not the continuous lipid dielectric required by the chemi-osmotic hypothesis for the submitochondrial particles, as contrasted with the intact mitochondria, where bulk Ca^{2+} transport is readily observed (ref 29, Fig. 5).

SUMMARY

The following conclusions may be drawn from the experiments reported here:

1. Energy coupling Site I is located in the span of the flavoprotein chain between the low and high potential flavoproteins. The roles of non-heme iron and the mersalyl site are discussed; however, they cannot be definitely identified as essential components of electron transport or energy coupling.
2. Direct observation of ubiquinone kinetics eliminates this component as a potential hydrogen carrier functional in electron transport or energy coupling in the region of the respiratory chain between flavoprotein and oxygen. At present, there is no hydrogen carrier identified with energy coupling at Sites II and III.
3. Direct observations using aequorin to indicate the lability of the bound calcium of the mitochondrial membrane suggest that any process which interrupts electron transport will release the endogenous calcium of mitochondria in a matter of seconds. Thus, subsequent activation of forward or reversed electron transport will necessarily cause movement of these, and possibly other, cations. These observations render dubious the available values for proton/electron ratios in oxygen pulse experiments.
4. Estimations of the energy storage in the mitochondrial membrane, using calcium release as a criterion, indicate that the value is non-zero, and is approximately 0·3 nmoles of ATP-equivalents per mg protein.
5. Observations of the failure to equilibrate ionic gradients in submitochondrial particles under intense sonic vibration suggest that the ion gradients observed in these particles are not a property of their "closedness", in the chemi-osmotic sense. Controls with intact mitochondria indicate the rapidity of ionic equilibrations with these closed structures.
6. A fluorescence decrease of the fluorochrome, DNS, indicates an opening of the electron transport structure in the transition from the reduced to the oxidized state, with a half-time of about 2 ms. This suggests that the membrane may shift its conformation as rapidly as cytochrome *a* may be oxidized. The re-establishment of the conformation which confers upon DNS a high fluorescence is slower than the reduction of cytochrome *a*. This is the first direct

observation of a conformation change which would at least participate in control of electron transfer in the mitochondrial membranes. The fluorescence of aurovertin, bound to the F_1 site in the membrane, is not sensitive to the oxidation-reduction state of the membrane; presumably, the change reported by DNS is highly localized.

7. The relationship of these small-scale and rapid changes to the larger and slower changes observed electron micrographically is discussed. It is unlikely that the large-scale changes are functional in the same sense as the smaller changes, but may represent the establishment of minimum entropy configurations of the mitochondrial membrane, by analogy with the quaternary structure of polymeric enzymes.

REFERENCES

1. Chance, B., *in* "Biochemistry of Mitochondria" (edited by E. C. Slater, Z. Kaniuga and L. Wojtczak), Academic Press, London, 1967, p. 93.
2. Klingenberg, M. and Kröger, A., *in* "Biochemistry of Mitochondria" (edited by E. C. Slater, Z. Kaniuga and L. Wojtczak), Academic Press, London, 1967, p. 11.
3. Schachinger, L., Eisenhardt, R. H. and Chance, B., *Biochem. Z.* **333** (1960) 182.
4. Hackenbrock, C. R., *J. Cell Biol.* **30** (1966) 269.
5. Watson, H. C. and Chance, B., *in* "Hemes and Hemoproteins" (edited by B. Chance, R. W. Estabrook and T. Yonetani), Academic Press, New York, 1966, p. 149.
6. Dickerson, R. E., Kopka M. L., Weinzierl, J., Varner, J., Eisenberg, D. and Margoliash, E., *J. biol. Chem.* **242** (1967) 3015.
7. Dickerson, R. E., Kopka, M. L., Weinzierl, J., Varner, J., Eisenberg, D. and Margoliash, E., *in* "Structural and Chemical Aspects of Cytochromes" (edited by K. Okunuki and M. Kamen), University of Tokyo Press, Tokyo, 1969, in press.
8. Chance, B., *Biochem. J.* **103** (1967) 1.
9. Chance, B., *Discuss. Faraday Soc.* **20** (1955) 205.
10. Chance, B., DeVault, D., Legallais, V., Mela, L. and Yonetani, T., *in* "Fast Reactions and Primary Processes in Chemical Kinetics", Nobel Symposium 5 (edited by S. Claesson), Almqvist and Wiksell, Stockholm, 1967, p. 437.
11. Wong, D., *Fedn Proc. Fedn Am. Socs exp. Biol.* **27** (1968) 527.
12. Chance, B., Ernster, L., Garland, P. B., Lee, C. P., Light, P. A., Ohnishi, To., Ragan, C. I. and Wong, D., *Proc. natn. Acad. Sci. U.S.A.* **57** (1967) 1498.
13. Garland, P. B., Chance, B., Ernster, L., Lee, C. P. and Wong, D., *Proc. natn. Acad. Sci. U.S.A.* **58** (1967) 1696.
14. Chance, B., *Fedn Proc. Fedn Am. Socs exp. Biol.* **27** (1968) 298.
15. Chance, B. and Pring, M., *in* "Biochemistry of Oxygen", 19 Mosbacher Colloquium (edited by B. Hess), Springer-Verlag, Heidelberg, 1969, in press.
16. Lee, C. P. and Ernster, L., *J. Eur. Biochem.* **3** (1968) 391.

17. Chance, B., Bonner, Jr., W. D. and Storey, B. T., *A. Rev. Pl. Physiol.* **19** (1968) 295.
18. Lee, C. P., Ernster, L. and Chance, B., *J. Eur. Biochem.* **8** (1969) 153.
19. Kröger, A., *in* "Biochemistry of Oxygen", 19 Mosbacher Colloquium (edited by B. Hess), Springer-Verlag, Heidelberg, 1969, in press.
20. Chance, B., Williams, G. R., Holmes, W. F. and Higgins, J., *J. biol. Chem.* **217** (1955) 439.
21. Chance B. and Mela, L., *J. biol. Chem.* **242** (1967) 830.
22. Lee, I. Y., *Fedn Proc. Fedn Am. Socs exp. Biol.* **27** (1968) 297.
23. Lee, I. Y. and Chance, B., *Biochem. biophys. Res. Commun.* **32** (1968) 547.
24. Hassinen, I. and Chance, B., *Biochem. biophys. Res. Commun.* **31** (1968) 895.
25. Ohnishi, T., Racker, E., Schleyer, H. and Chance, B., *in* "Flavins and Flavin Enzymes" (edited by K. Yagi), University of Tokyo Press, Tokyo, 1969, in press.
26. Sharp, C. W., Mackler, B., Douglas, H. C., Palmer, G. and Felton, S. P., *Archs Biochem. Biophys.* **122** (1967) 810.
27. Light, P. A., Ragan, C. I., Clegg, R. A. and Garland, P. B., *FEBS Lett.* **1** (1968) 4.
28. Imai, K., Asano, A. and Sato, R., *J. Biochem., Tokyo* **63** (1968) 219.
29. Chance, B. and Yoshioka, T. I., *Biochemistry* **5** (1966) 3224.
30. Ohnishi, T. and Ebashi, S., *J. Biochem., Tokyo* **54** (1963) 506.
31. Jöbsis, F. F. and O'Connor, M., *Biochem. biophys. Res. Commun.* **25** (1966) 246.
32. Geier, G., *Helv. Chim. Acta* **51** (1968) 94.
33. Mela, L. and Chance, B., *Biochemistry* **7** (1968) 11.
34. Chance, B. and Mela, L., *J. biol. Chem.* **241** (1966) 4588.
35. Chance, B. *in* "Abstracts of the Third FEBS Meeting", Polish Scientific Publishers, Warsaw, 1966, p. 109.
36. Chance, B. and Azzi, A., *Ann. N.Y. Acad. Sci.* **147** (1969) 753-856.
37. Azzi, A. and Chance, B., *in* "Abstracts, Fifth FEBS Meeting", The Czechoslovak Biochemical Society, Prague, 1968, p. 165.
38. Mela, L., *Archs Biochem. Biophys.* **123** (1968) 286.
39. Chance, B., Mela, L. and Harris, E. J., *Fedn Proc. Fedn Am. Socs exp. Biol.* **27** (1968) 902
40. Scarpa, A. and Azzi, A., *Biochim. biophys. Acta* **150** (1968) 473.
41. Mitchell, P., this volume, p. 219.
42. Mela, L., *Ann. N. Y. Acad. Sci.* in press.
43. Mela, L., *Fedn Proc. Fedn Am. Socs exp. Biol.* **27** (1968) 828.
44. Chance, B., *J. biol. Chem.* **240** (1965) 2729.
45. van Dam, K., *Biochim. biophys. Acta* **92** (1964) 181.
46. Shimomura, O., Johnson, F. H. and Saiga, Y., *J. cell. comp. Physiol.* **59** (1962) 223.
47. Chance, B. and Mela, L., *Nature, Lond.* **212** (1966) 369.
48. Crofts, A. R. and Chappell, J. B., *Biochem. J.* **95** (1965) 387.
49. Eisenhardt, R. H. and Rosenthal, O., *Biochemistry* **7** (1968) 1327.
50. Chance, B. and Mela, L., *Nature, Lond.* **212** (1966) 372.
51. Chance, B. and Packer, L., *Biochem. J.* **68** (1958) 283.
52. Packer, L., *J. Cell. Biol.* **18** (1963) 487.

53. Chance, B., Lee, C. P. and Mela, L., *Fedn Proc. Fedn Am. Socs exp. Biol.* **26** (1967) 902.
54. Azzone, G. F. and Azzi, A., *in* "Regulation of Metabolic Processes in Mitochondria", BBA library, Vol. 7 (edited by J. M. Tager, S. Papa, E. Quagliariello and E. C. Slater), Elsevier, Amsterdam, 1966, p. 332.
55. Hackenbrock, C. R., *J. Cell Biol.* 37 (1968) 345.
56. Green, D. E., *in* "Mitochondrial Structure and Compartmentation" (edited by E. Quagliariello, S. Papa, E. C. Slater and J. M. Tager), Adriatica Editrice, Bari, 1967, p. 126.
57. Green, D. E., Asai, J., Harris, R. A. and Penniston, J. T., *Archs Biochem. Biophys.* **125** (1968) 684.
58. Weber, G., *Adv. Enzymol.* **8** (1953) 415.
59. Slater, E. C., this volume, p. 205.
60. Lardy, H. A., Connelly, J. L. and Johnson, D., *Biochemistry* **3** (1964) 1961.
61. Racker, E., Tyler, D. D., Estabrook, R. W., Conover, T. E., Parsons, D. F. and Chance, B., *in* "Oxidases and Related Redox Systems" (edited by T. E. King, H. S. Mason and M. Morrison), John Wiley, New York, 1965, p. 1077.
62. Chance, B., *Fedn Proc. Fedn Am. Socs exp. Biol.* **23** (1964) 265.
63. Pring, M., 5th Meeting Federation European Biochemical Societies, Prague, 1968.
64. Watson, H. C. and Chance, B., *in* "Hemes and Hemoproteins" (edited by B. Chance, R. W. Estabrook and T. Yonetani), Academic Press, New York, 1966, p. 149.
65. Schoenborn, B. P., "Division of Biological Chemistry", American Chemical Society (1968) Abstr. 30.
66. Schoenborn, B. P., *J. molec. Biol.* in press.
67. Mildvan, A. S., Rumen N. and Chance, B., "Division of Biological Chemistry", American Chemical Society (1968) Abstr. 32.
68. Chance, B. and Rumen, N., *VIIth Int. Congr. Biochem. Tokyo,* 1968, Abstr. A-85.
69. Perutz, M. F., Plenary Lecture, 5th Meeting Federation European Biochemical Societies, Prague, 1968.
70. Perutz, M. F., Muirhead, H., Cox, J. M. and Goaman, L. C. G., *Nature, Lond.* **219** (1968) 131.
71. Kirschner, K., Eigen, M., Bittman, R. and Voigt, B., *Proc. natn. Acad. Sci. U.S.A.* **56** (1966) 1661.
72. Chance, B. and Park, J. H., *J. biol. Chem.* **242** (1967) 5093.

FEBS Symposium, Volume 17, 1969, pp. 275-284

Five Types of Uncouplers for Oxidative Phosphorylation

V. P. SKULACHEV, A. A. JASAITIS, V. V. NAVICKAITE and L. S. YAGUZHINSKY

Department of Bioenergetics, Laboratory of Bioorganic Chemistry, Moscow State University, Moscow, U.S.S.R., and

E. A. LIBERMAN, V. P. TOPALI and L. M. ZOFINA

Institute of Problems of Information Transfer, U.S.S.R. Academy of Sciences, Moscow, U.S.S.R.

At the present time several hundreds of agents are known to possess the ability to uncouple processes of oxidation and phosphorylation.

Such a variety of uncoupling agents calls for a classification of this group of chemical compounds according to their properties and to the mechanism of their action. This paper will deal with uncouplers which act closer to the respiratory chain than the oligomycin-sensitive point.

We studied the action of uncoupling agents using experimental systems of different degrees of complexity, from mitochondria to artificial phospholipid membranes (for materials and methods see [1, 2]). It was established that uncouplers belonging to the class of weak acids and bases are capable of inducing proton conductivity in artificial membranes. The effectiveness of uncouplers in mitochondria correlates with that in membranes [1, 3-6]. The study of such a simple model as artificial membrane leads to a better understanding of the uncoupling phenomenon. We revealed the uncoupling ability for two new classes of chemical compounds, i.e. lipid-soluble cations and anions, and differentiated between their action and that of previously known types of uncouplers such as lipid-soluble acids and bases transporting protons.

At the same time, development of the chemical scheme of energy-coupling, postulating the existence of a nucleophilic centre in the mechanism of coupling [7], allowed us to disclose one more type of uncoupler, i.e. electrophilic alkylating agents. This paper will present the results of a study of the five above-mentioned classes of uncouplers. The following substances were used: the weak acid tetrachlorotrifluoromethylbenzimidazole (TTFB) and its inactive derivative, N-methyl TTFB; the weak base tributylamine (TBA); the lipid-soluble cation dimethyldibenzylammonia (DDA^+); the lipid-soluble anion tetraphenyl-

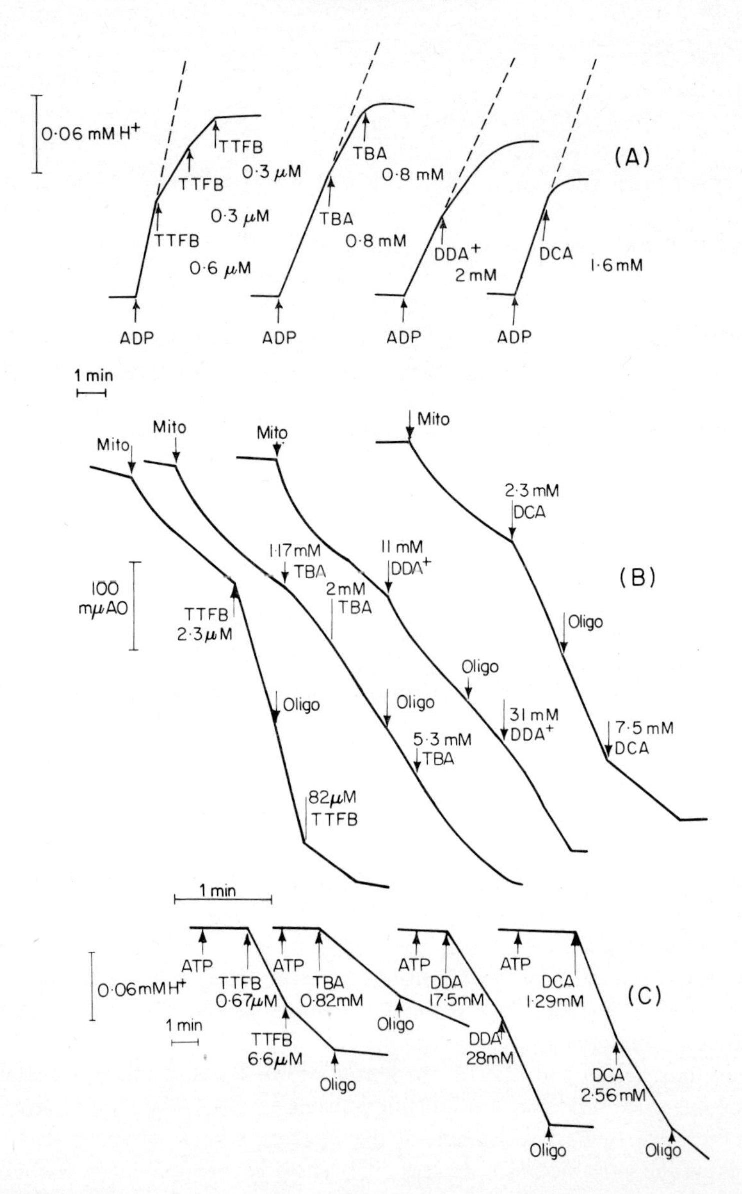

Figure 1. Effects of TTFB, TBA, DDA^{+} and DCA on mitochondrial functions. (A) Oxidative phosphorylation. Incubation mixture: 0·26M sucrose, 6 mM sodium phosphate, 16 mM sodium succinate, 0·5 mM ADP, 10^{-7}M rotenone, rat liver mitochondria (0·4 mg of protein per ml), pH 7·5, room temperature. Phosphorylation was measured using pH-meter techniques. (B) Respiration. Conditions as in (A). Concentration of mitochondrial protein 1·4 mg/ml. Polarographic measurement. (C) ATPase. Conditions as in (A), 0·6 mM ATP instead of ADP, succinate and rotenone.

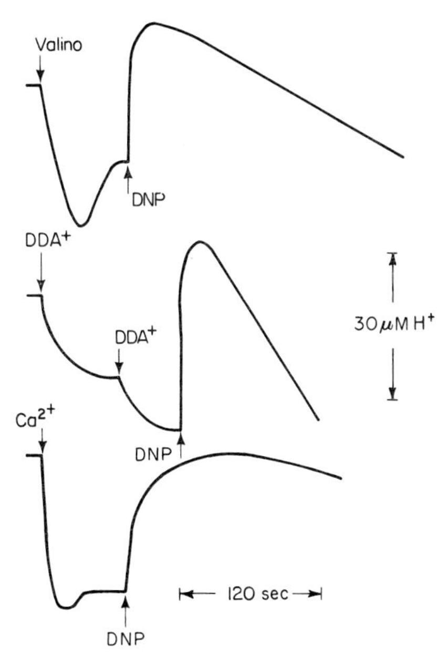

Figure 2. H^+-ejection induced by Ca^{2+}, K^+ + valinomycin, and DDA^+ in heavy beef heart mitochondria. Incubation mixture: 0·26M sucrose, 23 mM potassium succinate, 15 mM KCl, mitochondria 3 mg of protein per ml. Additions: 10^{-7}M valinomycin, 0·24 mM $CaCl_2$, 1 mM DDA^+, 0·1 mM DNP.

boron (TPB^-); alkylating agent N,N-dichloroethyl-*p*-aminophenylacetic acid (DCA) and its non-alkylating analogue, N-acetyl-N-chloroethyl-*p*-aminophenylacetic acid (ACA). Decylamine and picric acid (trinitrophenol, TNP) were also used in some experiments.

Figure 1 shows the effects of TTFB, TBA, DDA^+ and DCA on rat liver mitochondria. It can be seen that all the four compounds have a similar effect on mitochondrial functions. They inhibit phosphorylation, stimulate ATPase and respiration in State 4. Oligomycin does not inhibit uncoupler-stimulated respiration and decreases ATPase activation. An excess of TTFB, TBA and DCA inhibits both respiration and ATPase, whereas an excess of DDA^+ does not. The effect of DDA^+ is accompanied by an acidification of the medium in a fashion similar to that of K^+ + valinomycin or Ca^{2+} (Fig. 2).

Study of the same compounds in the experiments with artificial phospholipid membranes showed (Fig. 3) that the proton conductors TTFB and TBA sharply increase the electrical conductivity of these membranes. DDA^+ also increases electrical conductivity of the membranes, but proton conductivity remains constant. This effect is due to the transfer of DDA^+ cation across the membrane. A sharp increase in electrical conductivity was observed in samples with the TPB^-

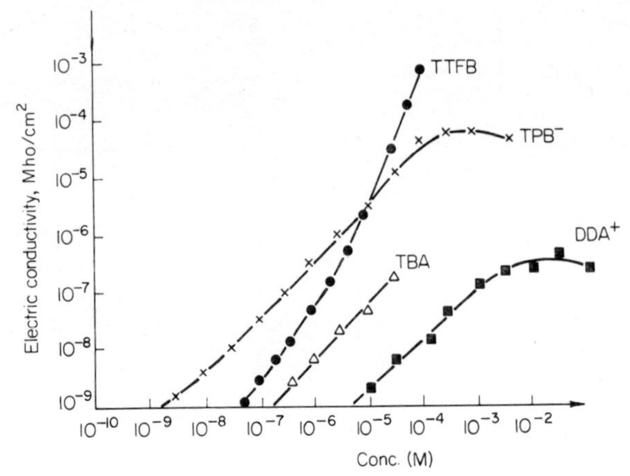

Figure 3. Effects of uncouplers on electrical conductivity of artificial phospholipid membranes.

anion. Small doses of this agent (less than 50μM) did not cause any appreciable increases in proton conductivity. Similar quantities of TPB^- did not influence the functions of mitochondria. High TPB^- doses induced proton conductivity in membranes and led to the appearance of uncoupling in mitochondria followed by strong inhibition of respiration (see also [8]). TNP displayed an effect similar to that of TPB^-. It is interesting that the action of TNP in both membranes and mitochondria was greatly potentiated by decylamine, an uncoupler of the base

Table 1. Effectiveness of different uncouplers in mitochondria and submitochondrial fragments

Uncoupler	Conc. causing complete inhibition of reversed electron transfer, M.	
	Mitochondria	"Sonic" fragments
TTFB	5×10^{-7}	5×10^{-7}
TBA	2×10^{-3}	10^{-4}
DDA^+	2×10^{-3}	4×10^{-2} is ineffective
TPB^-	5×10^{-4}	$1{\cdot}5 \times 10^{-5}$
DCA	$1{\cdot}5 \times 10^{-3}$	5×10^{-4}

Reaction mixture for mitochondria: 0·3M sucrose, 0·01M sodium phosphate, 0·012M sodium succinate, 1 mg protein per ml of rabbit heart mitochondria, pH 7·5.

Reaction mixture for fragments: 0·3M sucrose, 0·01M sodium phosphate, 0·012M sodium succinate, 0·005M NaCN, 10^{-3}M NAD^+, $1{\cdot}25 \times 10^{-3}$M ATP, 0·012M $MgCl_2$, 0·7 mg protein per ml of "sonic" beef heart mitochondria fragments, pH 7·5.

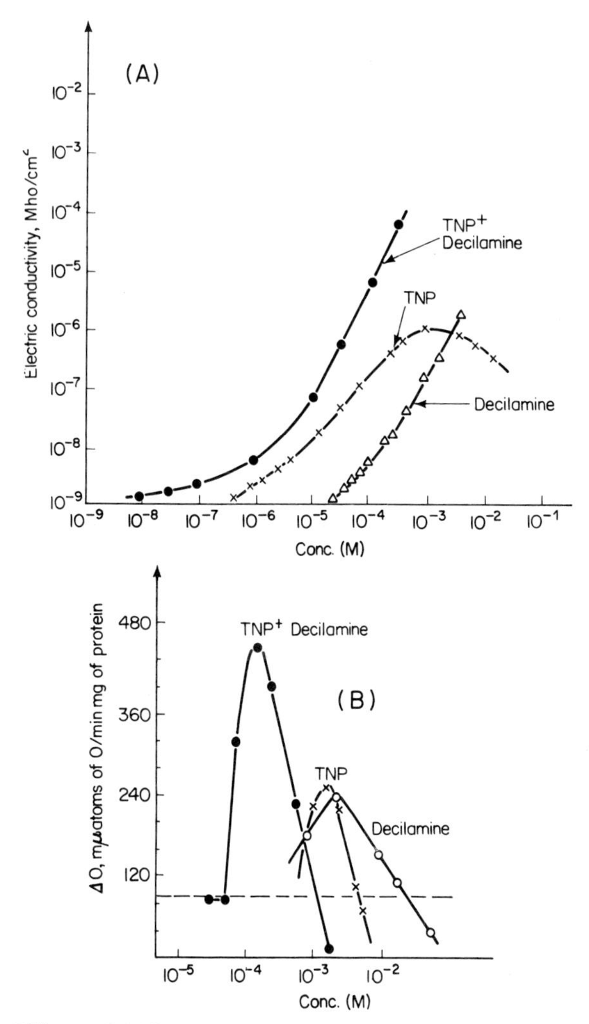

Figure 4. Effects of TNP and decylamine on electrical conductivity of phospholipid membranes (A), and mitochondrial respiration in State 4(B). Conditions as in Fig. 1(A).

type (Fig. 4). Alkylating agent slightly increased electrical conductance of membranes.

The effect of uncouplers mentioned above was then studied on "sonic" submitochondrial fragments. Electron microscopic data and results of pH measurements suggest that the membrane of these fragments, as compared to mitochondria, is orientated in the opposite direction towards the environment.

Table 2. Five types of uncouplers for oxidative phosphorylation

NN	Type of uncoupler	Example	Non-uncoupling analogue	The leading feature	The effect on the electrical conductivity of artificial membranes	Uncoupling activity	
						Mito-chondria	Frag-ments
I	Weak acid	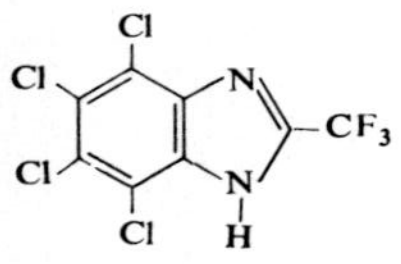 Cl, Cl, Cl, Cl, N, N–H, $-CF_3$	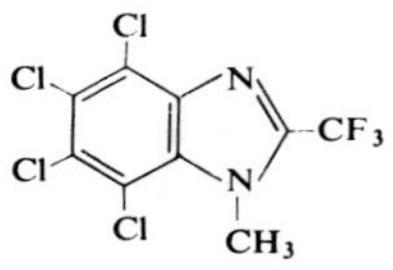 Cl, Cl, Cl, Cl, N, $N-CH_3$, $-CF_3$	Capability to donate H^+	Increases by the H^+-ion transfer	+	+
II	Weak base	$H-(CH_2)_4-N\langle^{(CH_2)_4-H}_{(CH_2)_4-H}$	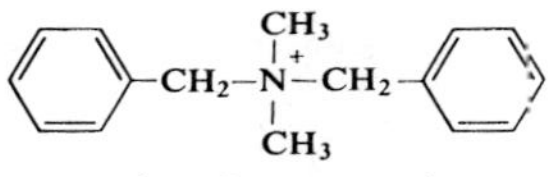 $-CH_2-\overset{+}{N}(CH_3)_2-CH_2-$ (on fragments)	Capability to accept H^+	Increases by the H^+-ion transfer	+	+

III	Cation	$C_6H_5-CH_2-N^+(CH_3)_2-CH_2-C_6H_5$; $(C_6H_5)_4B^-$ (on mitochondria)	Presence of (+)	Increases by the uncoupler cation transfer	+	–
IV	Anion	$(C_6H_5)_4B^-$; $C_6H_5-CH_2-N^+(CH_3)_2-CH_2-C_6H_5$ (on fragments)	Presence of (–)	Increases by the uncoupler anion transfer	In high conc. only	+
V	Alkylating agent	$(Cl-(CH_2)_2)_2N-C_6H_4-CH_2-COOH$; $Cl-(CH_2)_2(CH_3-CO)N-C_6H_4-CH_2-COOH$	Capability to electrophilic attack	Slightly increases	+	+

Reversed electron transfer from succinate to NAD^+ turned out to be a convenient system for comparison of the effects of various uncouplers in mitochondria and fragments (Table 1).

As is clear from Table 1, uncoupling effects of TTFB and DCA were similar in both mitochondria and fragments, TBA being more effective in fragments. The response to lipid-soluble ions was opposite in two experimental systems; DDA^+ causes uncoupling in mitochondria but not in fragments, while the uncoupling ability of TPB^- was found to be much higher in fragments than in mitochondria.

Analysis of all these data allows us to make a number of conclusions as to the mechanism of action of uncouplers belonging to different classes of chemical compounds.

The orientation of mitochondrial membrane is not critical for the uncoupling action of those compounds which induce proton conduction. Substitution of a mobile proton by an alkyl group in the molecule of such an uncoupler deprives it of its uncoupling ability. For example, N-methyl TTFB had no uncoupling activity in our experiments (see also [9]); the quarternary amine DDA^+ was ineffective in fragments where the tertiary amine TBA completely uncoupled oxidative phosphorylation.

Orientation of the mitochondrial membrane is essential for the uncoupling effect of lipid-soluble ions which do not transfer protons across the membranes. The sign of the charge of the ionized centre is evidently decisive for the above phenomenon; DDA^+ cation was effective in mitochondria, TPB^- anion in fragments.

The alkylating agent DCA was effective both in mitochondria and fragments. Activity of DCA was ensured by its ability for electrophilic substitution. Its non-alkylating analogue ACA did not possess the uncoupling ability (see also [10]). The above considerations are summarized in Table 2.

The mechanism of action of the first four types of uncouplers may be explained within the framework of the concept which states that uncoupling is the result of collapse of the membrane potential. Let us assume, together with Mitchell [11], that mitochondrial membrane has the difference of electric potential with the plus outside and the minus inside. In this case the lipid-soluble DDA^+ cation should move in the electric field inwards with respect to the mitochondria (from the plus towards the minus) and the lipid-soluble TPB^- anion outwards. The DDA^+ cation will be transferred from the outer space into the mitochondria and uncouple, whereas such "electrophoresis" will be impossible for TPB^- anion. With fragments the situation will be the opposite. The TPB^- anion added into the incubation medium will go inside, towards the plus, while the DDA^+ cation will not be taken in.

As for the uncoupler of the TTFB and TBA type, the electric field is no obstacle for them on their way into mitochondria because one of the two forms (protonated for TTFB and non-protonated for TBA) is electrically neutral. This is probably why proton carriers uncouple both in mitochondria and in fragments, while lipid-soluble ions do so only in one of these systems.

The membrane potential conception does not explain the action of the alkylating agents so well as in the case of the first four groups. Alkylating agent itself can hardly be current conductor.

Comparison of the chemical structure of the alkylating and non-alkylating agents shows that it is the alkylating ability that is responsible for the uncoupling effect. If alkylators also act by membrane potential collapse, this effect should be secondary (for example, increase in electrical conductivity of the membrane under alkylation of some components of the membrane).

The alternative explanation of the uncoupling mechanism may be that one of the intermediates of the energy transfer chain is attacked by the H^+ ion or an alkylator. In this case, the role of the proton carriers will be to carry H^+ ions to the sites of energy coupling localized in the hydrophobic phase of mitochondria. However, one cannot explain by this method the mechanism of uncoupling by penetrating ions, DDA^+ and TPB^-.

ACKNOWLEDGEMENT

The authors wish to thank Miss N. Asarenkova, Miss S. Smirnova and Mr. G. Pushkash for participation in some experiments, and Dr. T. Kheifez for translation of the article into English.

REFERENCES

1. Skulachev, V. P., Sharaf, A. A., Yaguzhinsky, L. S., Jasaitis, A. A., Liberman, E. A. and Topali, V. P., *Currents Modern Biol.* 2 (1968) 98.
2. Liberman, E. A., Mokhova, E. N., Skulachev, V. P. and Topali, V. P., *Biofizika* **13** (1968) 188.
3. Skulachev, V. P., Sharaf, A. A. and Liberman, E. A., *Nature, Lond.* **216** (1967) 718.
4. Skulachev, V. P., *VIIth Intn. Congr. Biochem,* Tokyo, 1967, Abstr. 2.
5. Skulachev, V. P., Yaguzhinsky, L. S., Jasaitis, A. A., Liberman, E. A., Topali, V. P. and Zofina, L. M., *in* "Energy Level and Metabolic Control in Mitochondria", Adriatica Editrice, Bari, 1969, p. 263.
6. Ratnikova, L. A., Sharaf, A. A., Skulachev, V. P., Yaguzhinsky, L. S., Liberman, E. A., Topali, V. P. and Zofina, L. M., Abstracts, 5th FEBS Meeting (1968).
7. Skulachev, V. P., "Energy Accumulation in the Cell", Nauka, Moscow (1969).
8. Utsumi, K. and Packer, L., *Archs Biochem. Biophys.* **122** (1967) 509.

9. Büchel, H., Corte, F. and Beechey, R. B., *Angew. Chem.* **17/18** (1965) 814.
10. Belousova, A. K., Romanova, I. N., Kusmina, S. V. and Sefirova, L. I., *Biokhimiya* **31** (1966) 13.
11. Mitchell, P., "Chemiosmotic Coupling in Oxidative and Photosynthetic Phosphorylation", Glynn Research Ltd., Bodmin, Cornwall (1966).

FEBS Symposium, Volume 17, 1969, pp. 285-292

Energy-dependent Functions in Cytochrome *c*-depleted Rat Liver Mitochondria

Y. AVI-DOR and S. STREICHMAN

Department of Chemistry, Technion–
Israel Institute of Technology,
Haifa, Israel

According to the chemical hypothesis of oxidative phosphorylation (cf. Ref. 1), a respiratory carrier can have a dual role; (a) in electron transfer, and (b) in energy transformation. To support this hypothesis, attempts were made to isolate respiratory carriers in a high-energy form or to identify the energy-rich form spectrophotometrically. Another approach in assessing the direct role of a respiratory carrier in energy transformation is to show that, under certain conditions, it affects differently the electron and energy transfer. Experiments in this line were carried out in the case of cytochrome *c* [2-4], since the latter can be removed and rebound [5] without excessive damage to the mitochondria. It was found that added cytochrome *c* increases the Ca/O ratio in cytochrome *c*-deficient beef heart particles [2] and the P/O ratio in deficient beef heart mitochondria [3]. Several energy-linked functions were stimulated by cytochrome *c* in deficient beef heart mitochondria [3] and in a sub-mitochondrial preparation [4] when supported by substrate. In the case of the latter functions, however, it was not established whether cytochrome *c* causes preferential increase in the rate of energy transfer.

An earlier study by the authors [6] dealt with the pattern of spontaneous swelling in cytochrome *c*-depleted rat liver mitochondria (CDM) prepared by diluting stock mitochondria suspended in 0·25M sucrose with four volumes of ice-cold water and transferring an aliquot of the osmotically "shocked" mitochondria into the reaction medium. Because of the electrolyte content of the latter, the "shocked" mitochondria immediately lost the major part of their cytochrome *c* (cf. Ref. 6). CDM did not swell spontaneously, but swelling could be induced either by cytochrome *c* or by ATP (Fig. 1). Cytochrome *c*-induced swelling, like spontaneous swelling, is supported by the endogenous respiration, and the rate of the latter is limited by the low concentration of the endogenous

Abbreviations. *In text*: EDTA, ethylenediaminetetracetate; TMPD, tetramethyl-*p*-phenylenediamine; CDM, cytochrome *c*-depleted rat liver mitochondria. *In figures*: Cyt. c, cytochrome *c*; M, intact rat liver mitochondria.

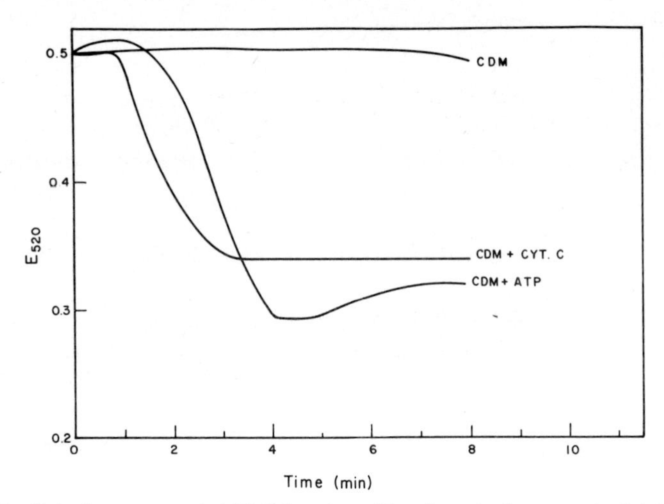

Figure 1. Cytochrome *c* and ATP-induced swelling in cytochrome *c*-depleted rat liver mitochondria (CDM). The assay medium contained 130 mM tris chloride (pH 7·4) and 6·6 μM cytochrome *c* or 80 μM ATP as indicated. CDM was prepared as described in the text and an aliquot equivalent to 0·51 mg protein was added to the assay medium at zero time. Total volume 3·0 ml, temp. 29°C. The time course of the turbidity changes was followed at 520 mμ or 540 mμ wavelength on a Beckman DB spectrophotometer supplemented with a strip-chart recorder.

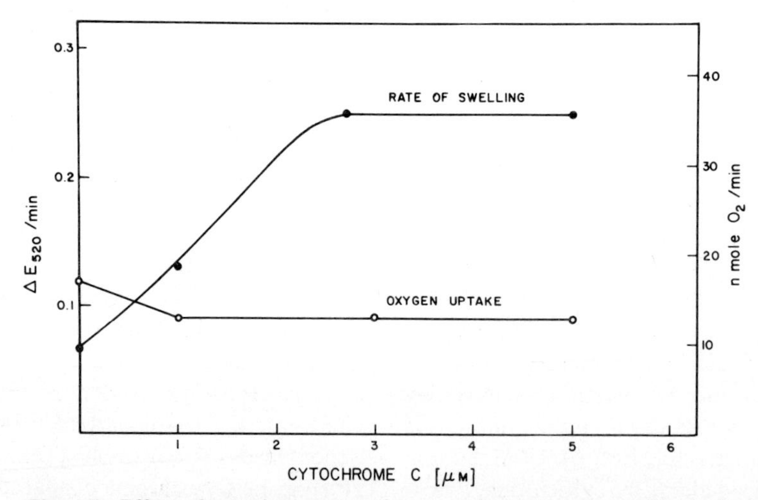

Figure 2. Effect of cytochrome *c* on the rates of spontaneous swelling and endogenous respiration in CDM. The assay medium for both swelling and respiration contained 130 mM tris chloride (pH 7·4) and cytochrome *c* as indicated. The mitochondria (CDM) concentration was equivalent to 0·24 mg protein/ml in the swelling assay and 1·56 mg protein/ml in the respiratory assay. Swelling was measured as in Fig. 1, oxygen uptake was determined polarographically in a Gilson Oxygraph. Total volume for swelling, 3 ml, and for respiration, 2 ml.

substrates. In order to learn more about the role played by cytochrome *c* in energy-dependent functions, its effect on spontaneous swelling was now studied in greater detail. In the experiment shown in Fig. 2, the effect of graded amounts of cytochrome *c* on spontaneous swelling on the one hand, and on the rate of endogenous respiration on the other, was compared. It can be seen that the rate of swelling was enhanced by increasing the cytochrome *c* concentration, while the rate of endogenous respiration remained practically unchanged. The maximum rate of swelling was reached when the cytochrome *c* concentration in the medium was approx. 3 μM. The important conclusion from this experiment is that under the above conditions cytochrome *c* was apparently able to exercise a selective effect on energy transfer, since its stimulatory effect on the energy-dependent function was not accompanied by simultaneous acceleration of electron transport.

In the experiment shown in Figs. 3a and b, an attempt was made to demonstrate a preferential effect of cytochrome *c* on energy transformation in the case of swelling linked to exogenous substrate. Since, according to MacLennan *et al.* [3, 4], only site III is affected by cytochrome *c*, ascorbate + TMPD was used as substrate. Antimycin A was added to the assay medium. It can be seen from the figure that the difference in rate of swelling between intact and depleted mitochondria exceeds that in rate of respiration. Even when respiration of the intact mitochondria was partially inhibited by KCN, so that the rate was lower than in CDM, they still swelled much more rapidly than their depleted counterparts. The responsibility of the higher cytochrome *c* content of the intact mitochondria, *per se*, for the above difference, was indicated by the results shown in Fig. 3b: namely, cytochrome *c* added to CDM accelerated swelling even when increase of the respiratory rate was suppressed by KCN.

Another energy-dependent reaction included in the present investigation was succinate-linked reduction of intramitochondrial pyridine nucleotides. As is seen from the results in Fig. 4, this reduction was suppressed in CDM and restored almost completely when cytochrome *c* was added to the medium. ATP was much less effective than cytochrome *c* in supporting the reversal of electron transport, either in CDM or (in the presence of KCN) in the intact mitochondria. It is noteworthy that ATP added in the absence of KCN induced a larger reduction of pyridine nucleotides in CDM than when added in the presence of KCN, unless cytochrome *c* was also added in the anaerobic experiment. This stronger effect of ATP in the absence of KCN is attributable to its capacity (in addition to being a source of energy) to improve the effectiveness of coupling by stabilizing the mitochondrial structure (cf. Ref. 7). (The EDTA effect observed is probably exclusively due to such stabilization.) As for the synergism between ATP and cytochrome *c* in the presence of KCN, it is probably due to the fact that under anaerobic conditions the added cytochrome *c* was reduced by succinate

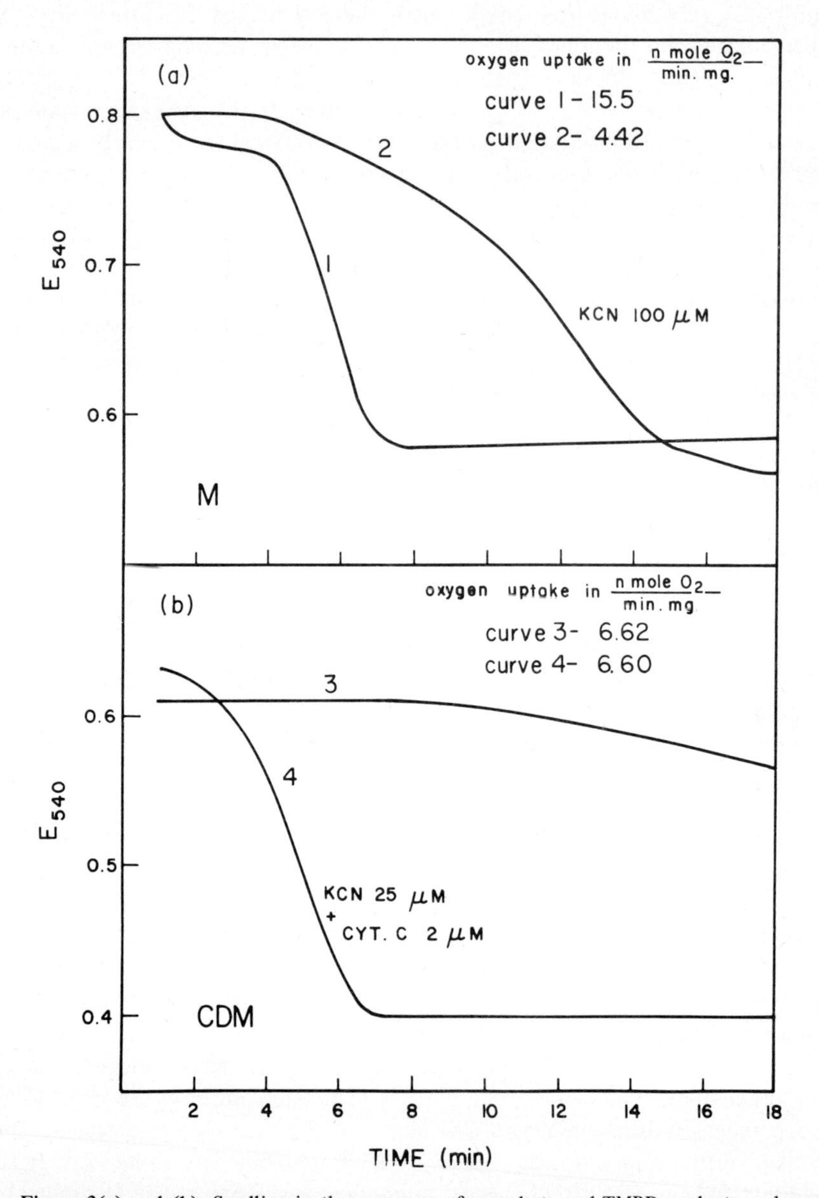

Figure 3(a) and (b). Swelling in the presence of ascorbate and TMPD as electron donor. The assay medium contained 130 mM tris chloride (pH 7·4) 0·016 μg/ml antimycin A and 3·3 mM ascorbate + 0·066 mM TMPD in both the swelling and respiratory assays. KCN or cytochrome *c* was added as indicated. Mitochondrial concentration was equivalent to 0·23 mg protein/ml in the swelling assay and to 1·29 mg protein/ml in the respiratory assay. Other conditions and the methods of assay as in Fig. 2.

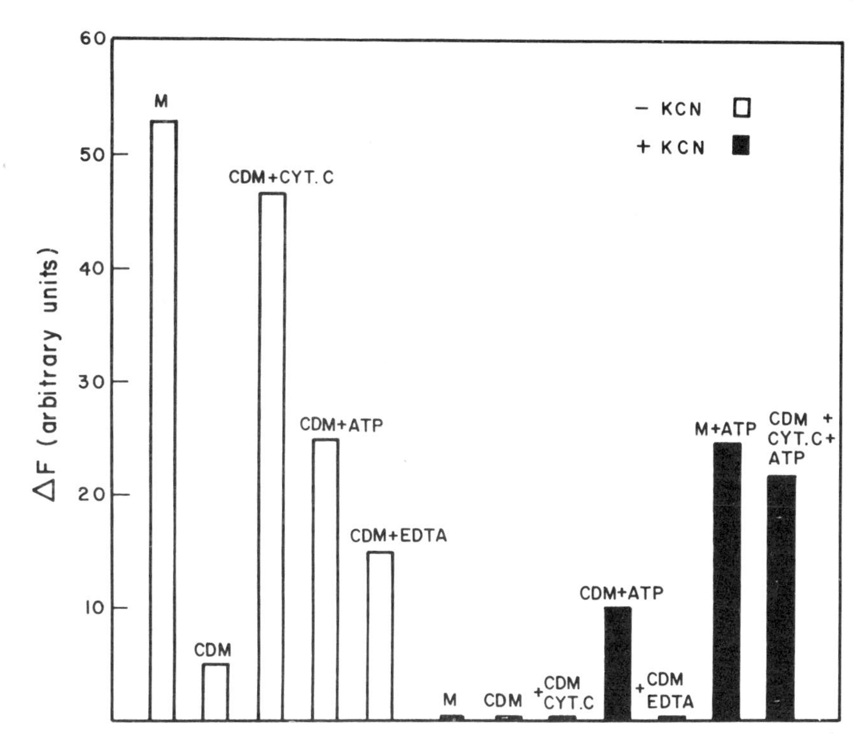

Figure 4. Succinate-linked reduction of intramitochondrial pyridine nucleotides. The reaction mixture contained 16 mM tris chloride (pH 7·4), 150 mM KCl and intact mitochondria or CDM equivalent to 1 mg protein/ml. 2 mM KCN was included in the medium when indicated. The reaction was started by injecting 6·6 mM succinate to the medium and when indicated also 10 μM cytochrome *c*; 26 μM ATP or 0·6 mM EDTA were injected into the medium immediately after the succinate. Total volume 3·0 ml, temp. 25°C. Change in fluorescence was measured according to Estabrook [8] in an Eppendorf spectrophotometer equipped with fluorimetric attachment and connected to a Varian Model G-14A-1 graphic recorder.

and served as an alternative electron donor for reduction of pyridine nucleotides.

The cytochrome *c*-dependent restoration of pyridine nucleotide reduction in CDM was further studied with the rate of electron transport limited by using low (0·66 mM) succinate concentration (Fig. 5). Under such conditions addition of cytochrome *c* at low concentrations stimulated the extent of reduction as well as respiration. Above a certain concentration of cytochrome *c*, however, when the substrate concentration became rate-limiting, only the extent of reduction continued to increase, whereas the oxygen uptake remained constant.

Further information on the relationship between cytochrome *c* concentration, rate of respiration and rate of generation of high-energy compounds

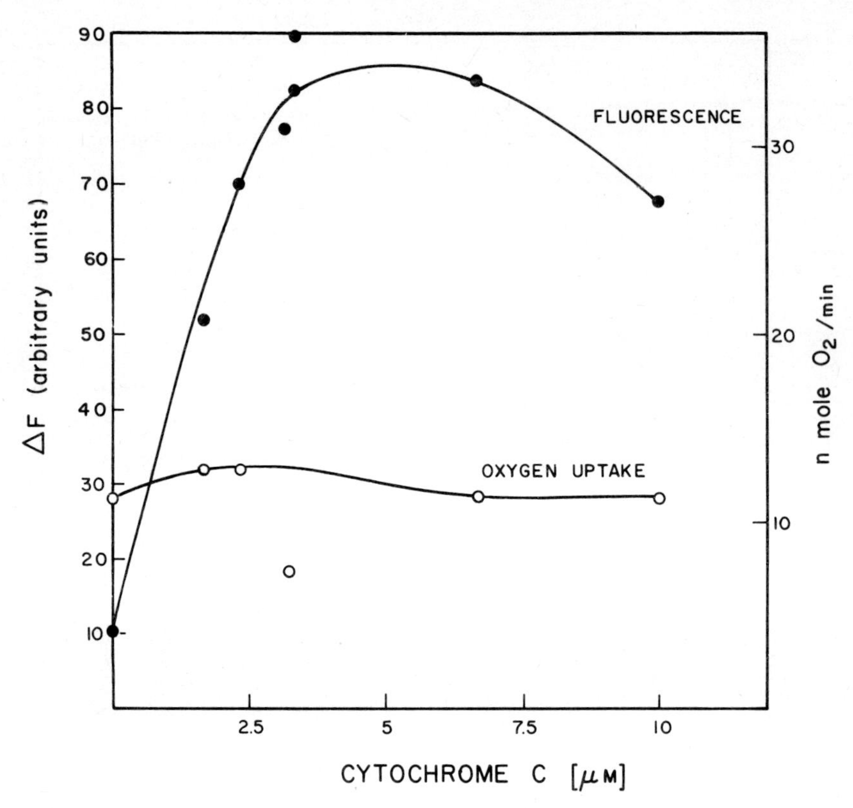

Figure 5. Effect of cytochrome *c* on succinate oxidation and energy-dependent reduction of intramitochondrial pyridine nucleotides in CDM. The assay medium contained 16 mM tris chloride (pH 7·4), 150 mM KCl, 0·66 mM succinate and mitochondria (CDM) in a concentration equivalent to 1·26 mg protein/ml. Temp. 18°C. Other conditions and methods of measurement as in Figs. 2 and 4.

was sought by measuring the P/O ratio in CDM at varying cytochrome *c* levels. The phosphorylating site studied was site III, with ascorbate + TMPD and antimycin A present in the reaction mixture. From the results shown in Fig. 6 it can be seen that added cytochrome *c* in the lower range of concentrations increased the respiratory rate more than that of phosphate esterification, whereas at higher concentrations its stimulating effect on phosphorylation was stronger. In particular, note the steep rise in the curve representing phosphate esterification in the intermediate range of cytochrome *c* concentration.

The results presented in this paper with regard to the effect of cytochrome *c* in CDM on swelling, reversed electron transport and phosphorylation are in accordance with the view that, under certain conditions, cytochrome *c* acts as a coupling factor. This coupling activity, in addition to its role in electron

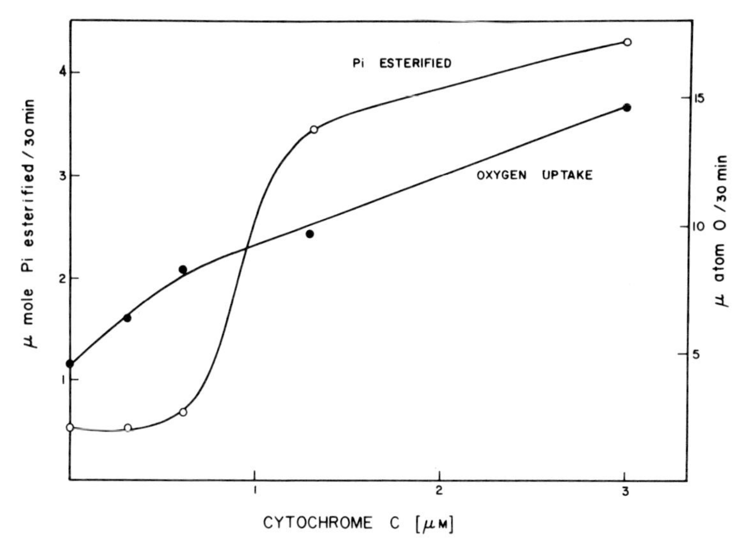

Figure 6. Effect of cytochrome *c* on respiration and phosphate esterification in CDM with ascorbate + TMPD as the substrate. The assay medium contained 24 mM tris chloride (pH 7·4), 48 mM KCl, 7 mM $MgCl_2$, 0·25 mM EDTA, 0·7 mM ATP, 10 mM inorganic phosphate (pH 7·4), 6·6 mM ascorbate, 0·13 mM TMPD, 0·016 μg/ml Antimycin A, 0·5 mg/ml hexokinase, Type III (Sigma Chemical Co.), 20 mM glucose, mitochondria (CDM) equivalent to 1·1 mg protein/ml and cytochrome *c* as indicated. Total volume 3 ml, temp. 29°C. Respiration and phosphorylation were measured as described by Slater [9], inorganic phosphate according to Fiske and SubbaRow [10].

transport, is attributable to several mechanisms. One could visualize that, when cytochrome *c* is released from the mitochondria, not only the respiratory activity but also the coupling mechanism is affected (probably at site III). At low external concentrations of cytochrome *c* a depleted chain binding would only participate in electron transport. Because of factors associated with the structure of the mitochondrial membrane, the conformational requirements for coupling at site III are only satisfied when, at higher cytochrome *c* concentrations, a larger number of chains rebind the respiratory pigment. In accordance with the assumption is the finding that at a fixed rate of electron flow, more energized intermediates are generated at high than at low cytochrome *c* concentrations, as manifested by the rate of swelling or by reversed electron transport. This hypothesis also accounts for the preferential effect of the higher concentrations on phosphorylation (cf. Fig. 6).

It is noteworthy that a similar pattern would result if different segments of the mitochondrial membrane (or different respiratory assemblies) were damaged to a different degree by the removal of cytochrome *c* and at the same time, for some reason, the damaged units retained their cytochrome *c* more firmly, and rebound it more easily, than the intact chains. Because of possible redistribution of the electron flow between "phosphorylating" and "non-phosphorylating" chains, the ~/O ratio may be improved in such a case by addition of external cytochrome *c*, even if the overall respiratory rate remains constant. It is, however, hard to see any reason for damaged binding sites having higher affinity for cytochrome *c* than intact ones; moreover, this latter hypothesis is hard to reconcile with the abrupt rise in the rate of phosphorylation observed in a titration experiment like the one in Fig. 6.

Heterogeneity of the mitochondrial population could be an additional possibility, but this was ruled out, in beef heart mitochondria [3], as a factor responsible for variation of the P/O ratio in view of the similar response by "light" and "heavy" mitochondrial fractions in this respect.

REFERENCES

1. Schatz, G., *Angew. Chem.* (International Edition) **6** (1967) 1035.
2. Penniston, J. T., Vande Zande, H. and Green, D. E., *Archs Biochem. Biophys.* **113** (1966) 512.
3. MacLennan, D. H., Lenaz, G. and Szarkowska, L., *J. biol. Chem.* **241** (1966) 5251.
4. Lenaz, G. and MacLennan, D. H., *J. biol. Chem.* **241** (1966) 5260.
5. Jacobs, E. E. and Sanadi, D. R., *J. biol. Chem.* **235** (1960) 531.
6. Streichman, S. and Avi-Dor, Y., *Biochem J.* **104** (1967) 71.
7. Avi-Dor, Y., Lamdin, E. and Kaplan, N. O., *J. biol. Chem.* **238** (1963) 2518.
8. Estabrook, R. W., *Analyt. Biochem.*, **4** (1962) 231.
9. Slater, E. C., *in* "Methods in Enzymology", Vol. 10 (edited by R. W. Estabrook and M. E. Pullman), Academic Press, New York (1967), p. 19.
10. Fiske, C. H. and SubbaRow, Y., *J. biol. Chem.* **66** (1925) 375.

FEBS Symposium, Volume 17, 1969, pp. 293-300

Piericidin A—a Tool for the Study of the Mechanism of Oxidative Phosphorylation

I. VALLIN AND H. LÖW

The Wenner-Gren Institute, University of Stockholm, Sweden

Piericidin A, originally intended to be used as an insecticide [1], has recently attracted considerable interest as a useful tool for the study of the oxidative phosphorylation mechanism [2-6]. The chemical structure of piericidin A is similar to that of coenzyme Q [1], the main difference being the ring structure, which is a pyridine ring in the insecticide instead of the benzene ring of the coenzyme. This similarity in structure initiated the study of piericidin action on mitochondrial respiratory activities performed by Crane and his group in Lafayette [7]. The piericidin A-inhibited succinic oxidase activity could be partially restored by the addition of coenzyme Q_2, in contrast to the NADH oxidase activity, which was sensitive to a concentration of piericidin about three orders of magnitude less.

The extreme potency of the inhibitor towards NADH oxidase, which was affected by concentrations lower than those reported for any compound so far, and the possibility of an interaction with the phosphorylation mechanism, made us investigate the effect of piericidin A on energy-linked reactions in submitochondrial systems.

We have demonstrated [8] that the energy-dependent NAD^+ reductions, supported by either ATP or high energy intermediates generated in the cytochrome oxidase region, are equally sensitive to the inhibitor and the effective concentrations are about one order of magnitude less than those affecting the NADH oxidase of the same particles (Fig. 1). The inhibitory pattern of the NAD^+ reduction systems follows a straight line when plotted against the logarithm of the piericidin concentration, whereas the inhibition of the NADH and succinic oxidases both proved to be biphasic (Fig. 2). The initial inhibitory phases of the oxidases are characterized by an unchanged P/O quotient, whereas the phosphorylative capacity is decreased more than the respiration during the second phases. A decrease in P/O ratio is also directly demonstrable when the rate of respiration in the NADH oxidase is limited by the amount of substrate added as in Table 1.

We regard these findings, together with the uncoupler-like stimulation by piericidin of the ATPase, dealt with later, as an indication that piericidin A inter-

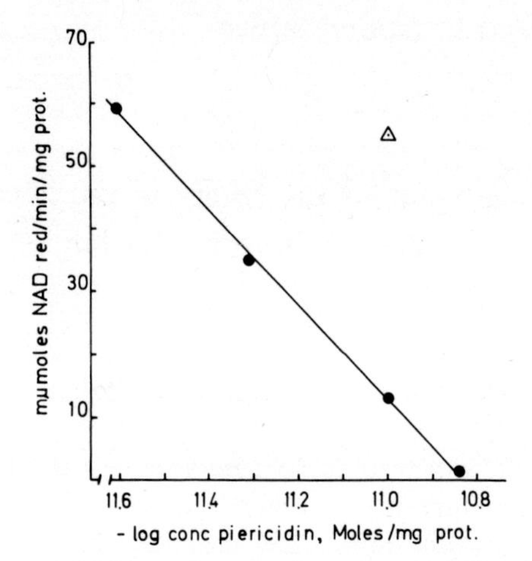

Figure 1. Piericidin A inhibition of energy-dependent NAD^+ reduction. Reduction of NAD^+ measured as an increase in fluorescence in an Eppendorf photometer at 30°C. Medium containing 50 mM Tris-HCl, pH = 8·0, 6 mM $MgCl_2$ and 0·25M sucrose. Particles corresponding to 0·15 mg protein per ml were added, followed by piericidin A as 3 μl of ethanolic solutions, 0·2 μg oligomycin and 0·67 μg antimycin A per mg protein, plus 0·3 mM TMPD, 5 mM ascorbate and the reaction was started by addition of NAD^+ in a final concentration of 1·5 mM.

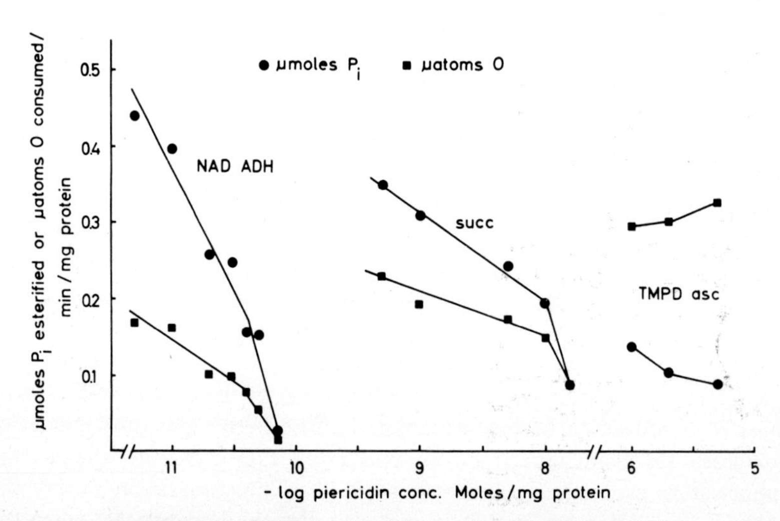

Figure 2. Piericidin A inhibition of oxidative phosphorylation. Medium containing 50 mM glycylglycine, pH = 7·5, 3 mM $MgCl_2$ and 0·25M sucrose at 30°C and particle concentration 0·33 mg of protein per ml. Substrates: 0·66 mM NAD^+ with a reducing ADH system; 3·3 mM succinate or 0·3 mM TMPD plus 5 mM ascorbate. Piericidin A added as 5 μl of ethanolic solutions except for the TMPD ascorbate system where 25 μl were added. ● denotes P_i (μmoles) and ■ 0 (μatoms).

feres with the generation of a primary high energy complex. The reversed electron transfer is directly dependent on the high energy state as evidenced by the need for an incubation of the particles in the presence of ATP prior to NAD^+ reduction [9]. In the NADH oxidase, however, the interference by piericidin seems not to be immediately rate limiting for the high energy complex

Table 1. Piericidin A inhibition of NADH oxidase

NAD^+ conc.	Piericidin A	Oxygen consumed	Phosphate esterified	P/O ratio
μM		μatoms/min/ mg prot.	μmoles/min/ mg prot.	
16·6	–	0·095	0·172	1·81
16·6	+	0·076	0·091	1·19
33·3	–	0·097	0·170	1·74
33·3	+	0·101	0·176	1·64
66·6	–	0·201	0·405	2·03
66·6	+	0·176	0·399	2·27

Respiratory rate limited by substrate concentration. Standard incubation at 30°C with particles corresponding to 0·33 mg of protein per ml except the amount of NAD^+ varied as indicated. Piericidin A added when indicated as 1 μl of a chloroform solution giving a final concentration of 6·6 pmoles/mg protein.

Table 2. Differential spectrum of reversed electron transfer

Additions	Flavoprotein 465-510	% Reduction		
		Cyt. *b* 564-575	Cyts. $c + c_1$ 550-541	Cyt. *a* 605-630
15 pmoles piericidin/mg prot.	–	–	–	–
1 mM ATP	10	–	–	–
10 mM succinate	40	–	–	–
1·5 mM NAD^+	67	–	–	–
1 mM ATP	–	3	–	–
10 mM succinate	60	84	26	12
1·5 mM NAD^+	100	81	38	2
15 pmoles piericidin/mg prot.	95	80	48	18

The numbers under each component refer to the wavelength pair used to measure the per cent of reduction as compared to a dithionite-treated sample. Each cuvette contained 50mM glycylglycine buffer pH 7·5, 0·25M sucrose, 1mM KCN and submitochondrial particles equal to 4 mg of particle protein. Additions to the sample cuvette as indicated in the table.

formation and the phosphorylative capacity is thus affected first when enough of the sensitive component is bound by the inhibitor. This would give rise to the two inhibitory phases found for the NADH oxidase and also explain the difference in sensitivity between the reversed electron transfer system and the oxidase.

The extreme sensitivity of the NAD^+ reduction systems to piericidin A can further be demonstrated by the differential spectrum obtained after addition of the inhibitor in a concentration which excludes the reversed electron transport but leaves the NADH oxidase unaffected (Table 2). The reduction of the cytochrome components after addition of succinate, as well as the oxidation-reduction change caused by the further addition of NAD^+, are completely abolished. A reduction of flavin components is the only effect obtained with piericidin A present.

The degree of piericidin A inhibition is also dependent on the oxidation-reduction state of the sensitive component. Addition of the inhibitor after anaerobic preincubation of the particles with ATP, or after succinate has been added to an aerobic system, decreases the extent of inhibition of the NAD^+ reductions. Under these conditions there is a change in the oxidation-reduction state of CoQ as measured by others [10, 11] and there may be a correlation so that the inhibition by piericidin A is less effective the more extensively the CoQ pool is reduced.

p–Chloromercuribenzoate has been demonstrated to react preferentially with dithiol groups and studies by Fluharty and Sanadi on rat liver mitochondria [12, 13] have revealed evidence for a dithiol site being located in the NADH oxidase flavin region between the oligomycin-sensitive site and the respiratory chain. Thus, it is of interest that a synergistic effect can be obtained when PCMB and piericidin A are added simultaneously to the NAD^+ reduction systems (Table 3).

It can further be demonstrated that 2, 3-dimercaptopropanol (BAL) releases the piericidin inhibition of the same system to a considerable degree, whereas glutathion and other –SH regenerating reagents are uneffective (Table 4). This constitutes strong evidence that proximal –SH groups are involved in the piericidin A-sensitive site.

By using homologs, Crane and collaborators [14] have recently studied how the inhibitory action is influenced by the chemical groups associated with the pyridine nucleus of the piericidin A molecule. They demonstrated that the inhibition of the NADH oxidase was highly dependent on both the hydroxyl group connected to the ring structure and on the length of the side chain. The inhibition of the succinic oxidase was, on the other hand, very little dependent on groups connected to the ring structure and even short fatty acid side chain homologs had some effect. These differences in strict specificity for the inhibitor in the two oxidases fit remarkably well with the reconstitution experiments recently published by Lenaz *et al.* [15]. Lyophilized beef heart mitochondria extracted with pentane lose both the NADH and the succinic oxidase activity.

Table 3. The synergistic effect of PCMB and piericidin A

PCMB concentration	Piericidin A concentration	Inhibition
μM	pmoles/mg prot.	%
—	3·6	32·2
1·0	—	22·5
1·0	3·6	71·0

NAD^+ reduction with high energy intermediates generated in the cytochrome oxidase region. Piericidin was added as 3 μl of an ethanolic solution.

The NADH oxidase activity can be restored by addition of CoQ_7–CoQ_{10}, CoQ_5 and CoQ_6 being half as effective. The succinic oxidase activity was, on the other hand, completely restored by CoQ homologs with two to ten isoprene units in the side chain. The authors conclude these findings as a support for the involvement of CoQ at two separate sites of electron transfer, one in the NADH oxidase and one in the succinic oxidase.

According to Horgan and Singer [2], the binding of C^{14}-labelled rotenone to submitochondrial particles is decreased in the presence of piericidin A or amytal, a finding favouring the concept of a common site of action of these inhibitors. In spite of this strong indication of a common binding site a few differences in action of the three inhibitors can be found. The piericidin action differs from that of rotenone by being more potent when tested on the NAD^+ reduction systems than when inhibiting the NADH oxidase. Rotenone is as potent in the two systems. Furthermore, piericidin A is more potent when the sensitive component is oxidized, whereas a reduced sensitive site favours the rotenone inhibition.

Piericidin A is also different from the two other inhibitors in the sense that it affects the phosphorylation sequence. In concentrations which leave the NADH oxidase unaffected, piericidin A initiates a release of P_i from ATP under

Table 4. Release of piericidin A inhibition by BAL

Piericidin A concentration	Rate of NAD^+ reduction – BAL	Rate of NAD^+ reduction + BAL
pmoles/mg prot.	μmoles/min/mg protein	μmoles/min/mg protein
—	0·173	0·198
11	0·096	0·127
22	0·054	0·081

NAD^+ reduction dependent on generated high energy intermediates. When indicated BAL was added in a final concentration of 8·3 μM.

Table 5. Effect of piericidin A on the P_i-release

Experiment	Additions to		Piericidin conc.	Rate of P_i—release	Rate of control
	Preincubation	Incubation			
			pmoles/mg protein	μmoles/min/mg protein	%
Expt. 1	—	ATP	—	0·358	100
	—	ATP, succinate, NAD^+	—	0·351	97·9
	ATP	—	—	0·225	62·8
	ATP	succinate, NAD^+	—	0·292	81·6
Expt. 2	ATP	—	—	0·150	100
	ATP, piericidin	—	445	0·140	93·5
	ATP	succinate, NAD^+	—	0·224	149·2
	ATP, piericidin	succinate, NAD^+	445	0·148	98·8
Expt. 3	ATP	—	—	0·127	100
	ATP	piericidin	4·45	0·171	134·5
	ATP	piericidin	44·5	0·201	157·7
	ATP	piericidin	445	0·212	166·9
	ATP	succinate, NAD^+	—	0·292	229·6
	ATP	piericidin, succinate, NAD^+	4·45	0·278	291·0
	ATP	as above	44·5	0·257	202·1
	ATP	as above	445	0·260	202·8

Piericidin A added as 1 μl of ethanolic solutions either to the preincubation medium (Expt. 2) or after a 5 min preincubation prior to succinate and NAD^+. Conditions equal to the ATP-dependent NAD^+ reduction with 1·5 ml as a final volume.

appropriate conditions (Table 5). The piericidin-stimulated release requires a preincubation procedure and the effect is thus directly comparable to that of DNP or Cl–CCP. In μM concentrations piericidin uncouples oxidative phosphorylation in the cytochrome oxidase region, but it has no effect on P_i–ATP exchange in contrast to the inhibition exerted by high concentrations of amytal.

The involvement of proximal –SH groups in the piericidin A-sensitive site and the structural similarity between piericidin and CoQ, together with the findings

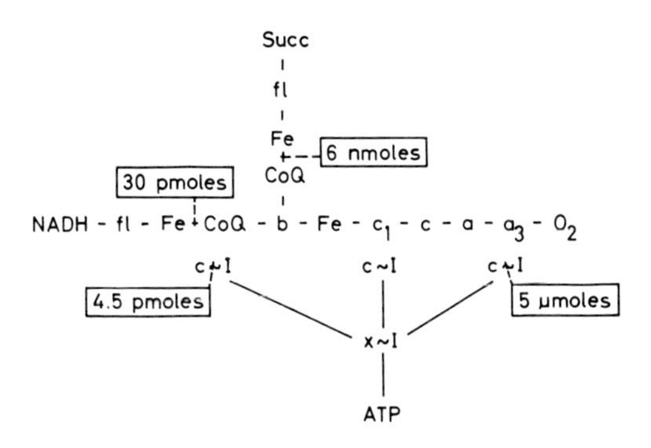

Figure 3. General scheme of piericidin A effects in energy-linked reactions.

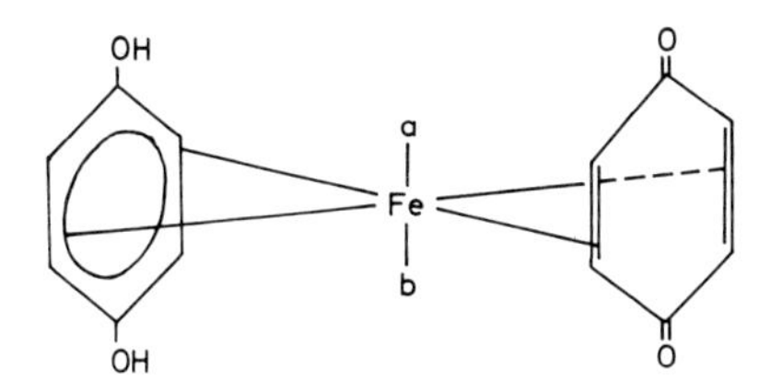

Figure 4. Planar chelate between non-heme iron and CoQ. According to Lenaz *et al.* [15].

presented above and summarized in Fig. 3, made us suggest a hypothesis for a possible mechanism of action of the agent. A planar chelate between CoQ and non-heme iron has earlier been suggested on organic chemical reasons by Folkers and Moore [16] and this hypothesis is further elaborated in the recent publication where the respiratory activity of pentane-extracted mitochondria was restored by CoQ [15]. We imagine the non-heme moiety to be structurally oriented in such a chelate (Fig. 4) by –S linkages in a way similar to the situation in the ferredoxin complex. A similar type of complex may also be possible between a flavin isoalloxazine molecule, non-heme iron and CoQ. An interference with the –S bond-fixed non-heme iron, where piericidin A substitutes one or two of the CoQ molecules, would cause the inhibition. The structural rigidity of the non-heme iron–CoQ complex might be a prerequisite for the formation of primary high energy bonds.

In recent EPR experiments by Palmer *et al.* [4], the action of piericidin A or rotenone on submitochondrial particles was investigated. After addition of NADH there was in both cases a decrease of 25-35 per cent of the g = 1·94 signal originating from the non-heme iron component in the NADH dehydrogenase. The signal related to the iron moiety in the cytochrome $b - c_1$ region was unchanged, whereas that emanating from the succinic dehydrogenase component was abolished. The authors regard the decrease in the NADH dehydrogenase iron signal as questionable. This may imply that this component can still be reduced with piericidin A bound to the complex. The entire signal pattern is interpreted as an indication of a second inhibitor site situated between the $b - c_1$ non-heme iron and cytochrome c_1, which would then be another possible example of the suggested non-heme iron–piericidin A interaction.

REFERENCES

1. Takahashi, N., Suzuki, A. and Tamura, S., *J. Am. chem. Soc.* **87** (1965) 2066.
2. Horgan, D. J. and Singer, T. P., *J. biol. Chem.* **243** (1968) 834.
3. Kosaka, T. and Ishikawa, S., *J. Biochem., Tokyo* **63** (1968) 506.
4. Palmer, G., Horgan, D. J., Tisdale, H. and Singer, T. P., *J. biol. Chem.* **243** (1968) 834.
5. Jeng, M. and Crane, F. L., *Biochem. biophys. Res. Commun.* **30** (1968) 465.
6. Horgan, D. J. and Casida, J. E., *Biochem J.* **108** (1968) 153.
7. Hall, C., Wu, M., Crane, F. L., Takahashi, N., Tamura, S. and Folkers K., *Biochem. biophys. Res. Commun.* **25** (1966) 373.
8. Vallin, I. and Löw, H., *Eur. J. Biochem.* **5** (1968) 402.
9. Löw, H., Vallin, I. and Alm, B., *in* "Energy-linked Functions of Mitochondria" (edited by B. Chance) Academic Press, New York, 1963, p. 5.
10. Storey, B. T., *Archs Biochem. Biophys.* **121** (1967) 265.
11. Kröger, A. and Klingenberg, M., *Biochem. Z.* **344** (1966) 330.
12. Fluharty, A. L. and Sanadi, D. R., *Proc. natn. Acad. Sci. U.S.A.* **46** (1960) 608.
13. Fluharty, A. L. and Sanadi, D. R., *Biochemistry,* 2 (1963) 519.
14. Jeng, M., Hall, C., Crane, F. L., Takahashi, N., Tamura, S. and Folkers, K., *Biochemistry* 7 (1968) 1311.
15. Lenaz, G., Daves, Jr, G. D. and Folkers, K., *Archs Biochem. Biophys.* **123** (1968) 539.
16. Moore, H. W. and Folkers, K., *J. Am. chem. Soc.* **86** (1964) 153.

FEBS Symposium, Volume 17, 1969, pp. 301-314

Sites and Mechanism of Ion Binding and Translocation in Liver Mitochondria

G. F. AZZONE, S. MASSARI, E. ROSSI and A. SCARPA

Institute of General Pathology, University of Padova, Padova, Italy

In the present paper we shall discuss some aspects of the mechanism of ion translocation in mitochondria. Attention will be focused on the following points:

1. The surface binding and the metabolism-independent Ca^{++} uptake;
2. The mechanism of the superstoichiometry ratios;
3. The stoichiometry between ion translocation and energy consumption;
4. The magnitude of the cation gradient;
5. The question of the membrane potential.

The surface binding and the metabolism-independent Ca^{++} uptake

The reaction steps for the translocation of Ca^{++} in liver mitochondria are indicated in Fig. 1 as: (1) the binding of Ca^{++} at the mitochondrial surface; (2) the energy-linked translocation of Ca^{++}; and (3) the coupling of the anion influx to the Ca^{++} uptake. We will first consider reaction (1). When added to a suspension of mitochondria, Ca^{++} binds to surface components of the mitochondrial membrane which have been shown to be phospholipids [1-3]. The binding is accompanied by a release of H^+ (or of other bound univalent cations) and does not require a supply of energy. It is necessary, however, to distinguish operationally the surface binding from the metabolism-independent binding [4]. The "surface" binding is a characterization of the binding of Ca^{++} according to a topographic criterion. The metabolism-independent binding is a characterization of the binding of Ca^{++} according to an energetic criterion. We have shown that Ca^{++} can be bound in the absence of energy supply both in the outer and in the inner mitochondrial spaces [4, 5]. A large and rapid translocation of Ca^{++} occurs when valinomycin is added to K^+-loaded mitochondria, and we have suggested that the uptake of Ca^{++} is coupled to the release of K^+ [6].

The role of the surface binding on the mechanism of Ca^{++} translocation has been studied in detail in our laboratory [2]. The primary question to answer was whether this surface binding was completely unspecific and independent or was a preliminary step of the process of Ca^{++} translocation. Both hypotheses

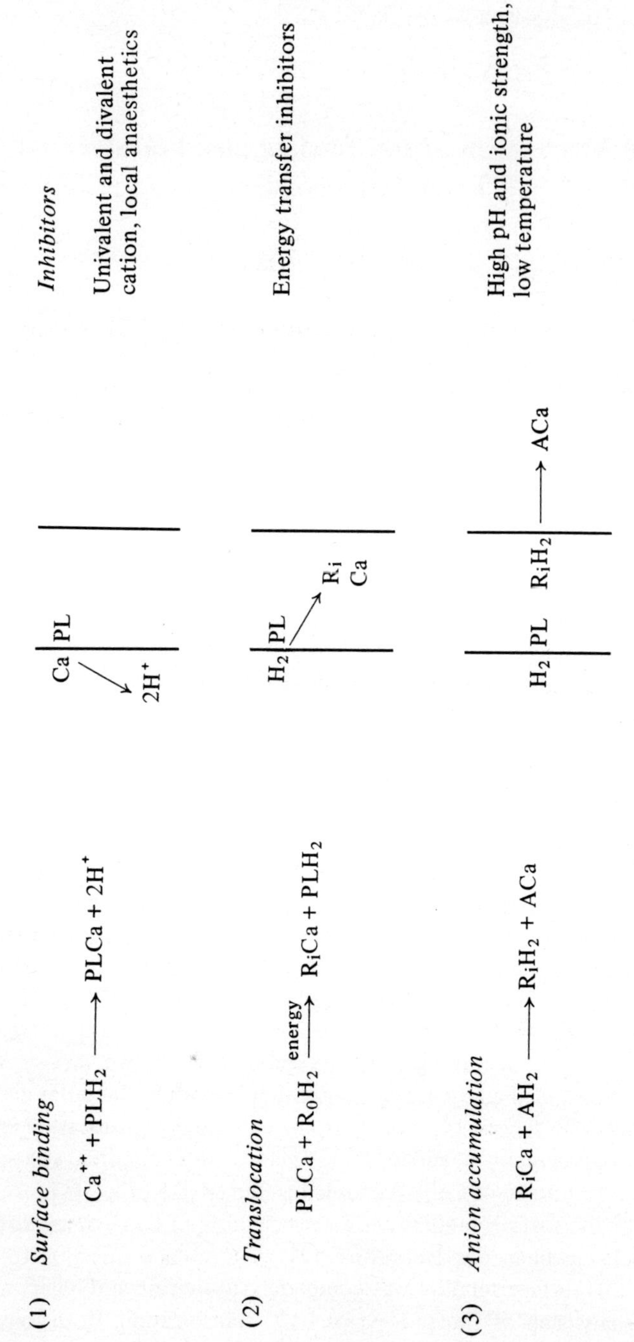

Figure 1. Mechanism of Ca^{++} translocation.

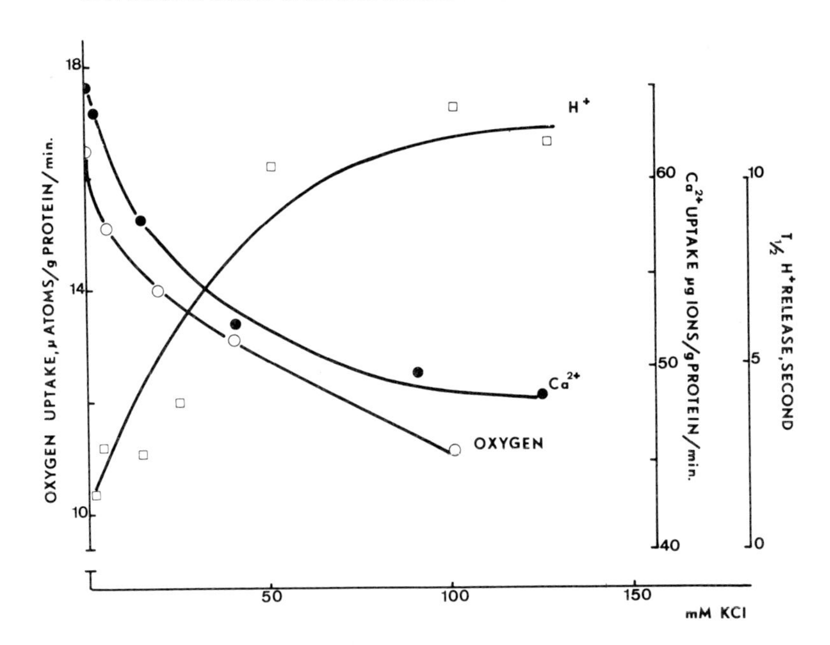

Figure 2. Effect of surface binding on the rate of Ca^{++} translocation. Experimental conditions as follows: the final osmolarity of 0·25 was obtained by adding various concentrations of sucrose and KCl as indicated in the figure, 8 mM Tris HCl at pH 7·4, 25 mg of mitochondrial protein. Final volume 3·2 ml, temp. 22°C. In the case of Ca^{++} uptake, 300 μM Ca^{++} was present in the medium and the reaction was started by the addition of 380 μM succinate. In the case of oxygen consumption, 380 μM succinate was present and the reaction was started with 300 μM Ca^{++}. When the H^+ release was measured the amount of mitochondrial protein was 18 mg.

have in fact been suggested [1, 7]. By testing the effect of several univalent and divalent cations which compete with Ca^{++} for binding to the membrane, we have found a close relationship between extent of surface binding and rate of Ca^{++} translocation.

Figure 2 shows the effect of KCl on the initial rate of Ca^{++} uptake and oxygen consumption. The rate of Ca^{++} uptake decreased from 65 μg ion/g protein per min in 0·25M sucrose to 48 μg ion/g protein per min in 0·125M KCl. Parallel to this there was also a decrease in oxygen uptake from 16 to 11 μatoms/g protein per min. Figure 2 also shows that the $t_{1/2}$ of the H^+ release was decreased several times when the sucrose medium was replaced with KCl medium. Inhibition of the rate of Ca^{++} translocation was observed with various univalent and divalent cations, the effectiveness of which is in the order $H^+ > Mg^{++} > K^+ > Na^+$.

The uptake of Ca^{++} can be measured in two steps, first the anaerobic binding, and then the aerobic translocation. Under these conditions, the majority of Ca^{++} which is bound at the surface in the anaerobic phase, is subsequently translocated into the mitochondria in the aerobic phase.

Our data are therefore compatible with a model of Ca^{++} translocation occurring through several steps, the first of which consists of a binding of Ca^{++} to some sites on the external surface. The energy-dependent binding of Ca^{++} to a membrane component is described by reaction (2), Fig. 1:

$$R_0H_2 + PLCa \xrightarrow{\text{energy}} R_iCa + PLH_2 \quad (2)$$

where R_0H_2 is a membrane component. Energy can be supplied in several ways for the translocation reaction, among which may be a potential created by the efflux of K^+. Reaction (2) will therefore be inhibited by agents which cause a de-energization of the mitochondria, primarily the uncouplers.

In the presence of anions the energy-dependent translocation is followed by an influx of anions, whereby the Ca^{++} bound to the membrane component R is transferred into the matrix and neutralized by the translocated anions (reaction (3), Fig. 1):

$$R_iCa + AH_2 \rightarrow R_iH_2 + ACa \quad (3)$$

The mechanism of the superstoichiometric ratios

When Ca^{++} is translocated in the absence of anions the mitochondria go into a state of inhibited respiration [8-12]. In this state the binding of Ca^{++} to the mitochondria requires a catalytic, but not a stoichiometric, energy supply. An interesting feature of the superstoichiometric ratios is that the plot of either the $Ca^{++}/\sim$ or the $H^+/\sim$ ratios versus the Ca^{++} concentration reveals a sigmoid curve [12, 13] (Fig. 3). Thus the $H^+/\sim$ ratio was about 2 at low Ca^{++} concentration and rose to 6 at higher Ca^{++} concentrations. The rise of the $Ca^{++}/\sim$ and $H^+/\sim$ ratios is due to the fact that above a certain Ca^{++} concentration, the mitochondria continue to take up Ca^{++} and release H^+ without a stoichiometric consumption of oxygen.

To explain this phenomenon we have proposed a mechanism which is summarized below. We assume that the membrane component R_i^* undergoes a conformational change:

$$R_1^* \xrightarrow{L} T_1^*$$

where L is the equilibrium constant for the transition and T^* is another form of the membrane component which may have 1, 2, ... n Ca^{++} bound and is therefore denoted as $T_1^*, T_2^*, \ldots T_n^*$ (the asterisk indicates that the membrane component is in an energized form). After setting the equilibrium condition $\alpha = (Ca^{++}/K_T)$ [13] and having written the equilibria of Ca^{++} with the various T^* forms we come to a final equation of the type:

$$Ca^{++}/\sim = \frac{1 + L(1+\alpha)^{n-2}(1+n\alpha)}{1 + L(1+\alpha)^{n-1}} \quad (4)$$

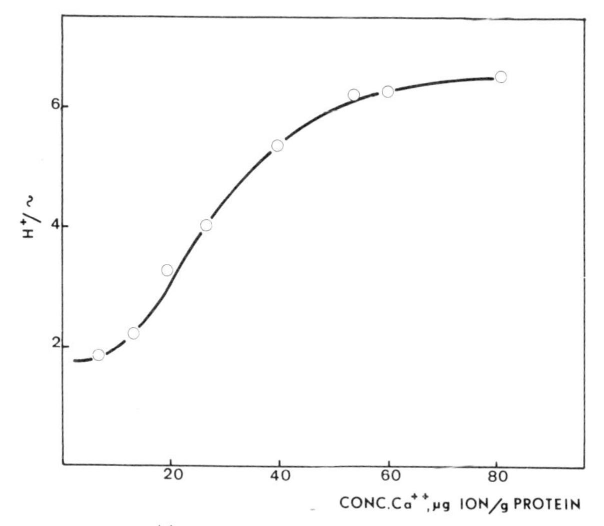

Figure 3. Effect of Ca^{++} concentration on $H^+/\sim$ ratio in the absence of anions. The incubation medium contained in 2·0 ml, 0·15M KCl, 5 mM Tris-HCl, pH 7·3, and various $CaCl_2$ concentrations, 15 mg mitochondrial protein.

Use of this equation gives a sigmoid curve for the $Ca^{++}/\sim$ versus Ca^{++} at equilibrium [13].

Another interesting feature of the superstoichiometric ratios is the sensitivity of the aerobically-bound Ca^{++} to releasing agents such as respiratory chain inhibitors or uncouplers.

Figure 4 shows that below 15 µg ion Ca^{++}/g protein, addition of rotenone at pH 7·4 induced practically no reuptake of H^+ (which corresponds stoichiometrically to the release of Ca^{++}), whereas dinitrophenol caused a reuptake of H^+ corresponding to about 70% of the H^+ released during the Ca^{++} uptake. Above 15 µg ion Ca^{++}/g protein there was a progressive increase in the extent of reuptake of H^+ due to rotenone. Thus, it seems that two aliquots of Ca^{++} can be distinguished; one which can be released whenever the mitochondria become de-energized and another which requires the specific addition of dinitrophenol.

Of great interest is the comparison between the amount of dinitrophenol-sensitive Ca^{++} release and the extent of stimulation of the respiration. Figure 4 shows that the respiration was stimulated only in correspondence with the dinitrophenol-sensitive part of H^+ reuptake and not in correspondence with the rotenone-sensitive part of H^+ release.

Since the release of the aerobically-bound Ca^{++} is presumably dependent on either the dissociation constant of Ca^{++} for the binding sites, or the rate of diffusion of the ions which must neutralize the potential originated during the Ca^{++} efflux, it follows that the differential effects of the respiratory chain

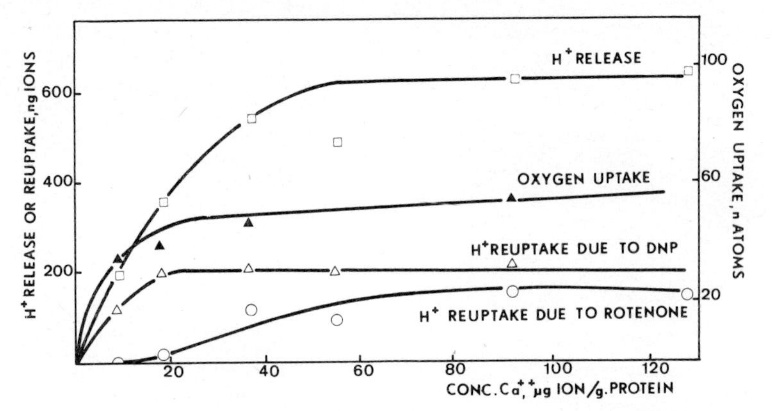

Figure 4. Effect of rotenone and dinitrophenol on the reuptake of H^+ at various Ca^{++} concentrations. The medium contained in 2·1 ml, 0·1M KCl, 2·4 mM Tris-HCl, pH 7·4, and various $CaCl_2$ concentrations. The reuptake of H^+ was started by the addition of 5 μM rotenone or of 250 μM dinitrophenol. 11 mg mitochondrial protein.

inhibitors and of the uncouplers must reside either in the dissociation constants or in the rates of diffusion of the counter ions.

According to the hypothesis expressed above, Ca^{++} can be bound to two types of sites: those defined as R_1^*, and T_1^* with the expenditure of a stoichiometric amount of energy, and those defined as T_2^*, T_3^*, . . . T_n^* with a catalytic amount of energy. Therefore, the differential releasing effects of rotenone and dinitrophenol on the bound Ca^{++} can be explained in two ways. According to one alternative, the R_1^* and T_1^* sites have a dissociation constant for Ca^{++} which is lower than the T_2^*, T_3^* sites. The Ca^{++} bound to sites with a very low dissociation constant cannot be released by rotenone but requires the presence of dinitrophenol. According to the other alternative, the sites R_1^* and T_1^* are located in a compartment to which H^+ ions have access only in the presence of dinitrophenol and not in the presence of rotenone. It is interesting to note that the dinitrophenol-sensitive sites can bind about 10 μg ion Ca^{++}/g protein which compares well with the buffering power of the inner space, as defined operationally by Mitchell and Moyle [14], which is about 13 μg ion H^+/g protein per pH unit.

The stoichiometric relationship between cation translocation and energy expenditure

In Table 1 are summarized the conclusions reached in our laboratories and in others in regard to the stoichiometry of cation translocation. The stoichiometry for divalent cations, as measured in the presence of permeant anions, is about 2 per energy conserving site [12, 15, 16]. The stoichiometry for univalent cations has been reported to be 3 by Cockrell *et al.* [17] with energy supplied by the respiratory chain. We have obtained results similar to those of Cockrell *et al.* The $K^+/\sim$ ratio was about 3 when measured on the total oxygen uptake and about 4 when measured on the Δ oxygen uptake. Since, except at very low K^+ concentrations, the reaction for cation translocation can compete efficiently with those causing the energy leak, we suggest taking the total oxygen uptake, in agreement with previous suggestions [8, 12].

Table 1. Stoichiometry of cation translocations

	Divalent	Univalent
(1) Number of ions translocated per energy conserving site (in presence of anions)	2	3–4
(2) Effect of ion gradients on the stoichiometry	none	none
(3) Difference between redox and chemical energy	none	–

The suggestion has been made [17, 18] that the stoichiometry of the translocation depends upon the gradient against which the ion is translocated. The effect of the ion gradient on the stoichiometry has therefore been investigated in great detail in our laboratory. In the case of Ca^{++} the stoichiometry is unchanged whether Ca^{++} is translocated into the mitochondria coupled to phosphate or to acetate, although it is precipitated in the former case and osmotically active in the latter case. Furthermore, the stoichiometry is independent of the amount of Ca^{++} added in the presence of acetate. In the case of K^+, the $K^+/\sim$ ratio was found to be independent of the gradient obtained by changing both the extra- and the intramitochondrial concentrations of K^+. Thus, we conclude that the energy expenditure is a function of the number of ions translocated and not of the work done against the gradient. No difference between redox and chemical energy has been found in regard to the stoichiometry of Ca^{++} translocation [19], whereas Cockrell *et al.* [17] have reported that twice as many K^+ ions are translocated per $\sim$ when energy is provided by ATP.

The magnitude of the cation gradient

For the determination of the maximal cation gradient there are two problems: (a) the measurement of the activity of the intramitochondrial cations and (b) the conditions of the steady state [20].

For the calculation of the activity of the intramitochondrial K^+ we have used the assumption of the osmotic equilibrium which implies that the sum of the activities of the intra- and the extramitochondrial solutes are equal:

$$\Sigma^{o}\gamma_j{}^{o}c_j = \Sigma^{i}\gamma_j{}^{i}c_j \tag{5}$$

where the superscripts o and i indicate the outer and inner (osmotically active) mitochondrial spaces, the suffix j indicates the ion species, γ the activity coefficient, and c the concentration. If we ignore the activity of the intramitochondrial Na^+, and the activities of the extramitochondrial Tris phosphate and succinate, (cf. Fig. 6) equation (5) can be written as:

$${}^{o}\gamma_{\text{choline}}\,{}^{o}c_{\text{choline}} + {}^{o}\gamma_{\text{ce}^-}\,{}^{o}c_{\text{ce}^-} = {}^{i}\gamma_{K^+}{}^{i}c_{K^+} + \sum_j {}^{i}\gamma_{A_j^-}\,{}^{i}c_{A_j^-} \tag{6}$$

where A_j^- indicates the soluble intramitochondrial anions. Charge neutrality, however, requires that:

$${}^{i}c_{K^+} = \Sigma z_j{}^{i}c_{A_j^-} \tag{7}$$

where z_j indicates the valence of the jth anion. In view of the large number of polyelectrolytes in the intramitochondrial spaces, with multiple charges and an activity coefficient presumably lower than that of K^+, it follows that:

$$\sum_j {}^{i}\gamma_{A_j^-}\,{}^{i}c_{A_j^-} < {}^{i}\gamma_{K^+}\,{}^{i}c_{K^+} \tag{8}$$

By combining equation (6) and equation (8) it follows that the total activity of the intramitochondrial K^+ must be comprised between (a's indicate activities):

$${}^{o}a_{\text{choline}} + {}^{o}a_{\text{ce}^-} > {}^{i}a_{K^+} > \tfrac{1}{2}({}^{o}a_{\text{choline}} + a_{\text{ce}^-}) \tag{9}$$

Equation (9) states that the intramitochondrial K^+ contributes more than 50 and less than 100% to the osmotic activity of the intramitochondrial solutes. Therefore the assumption that the activity of intramitochondrial K^+ is at least equal to 50% of the total activity of the extramitochondrial solutes is perfectly justified.

More recently equation (9) has received some experimental support. In the experiment shown in Table 2 we have measured the Cl^- accessible space, the osmotic space (cf. Bentzel and Solomon [21]), and the concentration of Rb^+ in the Cl^- space and in the osmotic space. We have assumed that the concentration of Rb^+ in the osmotic space is given by the Δ Rb^+ ± valinomycin. In the last column it is seen that the concentration of Rb^+ in the Cl^- space is higher than in the medium. This is presumably due to two factors: first, a binding of Rb^+ to the outer membrane and to the surface of the inner membrane; and second, an exchange of Rb^+ with the K^+ present in the osmotic space. This exchange, although slow, cannot be eliminated completely under our experimental conditions. It is seen in Table 2 that the concentration of Rb^+ was higher in the

Table 2. Mitochondrial spaces and ions

	ng ions/mg prot.	μl/mg prot.	Rb^+ conc. mM	
Cl^- space	67	0·67		
Rb^+	109			
Rb^+ + valinomycin	199			
Rb^+ ± valinomycin	90			
Osmotic space		0·53		
In outer medium			100	200
In Cl^- space			163	236
In osmotic space			170	344

osmotic space than in the Cl^- space and increased proportionally to the concentration of impermeant solutes in the medium. Indeed, the reduction of the osmotic space was proportional to the increase in concentration of impermeant solutes in the range between 0·25 and 1 osmolar (Fig. 5). If we now assume that the number of binding sites for Rb^+ per mg protein on the mitochondrial membrane is similar in the Cl^- and in the osmotic spaces we can calculate that about 30% of the internal Rb^+ is bound and that the concentration of unbound Rb^+ in the osmotic space is about 60% of the sum of the concentrations of impermeant solutes in the medium.

The evaluation of the conditions of the steady state is critical since there are several factors which do not permit attainment of the maximal cation gradient. For example, if the mitochondrial membrane becomes leaky, part of the cation gradient might be dissipated. Furthermore, the permeability of the membrane to water plays a role. The ion translocation mechanism would tend to increase the activity of the intramitochondrial solutes. However, it can do so only to the extent to which the activity of the internal ions does not exceed the activity of the extramitochondrial impermeant solutes. Due to the rapid equilibration of the water activity, part of the energy for ion translocation may be used for an osmotic transport of water.

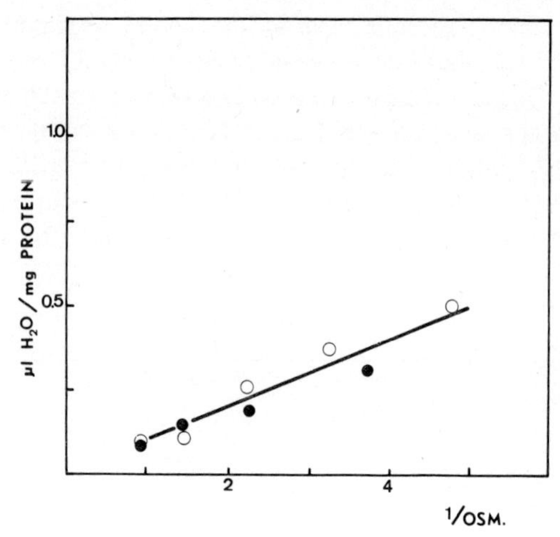

Figure 5. The water content in the osmotic space at various osmolarities. The osmolarity of the medium was changed by adding various KCl concentrations. The pellet volume was corrected for the Cl^- space as measured with $^{36}Cl^-$. The osmotic space is defined as total H_2O in Cl^- inaccessible space minus osmotically-inactive H_2O in Cl^- inaccessible space. The osmotically-inactive H_2O is calculated by extrapolating to the ordinate the regression line through the points of the H_2O content of the Cl^- inaccessible space at various osmolarities.

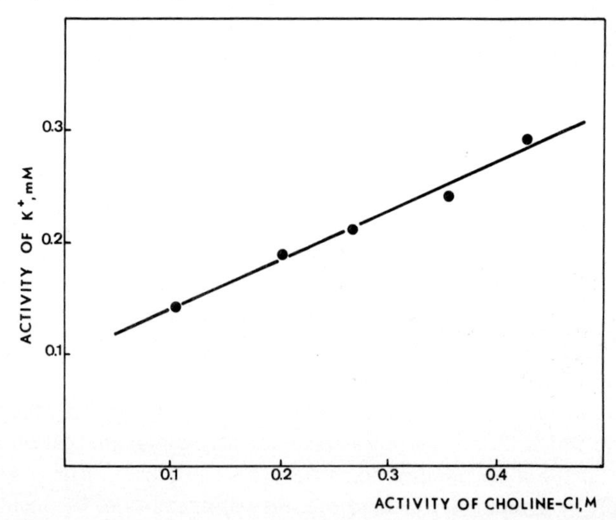

Figure 6. The activity of extramitochondrial K^+ at various activities of choline chloride. The medium contained variable amounts of choline-Cl, as indicated in the figure, 3·9 mM Tris-succinate, 7·8 mM Tris-HCl, pH 7·5, 3·1 mM Tris-phosphate, 6·2 μM rotenone, 0·5 μg valinomycin. Amount of mitochondrial protein was 5·0 mg/ml. Final volume, 3·2 ml, temp. 20°C. The values for the activities of the choline chloride media were obtained by multiplying the concentrations for the activity coefficients of choline chloride. The activity coefficients were obtained by measuring the depression of the freezing point of the choline chloride solutions.

Figure 6 shows an experiment where the activity of the extramitochondrial K^+ was measured at various choline chloride concentrations under steady state conditions. On the abscissa are indicated the activities of the choline media as obtained experimentally from the measurement of the depression of the freezing point. The activity of K^+ in the medium rose parallel to the increase of the activity of choline chloride. The experiment shown in Fig. 6 may be explained by assuming that the activity of the intramitochondrial K^+ increases parallel to the activity of choline because of osmotic equilibration (cf. Fig. 5).

The increase of the K^+ activity in the medium may thus be due to the presence of a constant gradient under steady conditions. Therefore an increase of the intramitochondrial K^+ activity must be followed by an increase of the extramitochondrial K^+ activity. By assuming that the activity of intramitochondrial K^+ is equal to the activity of the extramitochondrial choline, we can calculate a maximal gradient of K^+ across the membrane of about 1000 times [20], corresponding to an equilibrium potential of 180 mV.

When the K^+ gradient exceeds 1000 times, the accumulation of ions becomes inhibited. Addition of valinomycin to mitochondria, under conditions where the K^+ gradient has been increased beyond the above figures, causes a discharge of K^+ ions [20] under aerobic conditions.

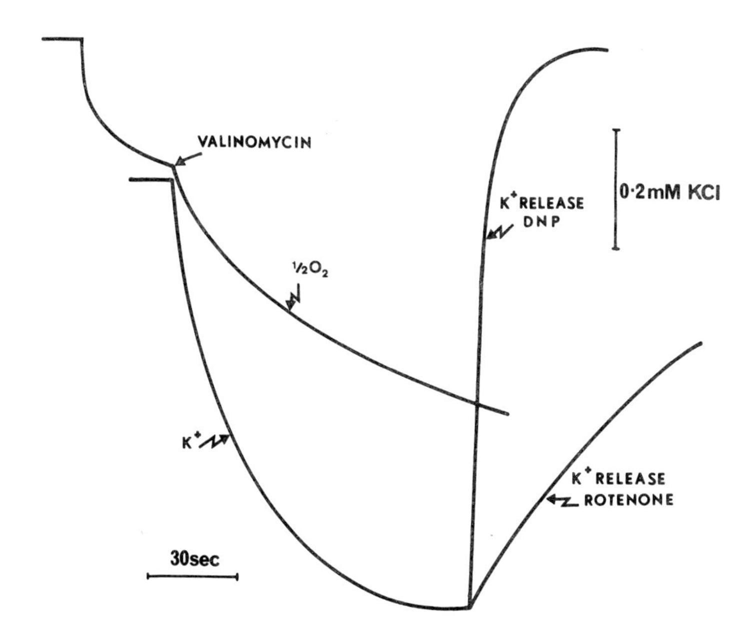

Figure 7. Energy demand for maintenance of K^+ gradient and releasing effects of rotenone and dinitrophenol. The medium contained, 0·25M sucrose, 1·5 mM KCl, 10 mM Tris-Acetate, 5 mM Tris-HCl, 1 mM β hydroxybutyrate. Final volume 2·1 ml, temp. 22°C.

The question of the membrane potential

Mitchell [22] has proposed a membrane potential as the basis for energy conservation in the respiratory chain. The magnitude of the membrane potential has been derived from thermodynamic calculations of the energy required to drive the synthesis of ATP.

Liver mitochondria were incubated in the presence of 1·5 mM KCl and 10 mM acetate. Addition of valinomycin caused a stimulation of the respiration which paralleled the uptake of K^+. After the uptake of K^+ was completed the respiration levelled off. Addition of rotenone caused a slow release of K^+. The release of K^+ was 20 times faster in the presence of dinitrophenol (Fig. 7). The essential feature of this experiment is that liver mitochondria were able to maintain a $[K^+]_i/[K^+]_o$ ratio of 350, corresponding to an equilibrium potential of 152 mV, without the expenditure of energy, although the membrane was highly permeable to K^+. Furthermore, K^+ was released at a slow rate unless dinitrophenol was added. A rate similar to that with dinitrophenol was obtained with Ca^{++}. One explanation is that K^+ is maintained within the mitochondria by an electrical field. This electrical field may either be created by a primary electrogenic reaction, as proposed by Mitchell, or by the efflux of K^+ down the chemical gradient. An alternative explanation for the low rate of respiration in valinomycin-treated mitochondria is that K^+ cannot leak out because the membrane carrier by which it is translocated becomes energized under State 4 conditions [23].

The following two postulates are formulated:

(1) Cations are translocated across the mitochondrial membrane and, on the expense of energy, from a lower to a higher thermodynamic potential. Protons are released in exchange with the cations and the translocation is thus electroneutral.

(2) The mitochondrial membrane has a low permeability to anions and cations, and possesses carriers for the translocation of protons, univalent and divalent cations.

It follows that the transmembrane potential is a function of the ion fluxes which are driven by the ion concentrations and potential gradients.

We have tried to calculate the membrane potential as a function of the ion leaks through the membrane.

The flux equations under steady state conditions give the membrane potential:

$$\frac{F\Delta\psi}{RT} P_{K^+} [K^+]_i - gJ_o + \left(gJ_o - \frac{F\Delta\psi}{RT} P_{K^+} [K^+]_o \right) e^{-F\Delta\psi/2\cdot 3RT} \tag{10}$$

where J_o is the flux of electrons through the respiratory chain and P_{K^+} is the permeability of the membrane to K^+. Since in equation (10) all values can be determined experimentally we can calculate the magnitude of the membrane potential under the various experimental conditions. According to equation (10) the membrane potential varies from values of 180 mV, under conditions when the respiration comes back to the initial rate after the uptake of K^+ is completed, to values of 20-40 mV when the rate of respiration remains very high.

ACKNOWLEDGEMENTS

The present studies have been aided by grants from the National Research Council and from NATO (No. 293).

The authors wish to thank Mr. Luciano Pregnolato and Mr. Paolo Veronese for expert technical assistance.

REFERENCES

1. Chappell, J. B. and Greville, G. D., *in* "Energy-linked Functions of Mitochondria" (edited by B. Chance), Academic Press, New York, 1963, p. 219.
2. Scarpa, A. and Azzone, G. F., *J. biol. Chem.* **243** (1968) 5132.
3. Scarpa, A. and Azzi, A., *Biochim. biophys. Acta* **135** (1968) 444.
4. Rossi, C., Azzi, A. and Azzone, G. F., *J. biol. Chem.* **242** (1967) 951.
5. Azzone, G. F. and Azzi, A., *in* "Regulation of Metabolic Processes in Mitochondria" (edited by J. M. Tager, S. Papa, E. C. Slater and E. Quagliariello), Elsevier, Amsterdam, 1966, p. 332.
6. Azzone, G. F., Rossi, E. and Scarpa, A., "Regulatory Functions of Biological Membranes" (edited by J. Järnefelt), Elsevier, Amsterdam, 1968, p. 236.
7. Chance, B. and Mela, L., *J. biol. Chem.* **241** (1966) 4588.
8. Chance, B., *J. biol. Chem.* **240** (1965) 2729.
9. Rossi, C. and Azzone, G. F., *Biochim. biophys. Acta* **110** (1965) 434.
10. Carafoli, E., Gamble, R. L., Rossi, C. S. and Lehninger, A. L., *Biochem. biophys. Res. Commun.* 22 (1966) 431.
11. Carafoli, E., Gamble, R. L., Rossi, C. S. and Lehninger, A. L., *J. biol. Chem.* **242** (1967) 1199.
12. Rossi, E. and Azzone, G. F., *J. biol. Chem.* **243** (1968) 1514.
13. Massari, S. and Azzone, G. F., in preparation.
14. Mitchell, P. and Moyle, J., *Biochem. J.* **104** (1967) 588.
15. Rossi, C. S. and Lehninger, A. L., *J. biol. Chem.* **239** (1964) 3971.
16. Chance, B., *J. biol. Chem.* **240** (1965) 2729.
17. Cockrell, R. S., Harris, E. J. and Pressman, B. C., *Biochemistry* **5** (1966) 2326.
18. Rottenberg, H. and Caplan, S. R., *Nature, Lond.* **216** (1967) 610.

19. Bielawsky, J. and Lehninger, A. L., *J. biol. Chem.* **241** (1966) 4316.
20. Rossi, E. and Azzone, G. F., *Eur. J. Biochem.* **7** (1968) 418.
21. Bentzel, C. J. and Solomon, A. K., *J. Physiol., Lond.* **50** (1967) 1547.
22. Mitchell, P., *Biol. Rev.* **41** (1966) 445.
23. Massari, S. and Azzone, G. F., in preparation.

Note added in proof. More recently we have further developed the concept of ion transport through carriers. According to the carrier mechanism the active uptake of cations does not require a membrane potential but involves only an increased affinity of the carrier for protons.

FEBS Symposium, Volume 17, 1969, pp. 315-333

Control of Mitochondrial Substrate Metabolism by Regulation of Cation Transport

B. C. PRESSMAN

University of Pennsylvania
Philadelphia, Pennsylvania,
U.S.A.

Rather than confine my presentation to the topic listed on the programme which has been discussed at several recent symposia [1-3], I wish to begin with the history of the discovery of the mechanism of action of the ionophorous antibiotics [4]. I will touch on various points of interest to this audience, including the chemiosmotic hypothesis, and finally will consider the relationship of cation and anion transport in mitochondria.

The story begins with earlier studies of the unique inhibition of energy transfer in mitochondria induced by guanidine and its derivatives [5]. We observed that the typical type of inhibition produced in State 3 mitochondria by alkylguanidines is a gradual decrease of respiration which can be partially restored by uncouplers such as DNP* [6, 7]. A group of uncoupling agents, octyl-DNP and dicumarol, however, fail to release alkylguanidine inhibition [7-9] (Fig. 1). These uncouplers are, however, equivalent in stimulating the respiration of mitochondria in which energy transfer is blocked by oligomycin. The figure shows that, even after the addition of sufficient dicumarol to mitochondria to release oligomycin inhibition, the subsequent addition of octylguanidine elicits its typical inhibition subject to reversal by DNP (Fig. 1, lower tracing). Thus the ability to release the respiratory inhibition by octylguanidine can differentiate two groups of uncoupling agents [8, 9]. Since alkylguanidines are much less effective in inhibiting succinate respiration than that of NAD-linked substrates, they apparently have a predilection to inhibit the first energy conservation site [6-9]. However, the related compound DBI effectively inhibits energy transfer with succinate as substrate, indicating that it inhibits the second energy conservation site [6, 7, 9]. The inferred site selectivity of the two types of guanidine compounds was confirmed by their effects on

*The following abbreviations have been used: DNP, 2,4-dinitrophenol; octyl-DNP, 2,6-dinitro-4-octylphenol; DBI (trade name, U.S. Vitamin), phenethylbiguanide; CCP, dicyanocarbonylphenylhydrazone; Cl-CCP, metachloro derivative of CCP.

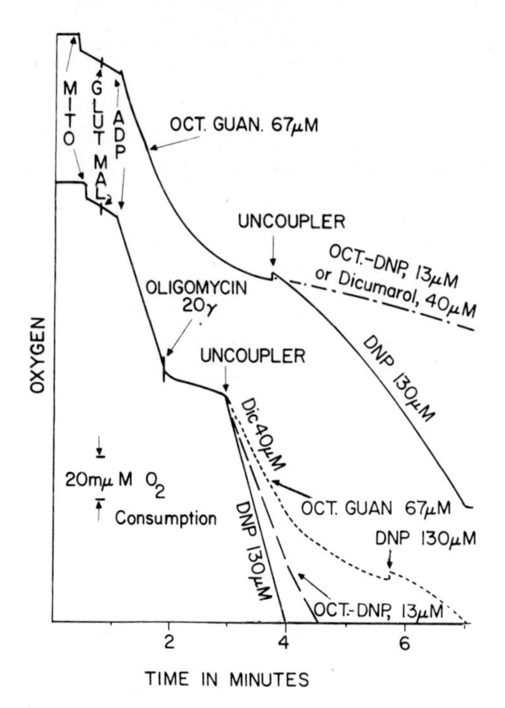

Figure 1. Release of energy transfer-blocked respiration of rat liver mitochondria by uncouplers. Reaction conditions are given in Reference 8; the apparatus is described in Reference 23.

components of the respiratory chain [6, 9]. Uncoupling agents such as dicumarol, which are unable to release the respiratory inhibition by alkylguanidines, can fully restore DBI-inhibited respiration.

These observations led to the scheme illustrated in Fig. 2, which features separate and distinct energy transfer pathways between the first and second energy conservation sites and the common intermediate which can energize ATP synthesis or drive ion transport [9]. By inference the pathway from the third energy conservation site is depicted as distinct from those emanating from the first two conservation sites. The site selective inhibitors also unmask the site specificity of uncoupling agents normally obscured by intercommunication between energy conservation pathways [7, 9]. These data have never been adequately reconciled with the chemiosmotic hypothesis which does not predict multiple energy transfer pathways between the electron transport chain and a common energized intermediate serving both ion transport and oxidative phosphorylation. An attempted reconciliation which ascribed the apparent site specificity of the guanidine inhibitions to a trivial phenomenon did not consider all of the reported experimental evidence [10].

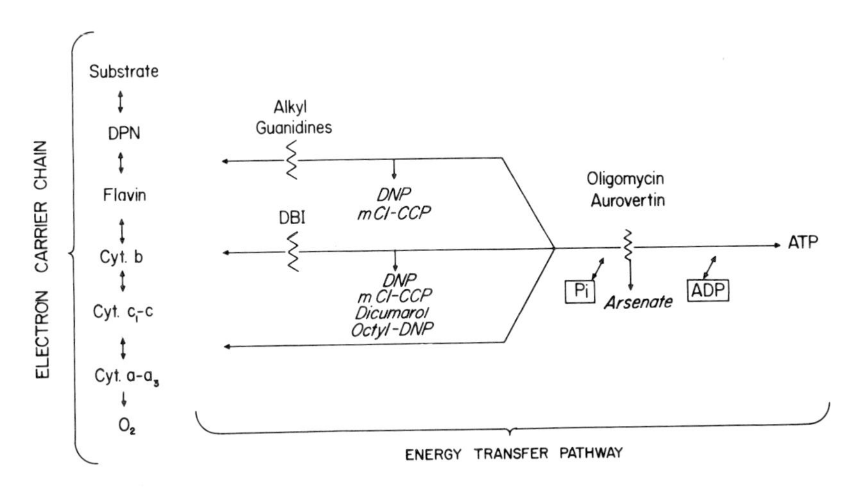

Figure 2. Interaction of site-specific reagents with mitochondrial energy transfer.

The site selective guanidines therefore offer a diagnostic test for grouping the known uncoupling agents into those like DNP, which release respiratory inhibitions by both types of guanidines, and those like dicumarol, which release only the inhibition induced by DBI but not by octylguanidine. In Table 1 four pairs of representative uncoupling agents are grouped. The first pair reverses the respiratory inhibition by DBI completely and the inhibition by octylguanidine about 50%. The second pair are also 100% effective against DBI but somewhat

Table 1. Comparison of the abilities of various uncouplers to release octylguanidine and DBI-inhibited respiration of mitochondria

Uncoupler	% Release Octylguanidine (100 μM)	% Release DBI (3 mM)
DNP	36	106
Penta-Cl-Phenol	54	86
CL-CCP	24	107
CCP	12·5	112
Dicumarol	3·2	140
Octyl-DNP	2·2	130
Gramicidin	16	35 (slow!)
Valinomycin	19	19 (slow!)

Glutamate-malate was used as substrate. Each uncoupler was added at five times the level required to stimulate a State 4 system to half maximal respiration.

less effective in releasing octylguanidine inhibition. This group is transitional between the typical DNP type of uncouplers and those of the third pair which releases DBI inhibition but not octylguanidine inhibition. The search for uncoupling agents with widely divergent structural characteristics led to valinomycin, which at the time appeared to be the most powerful uncoupling agent reported [11], and whose depsipeptide structure [12] differed from all the other tested agents. Surprisingly, it produced no extensive release of either type of guanidine inhibition, but only an atypical slow and partial release. Thus,

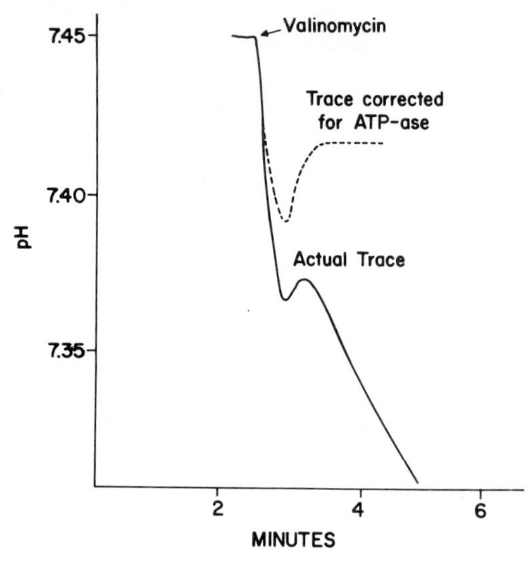

Figure 3. Effect of valinomycin plus K^+ on acid production by mitochondria. The reaction mixture contained: ATP, 2 mM; KCl, 25 mM; $MgCl_2$, 5 mM; Tris Cl, pH 7·4, 20 mM; sucrose, 200 mM, valinomycin added, 20 μg; mitochondrial protein, ca. 3 mg/3 ml.

the diagnostic test chosen revealed the unexpected existence of a new group of uncoupling agents, including not only valinomycin but the polypeptide gramicidin [8, 9].

Investigation of why the last pair of uncoupling agents differ from all the other agents in the series tested led us to examine their release of the latent ATPase of mitochondria [13]. At the time, I had just arrived at the Johnson Foundation and sensed some stigma attached to following the progress of an enzyme reaction in a biophysical laboratory by classical chemical colorimetric procedures. Accordingly I chose to track the hydrolysis of ATP by the production of acid which can be detected by means of a pH electrode and meter coupled to a recorder [14]. My first excursion into the realm of physical instrumentation produced a rather rude shock because, instead of tracing a linear rate of acid production, the apparatus produced the peculiar pattern shown in

Fig. 3. There was an initial rapid appearance of acid which abruptly reversed direction and then resumed at a slower rate. Despite my intuitive suspicion of instrumentation, the reproducibility of the data convinced me it was real. Historically, this pattern of acid production has been referred to as the "drain pipe" phenomenon [9].

When the reaction mixture was sampled at intervals and the liberated Pi measured colorimetrically, from the predetermined relationship between the hydrolysis of ATP and the liberation of protons [14] the amount of acid which should have appeared due to ATP hydrolysis was determined. On subtracting the chemically produced acid from the raw data curve the dotted curve resulted, representing the release of acid from mitochondria by some process other than the hydrolysis of ATP. This excess acid liberation seemed in many respects to resemble that which accompanies the accumulation of Ca^{++} by mitochondria [15]. The valinomycin-induced excess acid production could be energized either by ATP or by oxidizable substrates. It disappeared upon the addition of the detergent triton which destroys the mitochondrial membranes, hence this acid was due to a proton translocation rather than the net proton production by a chemical reaction. The antibiotics, however, functioned catalytically at extremely low concentrations, while in the case of Ca^{++} [15] and Mn^{++} [16] strict stoichiometries are obtained between the uptake of cations and the expulsion of protons. Despite our earlier work relating ion movements to gramicidin as well as valinomycin [7, 9] and detailed in a subsequent symposium [17], incomplete citations [18] have obscured the original discovery of the ionophorous properties of gramicidin.

Eventually we concluded that the antibiotic-evoked, energy-dependent, catalytic proton ejection represented the induced movement of one of the component ions of the medium. A further clue came from earlier work on the uncoupling effects and ATPase induction by fatty acids in mitochondria, which revealed a K^+ requirement not apparent in untreated mitochondria [19, 20]. This inspired us to omit K^+ from the reaction medium; the valinomycin response vanished, thereby confirming the deduction that K^+ was indeed the ion responding to valinomycin [21].

This was confirmed by applying ion-specific electrodes [22] to the study of mitochondrial ion metabolism [21, 23]. I wish to draw attention to the direct evidence for the movement of ions as sensed by the ion selective electrodes in Fig. 4. Figure 4A shows that addition of valinomycin to mitochondria changes the potential of the K^+ electrode, indicative of K^+ entering the mitochondria. This is accompanied by a stimulation of respiration responsible for the original discovery of the uncoupling effects of valinomycin [11]. Depriving the mitochondria of energy by permitting the medium to go anaerobic released the accumulated K^+; no evidence for ion movements appeared on the Na^+ selective electrode trace. In Figure 4B, valinomycin was added to mitochondria suspended in 10 mM Na^{++} in the absence of added K^+. After cessation of the return of the K^+ which had leaked out of the mitochondria, gramicidin was added and produced no further significant change of the K^+ electrode tracing, but now the Na^+ electrode indicated the uptake of Na^+ by the mitochondria. This is the first published record of the net movement of Na^+ into mitochondria under the influence of an antibiotic [24], not merely respiratory stimulation which could infer ion movement [18].

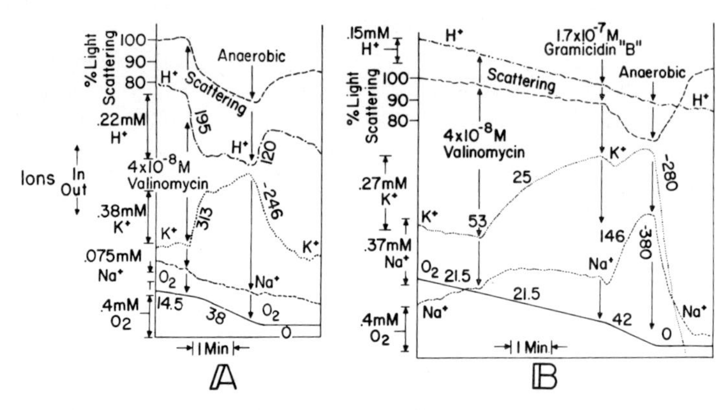

Figure 4. Multiparameter responses of mitochondria to valinomycin and gramicidin. Reaction conditions are given in Reference 24.

Specific ion electrodes were also used to measure the stoichiometry between the uptake of K^+ and the accompanying hydrolysis of ATP [25]. In the presence of acetate, $\Delta K/\Delta ATP$ values of an average of 6·5 to as high as 8 were obtained. Dr. Mitchell explained in detail in his paper (p. 219) how the chemiosmotic hypothesis calls for two charge separations per ATP [10, 26]. The observation of as many as 8 K^+ taken up per ATP split, equivalent to change separations, has not yet been explained by the chemiosmotic hypothesis.

Certain simple assumptions enabled us to compare the free energy necessary for K^+ accumulation with the free energy liberated during the hydrolysis of ATP. This calculation indicated that the ion accumulation process has a remarkably high thermodynamic efficiency [25]. If the energy conversion were really that efficient, ATP-driven K^+ accumulation ought to be demonstrably reversible. Accordingly we sought to synthesize ATP by permitting the intra- and extramitochondrial K^+ gradient to collapse by treatment with valinomycin. In Fig. 5 we see a typical record of an experiment carried out by Dr. Ron Cockrell, in which the addition of valinomycin to mitochondria leads to a rise in endogenous ATP and an equivalent disappearance of ADP, indicating the net formation of high energy phosphate. The time course of ATP synthesis correlates closely with the rapid phase of K^+ efflux from mitochondria. As soon

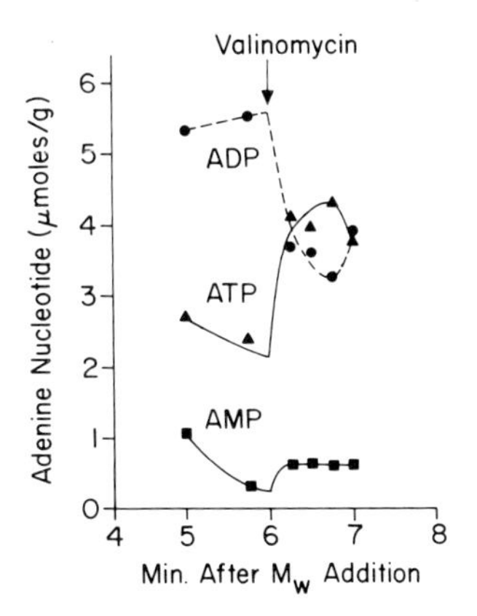

Figure 5. Net synthesis of ATP driven by collapse of the mitochondrial K^+ gradient with valinomycin. Reaction conditions and assay procedures are given in Reference 27.

as the rapid K^+ efflux phase culminates, the newly synthesized ATP resumes hydrolysis due to various exergonic reactions within the mitochondria. These experiments were carried out in the presence of respiratory inhibitors to prevent utilization of endogenous substrate to form ATP [27].

The Lardy group reported a second class of agents, typified by nigericin, which also affects mitochondrial ion transport [28]. In Fig. 6, upon the addition of valinomycin we observe the typical uptake of K^+; a subsequent addition of nigericin reverses the effects of valinomycin. Subsequent observations indicate, however, that there are more similarities than differences in the function of these two antibiotics.

The structure of valinomycin is shown in Fig. 7. It is rather chemically inert, containing the sequence L-lactate, L-valine, D-hydroxyisovalerate, D-valine,

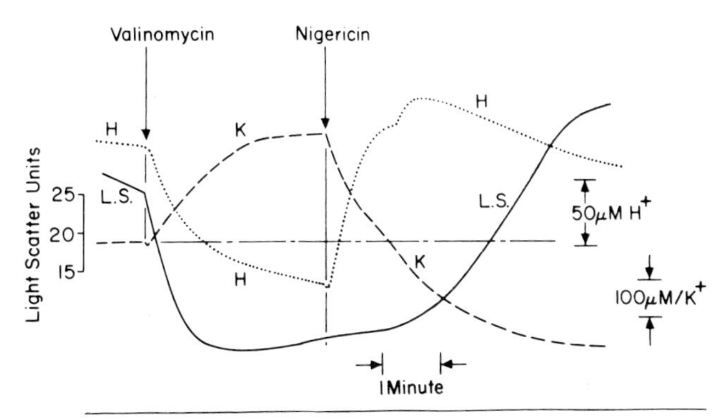

Figure 6. Opposing effects of valinomycin and nigericin on mitochondrial transport of K^+. Reaction conditions are given in Reference 4.

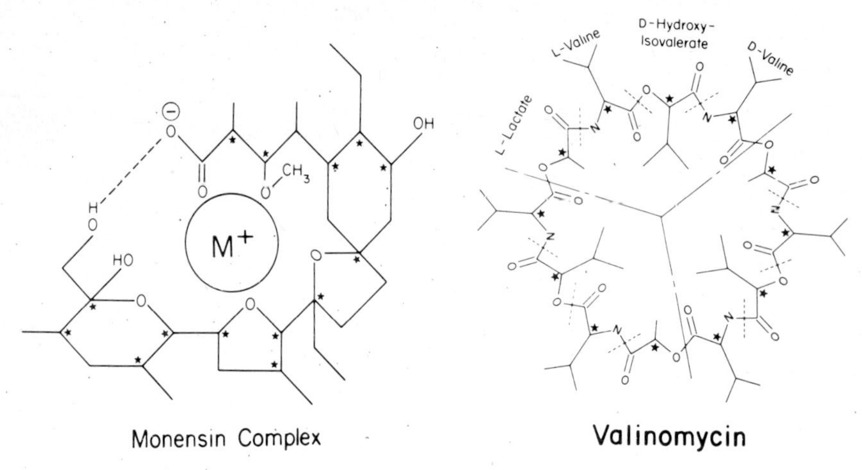

Figure 7. Structures of monensin and valinomycin.

repeated three times. The stars indicate the optically active centers. The structure of nigericin is still unknown, although we understand it is very close to being solved by the Steinrauf group. This group, however, has already obtained the structure of an antibiotic related to nigericin, monensin, by X-ray crystallography [29]. Monensin (Fig. 7) consists of a chain of heterocyclic rings containing ether oxygens and hydroxyl groups. The significant distinction between monensin and valinomycin is the existence in the former of a carboxyl group. All members of the nigericin series of antibiotics possess free carboxyl groups, while those of the valinomycin group have no charge or ionizable groups. The structure in Fig. 7 was determined by X-ray crystallography of the silver complex, although the ability of nigericin to form lipid soluble alkali ion complexes was known earlier [30]. Valinomycin and related antibiotics also form strong complexes with the alkali cations [27, 31].

For the purpose of analyzing the salient structural features of these antibiotics, the diagrammatic symbolizations in Fig. 8 will suffice. Valinomycin possesses neither charge nor ionizable groups, and the complexes it forms with

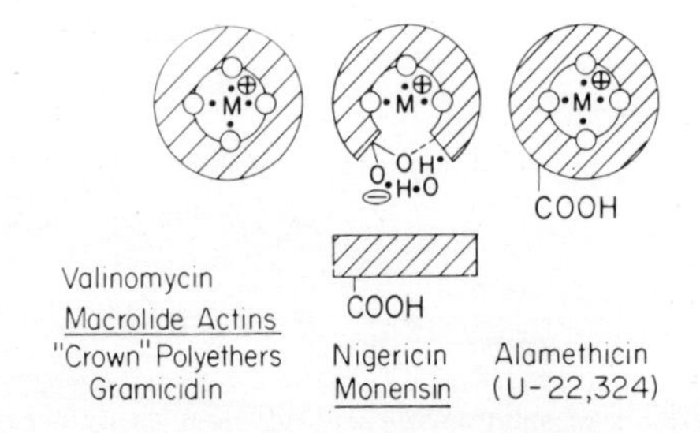

Figure 8. Schematic representations of ionophore complexes.

monovalent ions acquire the charge of the ion complexed, in this case represented by M^+. The small circles surrounding the complex ion in each compound represent oxygen atoms. Workers in Zurich [31, 32], Moscow [33], and Wilmington [34] have shown that the active regions of such complexing molecules are systematic arrays of oxygens, which form ion-dipole bonds with the metal ions complexed. The nigericin group of antibiotics exists in two forms. The protonated form which predominates at low pH cannot complex, and perhaps takes the conformation of an open chain. The deprotonated form produced at higher pH forms head to tail hydrogen bonds which stabilize a ring conformation in which the oxygens are appropriately arranged for complexing. The resulting complex is not a simple salt, since the charge on the carboxyl group is separated from the locus of the complexed metal ions. The charge of the cation complex is neutralized by the charge on the carboxyl group, so that as a whole it is electrically neutral. Due to the charge separation, this complex is a zwitterion rather than a simple salt. Such complexes have properties which differ from most salts, e.g. they dissolve in organic solvents.

The antibiotic alamethicin constitutes a distinct third class, since it has a covalently bound ring of L-amino acids plus a single free carboxyl group, as does the nigericin group [35]. The ability of alamethicin to complex with alkali ions is independent of the pH, i.e. it can form a complex either in the charged protonated form as depicted in Fig. 8, but when deprotonated it can also form a zwitterion complex. Hence, when the ring is held closed by covalent bonds, the state of ionization of a carboxyl group distal from the ring need not affect the complexing properties of an ionophore.

The essential difference for their opposing effects on mitochondria of the valinomycin-type and nigericin-type ionophores, both of which form analogous complexes with alkali ions, resides in the net charge of the complexed species. Valinomycin-type charged complexes can move ions down an electrochemical gradient, while the uncharged complexes of the nigericin type equilibrate proton with cation gradients uninfluenced by any existing membrane potential. More detailed explanations of how the difference in charge of the complexes explains their action on mitochondrial systems have been presented [1-3].

Let us now turn to the origin of the extreme ion specificity of the ionophores, e.g. valinomycin induces the mitochondrial uptake of K^+ but not Na^+, while gramicidin and the actins can bring about the net uptake of both these ions. An early explanation for the ion specificity of the ionophores, assuming that they were all cyclic (although we now know this is not so for the structure of gramicidin) [36], was that the specificity was determined by the relationship between the hole in the ring and the size of the hydrated or unhydrated alkali ions [37] (cf. also reference 24). However, in the two compounds illustrated in Fig. 9, the size of the planar ring representations differs by a factor of over two, yet they exhibit similar ion specificity [38]. Clearly, then, ionic specificity at the molecular level must reside elsewhere than in the dimensions of the planar ring.

The Zurich group have determined the structure of the K^+ complex of nonactin by X-ray crystallography, and they find that the backbone of the ring is not at all planar but rather puckers into the general shape of the seam of a tennis ball with the complexed ion in the center bonded to eight oxygens arranged cubically [32]. A profoundly significant aspect of this structure is that the polar groups of the ionophore gather about the centrally encaged K^+, while the exterior of the spheroidal complex is primarily aliphatic. This structure explains why such complexes are lipophylic and hence penetrate lipid membranes. This may well represent the general prototype for transport carriers which move polar substances through lipoidal membranes.

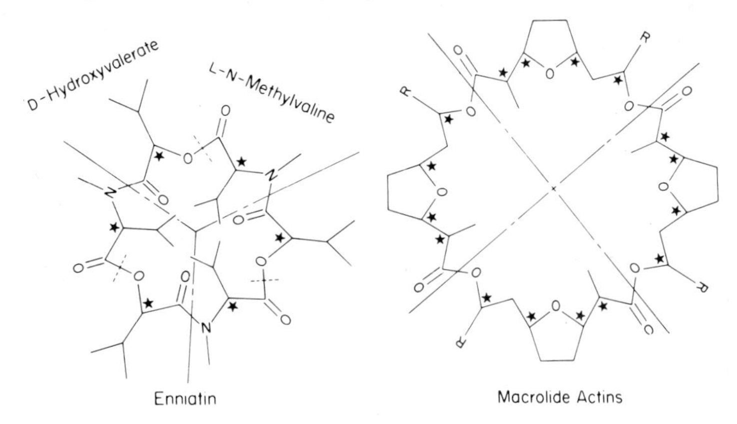

Figure 9. Comparison of structures of enniatin B and macrotetralide actins.

A space filled model of the K^+-monensin complex based on the Steinrauf structure [29] is analogous to the K^+-nonactin complex, except that the complexed ion is encaged by six oxygens. We were also able to extend the generalities observed in the known three-dimensional ionophore structures to construct a likely complex of valinomycin in which the cage consists of eight or nine oxygen atoms. If we insert the smaller Na^+ into the model before encaging it, in the case of those ionophores which do not form Na^+ complexes readily, if the model is shaken, the Na^+ rattles about. The ionophores which show high K^+/Na^+ selectivity, however, form good van der Waals contact with the larger anhydrous K^+. On the other hand, the six complexing oxygens in the monensin molecule can form a compact non-rattling cage about the smaller Na^+, consistent with its preference for complexing Na^+.

The partition of alkali ions between water and organic solvents in the presence of various lipophylic ionophores can measure the intrinsic ion selectivities of these compounds. The ionophores listed in Fig. 10 show a wide

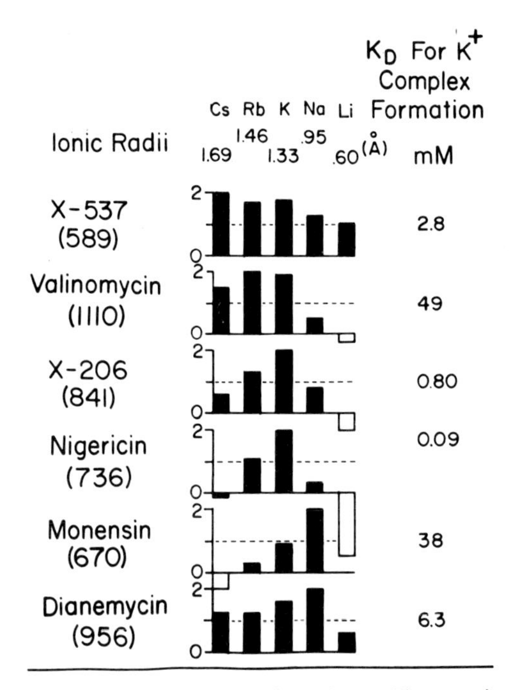

Figure 10. Ionic selectivities of various ionophores. The organic phase consisted of 30% *n*-butanol-70% toluene. For the valinomycin data the aqueous phase was buffered at pH 7·0 (glycin-tricine) and supplemented with 50 mM $Mg(CNS)_2$. In all other cases the pH was adjusted to 10·0 with tetramethylammonium hydroxide. The two-phase dissociation constants K_D are the aqueous $[K^+]$ necessary to half saturate the ionophore in the organic phase.

range of ion preferences ranging from Cs^+, Rb^+, and K^+, to Na^+ in the case of monensin. Note that the bars in the figure are logarithmic functions of the measured preferences, so that one unit in length represents a difference in ionic selectivity of 10 fold.

A new approach which we are using for studying the conformational changes involved during ionophore complex formation is protein NMR spectroscopy. The lower spectrum of Fig. 11 is that obtained with valinomycin in $CDCl_3$. The upper spectrum is that of the KCNS complex of valinomycin in the same solvent. The reference marker at zero p.p.m. is tetramethylsilane. The region between 1 and 2 p.p.m. contains the majority of the methyl protons which are shifted somewhat during complex formation with the emergence of a clear peak at about 1·3 p.p.m. The peak about 2·3 p.p.m. is shifted but little. The peak at 1·8 p.p.m. in the complex is due to a small quantity of adventitious H_2O. The bands at approximately 4 and 5 p.p.m. are shifted slightly upfield by complex formation and the fine structure of the 4 p.p.m. band is altered. At 7·3

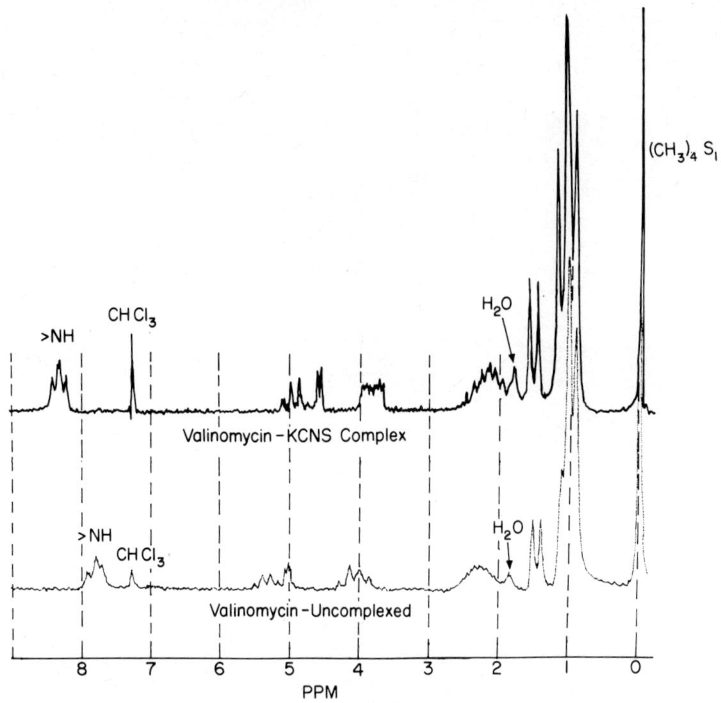

Figure 11. Proton NMR spectra of valinomycin and its K^+ complex. Determinations carried out on 60 megacycle Varian NMR instrument at a concentration of 50 mg valinomycin/ml in DCl/3. The complex was formed by a stoichiometric amount of KCNS.

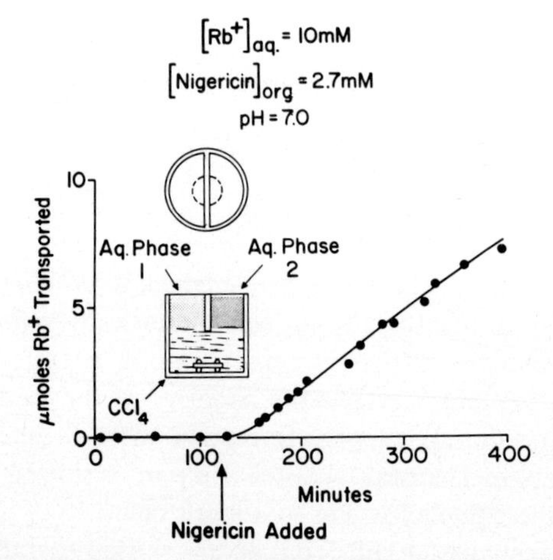

Figure 12. Translocation of $^{86}Rb^+$ across a bulk phase of CCl_4 by nigericin. Reaction conditions are described in the text and further detailed in Reference 4.

p.p.m. we see peaks of non-deuterated $CHCl_3$ present as an impurity which serves as a convenient reference marker. At 7·7 p.p.m. a triplet appears corresponding to the amide nitrogens of the valine which is shifted about 0·5 p.p.m. downfield during complex formation. This peak has been unambiguously identified as the amide proton, since it can be removed by exchange with deuterium in the presence of CD_3OD and DCl.

We hoped that the isolated amide proton peak would shift during complex formation. This anticipated shift was observed, but the entire proton spectrum was also altered. This indicates that, during complex formation in a solvent, the inter-relationship of all of the protons of valinomycin is affected, implying that a

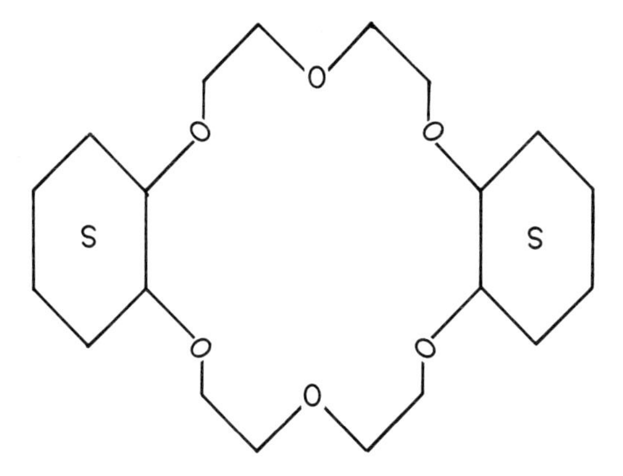

Figure 13. Structure of dicyclohexyl-18-crown-6.

radical conformational change occurs. We are currently pursuing this approach in greater detail because it is a powerful adjunct to X-ray crystallography which is limited to the solid state, and the conformations of these complexes in a lipophylic solvent are more relevant to transport phenomena.

In order to demonstrate that the ionophores can indeed move ions, we set up a Schulman-type model system by sealing a glass septum into the upper half of a small glass vessel. The upper part was separated into two compartments by adding enough CCl_4 to rise above the bottom of the septum. The upper compartments were then filled with aqueous buffer. $^{86}Rb^+$ was added to one aqueous compartment and its transfer to the opposite aqueous compartment measured as a function of time. The organic layer was stirred from below by a magnetic disc. During the control interval no transfer of $^{86}Rb^+$ was detected; however, following the addition of nigericin, a steady rate of $^{86}Rb^+$ transfer ensued (Fig. 12). This model experiment indicates that ionophores can transfer ions through a bulk phase of organic solvent equivalent to the lipid region of a biological membrane.

The actual procedure employed in obtaining the data in Fig. 10 was to measure the extent of formation of ion complex, i.e. the amount of ion migrating from the water phase to the organic phase at equilibrium under the influence of an ionophore, as a function of the ion concentration present in the water phase. It is possible to ascertain whether the complexing of ionophores follows a Langmuir saturation isotherm by plotting the reciprocal of complex formed against the reciprocal of the aqueous ion concentration. Ideally this would result in a straight line, the slope of which is the measure of the affinity between the ion and ionophore and the intercept on the Y-axis, indicating the stoichiometry between the ion and the ionophore in the complex. Representative data have been obtained with the synthetic ionophore illustrated in Fig. 13, a polyether which contains 6 complexing oxygen atoms [34] and shows a K^+/Na^+ preference of about 10. Its complexes resemble those of

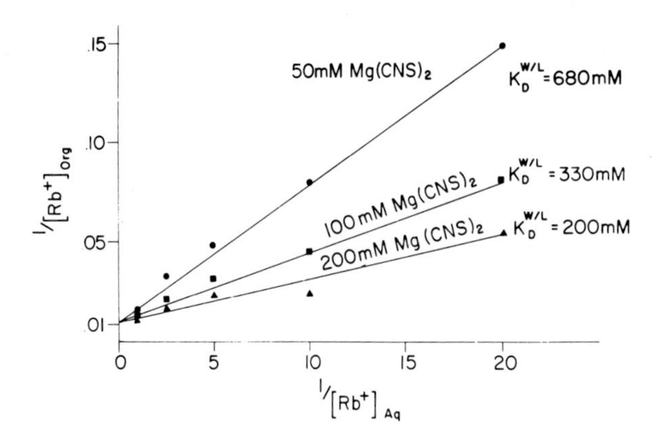

Figure 14. Saturation of polyether by Rb^+ as a function of $Mg(CNS)_2$ concentration. Reaction conditions are the same as in Figure 10 for valinomycin.

valinomycin in that they are charged. The migration of only small amounts of the complex into the organic layer would set up a strong interphase potential opposing further complex translocation unless a negative ion were available to accompany the complex. In our hands CNS^- has proved an ideal "gegenion." In Fig. 14 we see three ideal saturation curves for the Rb^+-polyether complex whose slopes differ with the $[CNS^-]$ present. All intersect at the same point, showing that, regardless of the $[CNS^-]$, at saturation a 1:1 complex forms; however, the higher the $[CNS^-]$, the greater the apparent complex affinity becomes.

Now we are coming to the heart of the matter, namely the inter-relationship and coupling of anion and cation transport. If we measure the migration of a labelled anion, e.g. $^{14}CNS^-$, into the organic phase as a function of the complexing cation in the aqueous phase, we get a very similar relationship (Fig. 15) to that exhibited in Fig. 14. Because of electrostatic interaction, anion and cation movements are so closely coupled that one can scarcely decide from the data (cf. Fig. 14) which ion is primarily complexed and which ion is being dragged into the organic solvent as a gegenion. In Fig. 16 we have carried out a similar sort of experiment with a Krebs-type substrate, malonate. Although this ion is much less effective than CNS^- as a gegenion, it demonstrates that polybasic carboxylic acid anions which have low intrinsic lipid solubility can be carried across lipid barriers by charged ionophore alkali ion complexes.

A neutral cation binding ionophore within a biological membrane could catalyze no fewer than five discrete types of transport: (1) net movement of a charged cation; (2) cation for cation exchange; (3) net movement of a charged anion; (4) anion for anion exchange; and (5) movement of an electrically balanced ion pair across the membrane. Ion translocations of the first and third

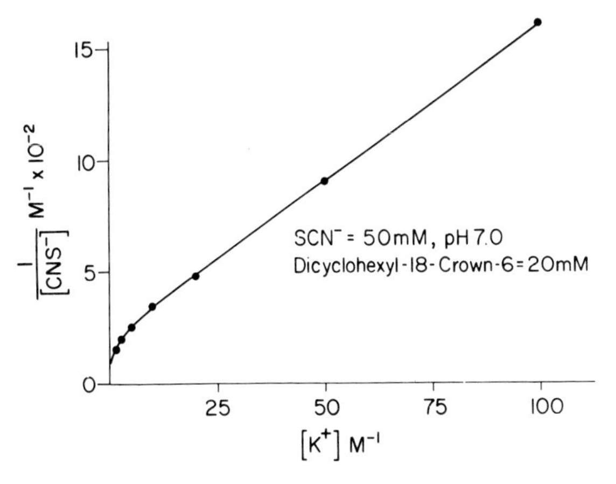

Figure 15. Association of $^{14}CNS^-$ with K^+-polyether complex. Reaction conditions were the same as in Figure 14.

class, if coupled to a source of energy so that the ions could be driven against an electrochemical gradient, would constitute electrogenic pumps, while the other types of transport would neither respond to, nor create, membrane potentials. Thus, if we succeed in isolating cation complexes from a biological pump, it would not be immediately apparent whether it normally functions to move anions, cations, or a combination of the two. In the mitochondria, although transport of the first class was recognized earliest in the cases of divalent ion transport and ionophore catalyzed monovalent ion transport, teleologically one could question what vital role such transport could play *in vivo* where the intra- and extramitochondrial $[K^+]$ are similar. It is much more likely that anion for anion exchange, i.e. the fourth type of transport, is more functional in mitochondria, playing an all important role in the metabolism of anionic substrates.

Returning to data obtained from mitochondrial systems, although the first recognized effects of ionophorous agents on mitochondria resulted from the spectacular changes in cation transport evoked, subsequent information implied that anion movement was affected as well. If mitochondrial phosphorylating systems are treated with carefully selected levels of these antibiotics, instead of uncoupling oxidative phosphorylation, K^+ can produce a large stimulation of net phosphorylation. Respiration keeps pace with the increasing phosphorylation

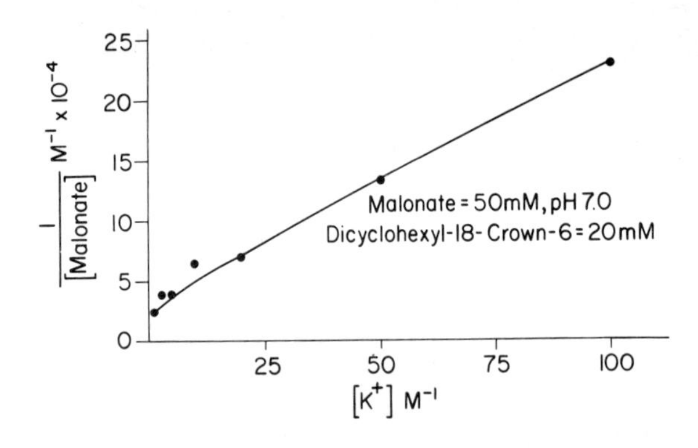

Figure 16. Association of malonate with K^+-polyether complex. Reaction conditions were the same as in Figure 14.

rates so that the P/O ratio remains constant below a certain threshold value of K^+; excessive K^+ produces true uncoupling [39]. In retrospect this data can be interpreted in terms of optimal cation transport increasing coupled anion transport, resulting in higher levels of rate limiting substrate entering the mitochondria. Subsequent experiments with labelled substrates indicated that the level of substrate in de-energized mitochondria is lower than that in normal mitochondria, and it is indeed possible with valinomycin and K^+ to increase the level of substrate within the mitochondria [40, 41]. Thus, the control of transport of cations in mitochondria can lead to an increased concentration, turnover, and metabolism of anions.

Our current view of the mitochondrial pump is depicted in Fig. 17. Mitochondria can be visualized as having an ion pump, which does not breach the membrane, in series with a barrier which must be traversed by ions in order to reach the pump. Thus, ion translocation not only depends on the intrinsic activity of the pump, but also on other factors which affect the barrier permeability. This might well constitute an important regulatory system of the mitochondria. Valinomycin would carry K^+ across the barrier in the manner suggested in the model experiments, such that the pump sees much higher $[K^+]$

than it does normally and thereupon transports it rapidly and electrogenically, i.e. against an electrochemical gradient. This is probably not a normal function of the pump, since in the cell it is difficult to visualize the biological value of this tremendous capacity for transporting K^+.

Valinomycin in effect unmasks the translocation capacity of an ionophore-containing, energy-driven anion exchange pump, transformed by the antibiotic into an electrogenic cation pump. In the absence of valinomycin the barrier is relatively impermeable to K^+ [42] and the pump drives the rapid turnover of metabolite anions. The pump probably contains an ionophore-K^+

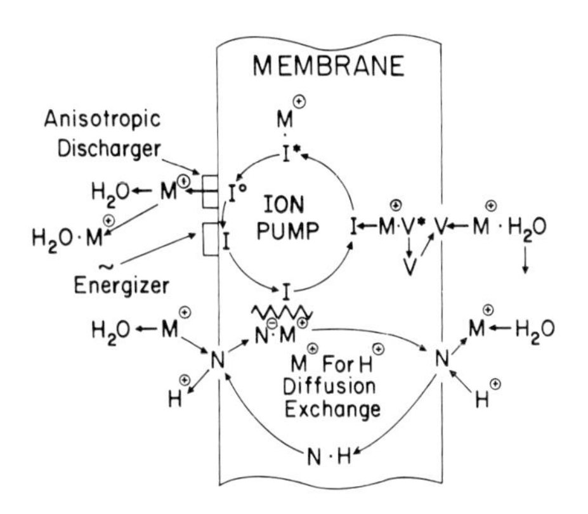

Figure 17. Hypothetical relationship of ion pump and barrier to mitochondrial transport. The following symbols are used: V, valinomycin; I, naturally occurring ionophore in pump; *, complexed conformation ionophore; N, nigericin; M^+, complexed cation. This scheme is described in further detail in Reference 4.

complex as the anion carrier. Normally the K^+ plays the role of a cofactor, the same cation remaining on the pump during the turnover of many anions. This would reconcile the normally low K^+ permeability of the mitochondrial membrane [42] and its intrinsically high permeability for substrate ions [40, 41].

This leads to the conclusion that anion and cation movements in the mitochondria are linked in a loosely coupled fashion. The same pump can carry out both the energy-linked electrogenic movement of cations and an energy-linked anion exchange, depending on what reaches the pump through a controlling barrier. Finally it should be pointed out that, if both anions and cations are subject to energy-driven transport, then it is invalid to assume that any given ionic species is in sole equilibrium with the membrane potential. It would therefore be impossible to calculate the transmembrane mitochondrial potential from the gradients of any particular ion without being certain it alone is in passive equilibrium with the potential (cf. Reference 43).

ACKNOWLEDGEMENTS

I wish to acknowledge the technical assistance of Genevieve Jones for the data of Figs. 10 and 14-16, and the collaboration of D. Haynes and A. Kowalsky in obtaining the NMR spectra. I am also grateful for the cooperation of the firms of Eli Lilly, CIBA, Hoffmann-La Roche, and du Pont for supplying the various ionophores employed. This work was supported by U.S. Public Health Service grant GM-12202 and carried out during the tenure of Career Development Award K3-GM-3626 from the same granting agency.

REFERENCES

1. Pressman, B. C., *in* "Symposium on Ion Transport and Intramitochondrial pH", New York, 1968 (edited by S. Addenki and J. F. Sotos), Annals of the New York Academy of Sciences, in press.
2. Pressman, B. C., *Fedn Proc. Fedn. Am. Socs exp. Biol.* **27** (1968) 1283.
3. Pressman, B. C., *in* "The Energy Level and Metabolic Control in Mitochondria" (edited by J. M. Tager, S. Papa, E. Quagliariello and E. C. Slater), Adriatica Editrice, Bari, p. 87.
4. Pressman, B. C., Harris, E. J., Jagger, W. S. and Johnson, J. H., *Proc. natn. Acad. Sci. U.S.A.* **58** (1967) 1949.
5. Hollunger, G., *Acta pharmac. tox.* **II** (1955) Suppl. 1.
6. Pressman, B. C., *Fedn Proc. Fedn Am. Socs exp. Biol.* **21** (1962) 55.
7. Pressman, B. C., *Fedn Proc. Fedn Am. Socs exp. Biol.* **22** (1963) 404.
8. Pressman, B. C., *J. biol. Chem.* **238** (1963) 401.
9. Pressman, B. C., *in* "Energy-linked Functions of Mitochondria" (edited by B. Chance), Academic Press, New York, 1963, p. 181.
10. Mitchell, P., "Chemiosmotic Coupling in Oxidative and Photosynthetic Phosphorylation", Glynn Research Ltd., Bodmin, Cornwall (1966).
11. Mcmurray, W. and Begg, R. W., *Archs Biochem. Biophys.* **84** (1959) 546.
12. Shemyakin, M. M., Vinogradova, E. I., Feigina, M. Yu. and Aldanova, N. A., *Tetrahedron Lett.* **(1963)** 1921.
13. Lardy, H. A. and Wellman, H., *J. biol. Chem* **201** (1953) 357.
14. Nishimura, M., *Biochim. biophys. Acta* **59** (1962) 183.
15. Saris, N.-E. L., Societas Scientiarum Fennica, Commentationes Physico-Mathematicae, *XXVIII, 11* (1963).
16. Bartley, W. and Amoore, J. E., *Biochem. J.* **69** (1958) 348.
17. Chance, B. and Estabrook, R. W., Report on Symposium on Mitochondrial Structure and Function, *Science, N.Y.* **146** (1964) 957.
18. Chappell, J. B. and Crofts, A. R., *Biochem. J.* **95** (1965) 393.
19. Pressman, B. C. and Lardy, H. A., *J. biol. Chem.* **197** (1952) 547.
20. Pressman, B. C. and Lardy, H. A., *Biochim. biophys. Acta* **18** (1955) 482.
21. Moore, C. and Pressman, B. C., *Biochem. biophys. Res. Commun.* **15** (1964) 562.
22. Eisenman, G., *in* "Glass Electrodes for Hydrogen and Other Cations (edited by G. Eisenman), Dekker, New York, 1967, p. 268.
23. Pressman, B. C., *Meth. Enzym.* **10** (1967) 714.
24. Pressman, B. C., *Proc. natn. Acad. Sci. U.S.A.* **53** (1965) 1076.

25. Cockrell, R. S., Harris, E. J. and Pressman, B. C., *Biochemistry* **5** (1966) 2326.
26. Mitchell, P., this volume, p. 219.
27. Cockrell, R. S., Harris, E. J. and Pressman, B. C., *Nature, Lond.* **215** (1967) 1487.
28. Graven, S. N., Estrada-O, S. and Lardy, H. A., *Proc. natn. Acad. Sci. U.S.A.* **53** (1965) 1076.
29. Agtarap, A., Chamberlin, J. W., Pinkerton, M. and Steinrauf, I., *J. Am. chem. Soc.* **89** (1967) 5737.
30. Harned, R. L., Harter, P. H., Corum, C. J. and Jones, K. L., *Antibiotics Chemother.* **1** (1951) 592.
31. Pioda, L. A. R., Wacter, H. A., Dohner, R. E. and Simon, W. *Helv. Chim. Acta* **50** (1967) 1373.
32. Kilbourn, B. T., Dunitz, J. D., Pioda, L. A. R. and Simon, W., *J. molec. Biol.* **30** (1967) 559.
33. Shemyakin, M. M., Ovchinnikov, Yu. A., Ivanov, V. T., Antonov, V. K., Shkrob, A. M., Mikhaleva, I. I., Evstratov, A. V. and Malenkov, G. G., *Biochem. biophys. Res. Commun.* **29** (1967) 834.
34. Pedersen, C. J., *J. Am. chem. Soc.* **89** (1967) 7017.
35. Meyer, C. E. and Reusser, F., *Experientia* **23** (1967) 85.
36. Sarges, R. and Witkop, B., *J. Am. chem. Soc.* **87** (1965) 2011.
37. Mueller, P. and Rudin, D. O., *Biochem. biophys. Res. Commun.* **26** (1967) 398.
38. Pressman, B. C. and Harris, E. J., Abstracts of Seventh International Congress of Biochemistry, Tokyo, August 1967, Vol. V, p. 900 (1967).
39. Höfer, M. P. and Pressman, B. C., *Biochemistry* **5** (1966) 3919.
40. Harris, E. J., van Dam, K. and Pressman, B. C., *Nature, Lond.* **213** (1967) 1126.
41. Harris, E. J., van Dam, K. and Pressman, B. C., *Fedn Proc. Fedn Am. Socs exp. Biol.* **26** (1967) 610.
42. Harris, E. J., Catlin, G. and Pressman, B. C., *Biochemistry, N. Y.* **6** (1967) 1360.
43. Mitchell, P., Moyle, J. and Smith, L., *Eur. J. Biochem.* **4** (1968) 9.

FEBS Symposium, Volume 17, 1969, pp. 335-346

Substrate Transport in Mitochondria and Control of Metabolism

E. QUAGLIARIELLO and S. PAPA

Department of Biochemistry, University of Bari, Bari, Italy and

A. J. MEIJER and J. M. TAGER

Laboratory of Biochemistry, B.C.P. Jansen Institute, University of Amsterdam, Amsterdam, The Netherlands

The penetration of mitochondria by substrates has recently been the subject of intensive investigation (see refs. 1, 2 for reviews). Chappell and co-workers [1, 3-7] have obtained evidence for the existence in rat liver mitochondria of a specific carrier for the transport of dicarboxylic acids and of another for that of tricarboxylic acids. The dicarboxylic acid carrier is activated by P_i (ref. 5) and the tricarboxylic acid carrier requires P_i and L-malate [3-7]. De Haan and Tager [8] have presented evidence that a third carrier exists, specific for α-oxoglutarate and differing from the tricarboxylic acid carrier in that it is activated not only by L-malate, but also by malonate. Additional support for the existence of a separate α-oxoglutarate carrier has been presented by Chappell and co-workers [1, 6, 7].

This paper concerns the transport of tricarboxylic acids and of α-oxoglutarate in rat liver mitochondria, attention being focused on the mechanism by which this is activated by dicarboxylic acids.

Robinson and Chappell [7] have proposed that malate must first enter the mitochondria before performing its stimulatory effect. It has also been found by Chappell [3, 9] that to demonstrate a malate requirement for the oxidation of tricarboxylic acids, mitochondria have to be preincubated with ADP plus P_i or an uncoupler. Ferguson and Williams [10], and Meijer and Tager [11] have in fact shown that freshly isolated rat liver mitochondria may contain sufficient malate to allow the aconitate hydratase reaction to proceed at a maximal rate.

To throw light on the mechanism by which intramitochondrial malate activates the entry of tricarboxylic acids into the mitochondria we have studied the effect of isocitrate on the efflux from mitochondria of endogenous intramitochondrial malate (Table 1). The effect of P_i and butylmalonate was also tested. Robinson and Chappell [7] have introduced the use of the latter compound and have proposed that it inhibits the transport of malate into mitochondria. This table shows that the addition of P_i promoted efflux of intramitochondrial malate and that this effect was counteracted by butylmalonate. Isocitrate also caused efflux of intramitochondrial malate. However, the isocitrate-induced efflux was butylmalonate insensitive. This experiment indicates that there are two ways of getting malate out of the mitochondria: in exchange with P_i on the dicarboxylic acid carrier, a process which is inhibited by

Table 1. Effect of P_i, butylmalonate and isocitrate on the efflux of intramitochondrial malate from rat liver mitochondria

Additions	Malate found (nmoles)		
	Intramit.	Extramit.	Total
None	22	36	58
P_i	5	53	58
Butylmalonate	21	30	51
Butylmalonate + P_i	21	29	50
Isocitrate	10	88	98
Isocitrate + P_i	5	81	86
Isocitrate + butylmalonate	5	51	56
Isocitrate + butylmalonate + P_i	6	54	60

Mitochondria (8·1 mg protein) were incubated at 22°C with 15 mM KCl, 5 mM $MgCl_2$, 2 mM EDTA, 50 mM Tris (pH 7·5), 1·6 μg rotenone, 2% ethanol, 47 mM sucrose, and (where present) 20 mM P_i (pH 7·5), 3 mM isocitrate and 5 mM butylmalonate. Final volume, 0·8 ml. After 1 min, the mitochondria were separated from the reaction mixture by the centrifugation-filtration technique [12] as described by Harris and van Dam [13]. Malate was determined with NAD^+, malate dehydrogenase, acetyl-CoA and citrate synthase as described by Davis [14], and based on the method of Ochoa *et al.* [15] for assaying citrate synthase.

butylmalonate, and in exchange for isocitrate. The insensitivity of the latter process to butylmalonate indicates that it is mediated by the tricarboxylic acid carrier. Further support for this is given below.

Table 2 shows an experiment on the effect of butylmalonate and citrate on the uptake of [^{14}C]malate by rat liver mitochondria in the presence of rotenone. Butylmalonate alone did not inhibit the accumulation of malate; citrate, on the other hand, did. However, when both were added together, there was a marked inhibition. The fact that butylmalonate alone does not inhibit malate uptake suggests that malate can penetrate the mitochondria in exchange for intramitochondrial citrate, always present in substantial amounts in freshly-isolated mitochondria [16-18]. In fact, we will see below that when mitochondria are depleted of substrates, butylmalonate inhibits the uptake of dicarboxylic acids. The entry of malate in exchange for intramitochondrial citrate is, however, inhibited by the addition of citrate outside the mitochondria.

Table 2. Effect of butylmalonate and citrate on the accumulation of $[^{14}C]$malate by freshly prepared rat liver mitochondria

Additions	$(\frac{^{14}C}{^{3}H})_{mit} / (\frac{^{14}C}{^{3}H})_{sup}$
None	3·1
Butylmalonate (10 mM)	4·8
Citrate (4 mM)	2·0
Butylmalonate + citrate	1·1

Mitochondria (7·6 mg protein) were incubated at 22°C with 15 mM KCl, 5 mM $MgCl_2$, 2 mM EDTA, 50 mM tris (pH 7·5), 5 mM P_i (pH 7·5), ^3H_2O, 0.4 mM $[^{14}C]$malate, 1·2 μg rotenone, 2% ethanol and 37 mM sucrose. Final volume, 1 ml. After 1 min the mitochondria were separated from the reaction mixture as described in Table 1.

When both butylmalonate and citrate are added, entry of malate on both the dicarboxylic acid- and the tricarboxylic acid-carrier is inhibited (Fig. 1). That added malate promotes the efflux of intramitochondrial citrate is directly shown by the experiment in Table 3. Mitochondria were incubated with $[^{14}C]$citrate in the presence of rotenone and antimycin. Then part of the mitochondria were transferred by layer-centrifugation [19] into $HClO_4$ for instantaneous determination of the $[^{14}C]$citrate accumulated inside the mitochondria; simultaneously, another part was transferred to a second layer free of citrate and the incubation continued. Citrate accumulated inside the mitochondria during the first phase and went out in the second. When malate was present during this second incubation, the efflux of citrate was greatly enhanced. In another experiment it was observed that the efflux of intramitochondrial citrate induced by malate was insensitive to butylmalonate.

In conclusion, the data presented provide direct evidence that the tricarboxylic acid carrier mediates an exchange-diffusion between a

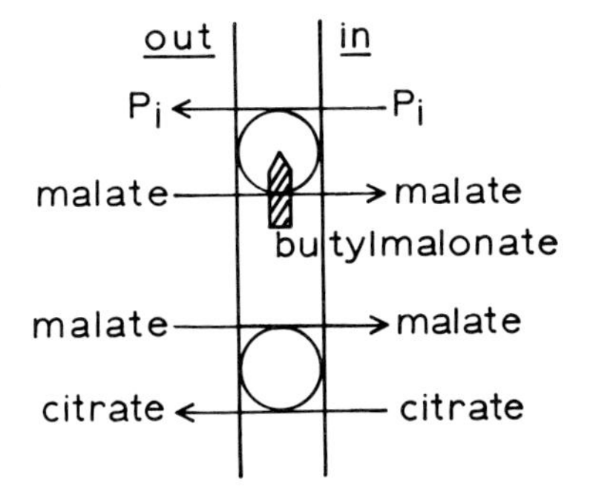

Figure 1. Scheme.

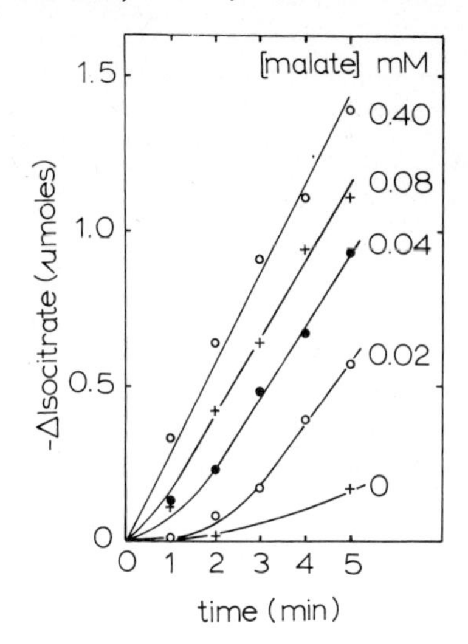

Figure 2. Effect of malate concentration on the aconitate hydratase reaction in preincubated rat liver mitochondria. Mitochondria (4 mg protein) were preincubated for 3 min at 25°C with 15 mM KCl, 5 mM $MgCl_2$, 2 mM EDTA, 50 mM Tris-HCl, 2 mM ADP, 20 mM potassium phosphate and 1 mM arsenite. Then 5 mM *threo* Ds (+) isocitrate was added. Final volume, 1 ml. Final pH, 7·5. The reaction was stopped with $HClO_4$ and isocitrate determined with isocitrate dehydrogenase on the neutralized extracts.

tricarboxylate anion and malate. As part of the collaborative programme between our laboratories, confirmatory evidence for this view has been obtained by Palmieri and Quagliariello [20].

There is another aspect of the transport of tricarboxylic acids which is illustrated by the experiment of Fig. 2. Here the effect of exogenous malate on the utilization of isocitrate in the aconitate hydratase reaction is shown. Mitochondria were preincubated with ADP, P_i and arsenite in order to deplete

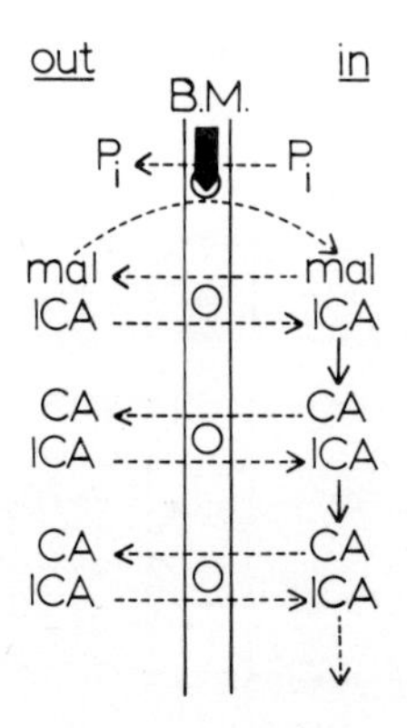

Figure 3. Scheme.

Table 3. Effect of L-malate on the efflux of [^{14}C]citrate accumulated inside rat liver mitochondria

Number of layers	Additions to		Intramitochondrial [^{14}C]citrate (nmoles)
	Layer I	Layer II	
1	[^{14}C]Citrate (1 mM)	–	150
2	[^{14}C]Citrate	None	70
	[^{14}C]Citrate	L-Malate (3 mM)	10

Mitochondria (18·2 mg protein) were incubated for 3 min at 29°C with 175 mM sucrose, 20 mM Tris-HCl (pH 7·5), 10 mM KCl, 3 mM $MgCl_2$, 3·5 μg rotenone, 1·75 μg antimycin and 1 mM [^{14}C]citrate. Final volume, 3·5 ml. Final pH, 7·5. After this incubation a part of the mitochondria were transferred by layer-centrifugation [19] to $HClO_4$ for instantaneous determination of the intramitochondrial content of [^{14}C]citrate, and another part was transferred by the same technique to a second incubation mixture containing all the components of the first except citrate and with dextran present to adjust the density of the mixture.

them of endogenous malate. After this the aconitate hydratase reaction was started by adding rotenone, isocitrate and different amounts of malate. There was a lag in the aconitate hydratase reaction and the length of this lag was an inverse function of the concentration of added malate. However, the final rate tended to approach that found with the highest concentration of malate. It is therefore clear that only the initial velocity is dependent on the malate concentration. A possible explanation of this is given in Fig. 3. In this scheme, presented by Meijer [21] in Polignano, malate is taken up by the mitochondria on the dicarboxylic acid carrier. Isocitrate is now able to enter the mitochondria in exchange for intramitochondrial malate. Once inside, isocitrate is converted to citrate and citrate leaves the mitochondria in exchange for a second isocitrate molecule. This process is repeated. According to this mechanism the intramitochondrial malate serves only as an initiator of an isocitrate-citrate exchange-diffusion. Furthermore, P_i activates only the entry of malate [22].

Let us now consider the transport of α-oxoglutarate. The experiment presented in Table 4 clarifies the mechanism by which dicarboxylic acids activate the uptake of α-oxoglutarate by mitochondria. Mitochondria were preincubated with ADP, glucose, hexokinase, P_i and arsenite. This converted endogenous tricarboxylic acid cycle intermediates into α-oxoglutarate. Rotenone and antimycin were then added. The two top lines show that [^{14}C]malate was actively taken up by mitochondria. Two minutes after malate addition, 77 nmoles were accumulated inside the mitochondria, and in the subsequent 2 min there was no further change in the intramitochondrial content of malate. It should be noted that after the incubation with malate no α-oxoglutarate was found inside the mitochondria. The third line shows that 35 nmoles α-oxoglutarate penetrated in 2 min on addition of the oxo-acid in the absence of malate. The last line shows that when α-oxoglutarate was added to mitochondria which had been allowed to accumulate malate, 76 nmoles α-oxoglutarate were taken up. Thus, there is an extra uptake of 41 nmoles α-oxoglutarate with respect to that observed in the absence of malate. If we now compare the intramitochondrial content of [^{14}C]malate in lines 2 and 4 we see that the extra uptake of α-oxoglutarate was coupled with the efflux of an equal amount of [^{14}C]malate. In other words, we have a 1:1 exchange-diffusion of extramitochondrial α-oxoglutarate with intramitochondrial malate. This experiment has therefore, clarified the sequence of reactions by which added malate activates the uptake of α-oxoglutarate: malate first enters the mitochondria, then it pulls α-oxoglutarate in by exchange-diffusion.

An exchange-diffusion between α-oxoglutarate and dicarboxylic acids can also occur in the reverse direction, viz., between intramitochondrial α-oxoglutarate and an extramitochondrial dicarboxylic acid (malate or malonate). This is shown in the experiment of Table 5. Mitochondria were preincubated with ADP, glucose, hexokinase and arsenite. In the first two lines glutamate was added and its oxidation generated α-oxoglutarate inside the

Table 4. Stoichiometric exchange between intramitochondrial malate and extramitochondrial α-oxoglutarate in preincubated mitochondria

First addition	Incubation time (min)	Addition	Incubation time (min)	Intramitochondrial content (nmoles) of	
				[^{14}C]-L-Malate	α-Oxoglutarate
[^{14}C]Malate	2	None	–	76·7	1·8
[^{14}C]Malate	4	None	–	77·4	0
None	2	α-Oxoglutarate	2	–	35·5
[^{14}C]Malate	2	α-Oxoglutarate	2	39·4	76·2
		ΔMalate		−38·0	
		Δα-Oxoglutarate			+40·7

Rat liver mitochondria (12·9 mg protein) were preincubated at 25°C with 125 mM KCl, 3 mM $MgCl_2$, 20 mM Tris-HCl (pH 7·5), 0·5 mM ADP, 0·5 mM P_i, 20 mM glucose, hexokinase, and 1 mM arsenite. Final volume, 2·5 ml. Final pH, 7·5. After 2 min, 2·5 μg rotenone and 1·25 μg antimycin were added, followed 10 s later by the additions given in the table. The incubation was then continued as specified in the table. Where indicated, 1 mM [^{14}C] L-malate and 3 mM α-oxoglutarate were added. The mitochondria were separated from the suspending medium as described by Pfaff [19], and α-oxoglutarate was determined as described earlier [24].

Table 5. Stoichiometric exchange between intramitochondrially-generated α-oxoglutarate and extramitochondrial malonate

Addition during		Intramitochondrial	
Preincubation	Incubation	α-Oxoglutarate	[^{14}C]Malonate
		(mM)	
Glutamate	None	1·68	–
Glutamate	[^{14}C]Malonate	0·53	3·73
None	None	0·20	–
None	[^{14}C]Malonate	0·17	2·42
Δα-Oxoglutarate		−1·15	
Δ[^{14}C]Malonate			+1·31

Rat liver mitochondria (16·0 mg protein) were preincubated for 2 min with 125 mM KCl, 20 mM Tris-HCl (pH 7·5), 3 mM $MgCl_2$, 1 mM ADP, 1 mM arsenite, 20 mM glucose and hexokinase. Final volume, 4 ml. Where indicated, 10 mM glutamate was added. After 3 min (± glutamate) 4 μg rotenone, 2 μg antimycin and 20 μg oligomycin were added, and (where indicated) 1·5 mM [^{14}C]malonate. The incubation was continued for 2 min and the mitochondria were separated by the suspending medium as described by Pfaff [19].

Table 6. Insensitivity of the α-oxoglutarate–malonate exchange to butylmalonate

Addition during		Intramitochondrial	
Preincubation	Incubation	α-Oxoglutarate	[^{14}C]Malonate
		(mM)	
Glutamate	None	0·66	–
Glutamate	[^{14}C]Malonate	0·27	1·69
Glutamate	[^{14}C]Malonate + butylmalonate	0·31	1·30
None	[^{14}C]Malonate	0·08	1·43
None	[^{14}C]Malonate + butylmalonate	0·00	0·85
− Butylmalonate	Δα-Oxoglutarate	−0·39	
	Δ[^{14}C]Malonate		+0·26
+ Butylmalonate	Δα-Oxoglutarate	−0·35	
	Δ[^{14}C]Malonate		+0·45

The experimental conditions are those described in the legend to Table 5, except that 20 mg mitochondrial protein were used. Concentration of the additions: 1·5 mM [^{14}C]malonate and 10 mM butylmalonate.

mitochondria [8, 23, 24]. Glutamate oxidation was then arrested with rotenone plus antimycin. Line 1 shows that on continuing the incubation, much α-oxoglutarate was retained inside the mitochondria. The addition of malonate (line 2) promoted efflux of α-oxoglutarate. Concomitantly malonate was actively taken up. When glutamate was absent during the preincubation, very little α-oxoglutarate was found inside the mitochondria, and less malonate was taken up. It can be seen that the amount of α-oxoglutarate driven out of the mitochondria by malonate was equal to the extra amount of malonate taken up by α-oxoglutarate-charged mitochondria. Thus, extramitochondrial malonate promoted the efflux of intramitochondrial α-oxoglutarate by entering the mitochondria in exchange for α-oxoglutarate. Again, the stoichiometry of the exchange-diffusion reaction was 1:1.

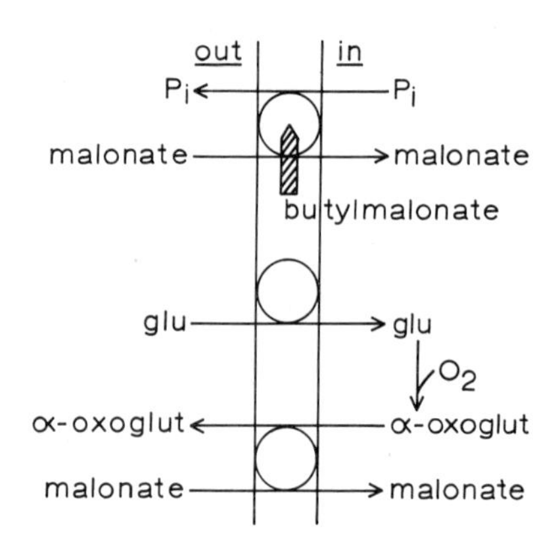

Figure 4. Scheme.

Table 6 shows the effect of butylmalonate. It inhibited the uptake of malonate both in mitochondria charged with α-oxoglutarate and in uncharged mitochondria. The extent of inhibition, however, was smaller in the first case. On the other hand, butylmalonate had practically no effect on the malonate-induced efflux of α-oxoglutarate. The lower part of the table shows that practically the same amount of intramitochondrial α-oxoglutarate was exchanged with extramitochondrial malonate, both in the absence and presence of butylmalonate. Thus, the latter had no effect on the exchange-diffusion between α-oxoglutarate and malonate.

It can be concluded (see Fig. 4) that when α-oxoglutarate is generated inside the mitochondria by means of glutamate oxidation, malonate enters on two carriers: the dicarboxylic acid carrier, which is inhibited by butylmalonate, and the α-oxoglutarate carrier in exchange-diffusion with intramitochondrial α-oxoglutarate, a process which is insensitive to butylmalonate.

The data presented are consistent with the observation of Robinson and Chappell [7] that butylmalonate prevents added malate from activating the entry of α-oxoglutarate into the mitochondria. Indeed it has been shown above that in this case malate has first to enter the mitochondria on its own carrier and then activates α-oxoglutarate entry by exchange-diffusion.

An experiment on the effect of malonate and butylmalonate on the Krebs-Cohen dismutation illustrates this (Table 7). It can be seen that malonate stimulated the Krebs-Cohen dismutation and butylmalonate largely prevented this effect. Furthermore, butylmalonate alone suppressed almost completely the dismutative synthesis of glutamate. This suggests that the entry of α-oxoglutarate into the mitochondria is tightly coupled to a counterflux of intramitochondrial dicarboxylic acids (see also ref. 25). When the intramitochondrial generation of these is blocked, as in this case, the entry of α-oxoglutarate on its carrier is dependent upon the continuous action of the dicarboxylic acid carrier. This will bring back into the mitochondria endogenous dicarboxylic acids which leave in exchange for α-oxoglutarate, so that a little amount of dicarboxylic acids can act catalytically.

Table 7. Effect of malonate and butylmalonate on the Krebs-Cohen dismutation in rat liver mitochondria

Additions	Δα-Oxoglutarate (μmoles) in	
	Expt. 1	Expt. 2
None	3·55	3·28
Butylmalonate (5 mM)	0·27	0·02
Malonate (3 mM)	4·68	5·14
Butylmalonate + malonate	1·90	2·73

Mitochondria (5·5 mg protein in Expt. 1 and 5·2 mg protein in Expt. 2) were incubated for 10 min at 25°C in 50 mM KCl, 50 mM Tris-HCl (pH 7·5), 5 mM $MgCl_2$, 2 mM EDTA, 10 mM P_i, 20 mM NH_4Cl, 10 mM α-oxoglutarate, 1 μg rotenone, and 0·5 μg antimycin. Final volume, 1 ml. Final pH, 7·5. The reaction was stopped with $HClO_4$ and α-oxoglutarate was determined on the neutralized extract as described in the legend to Table 4.

The data presented in this paper show the occurrence of exchange-diffusion reactions of tricarboxylic acids or α-oxoglutarate with dicarboxylic acids through the mitochondrial membranes of rat liver. The insensitivity of these antiports [26] to butylmalonate differentiate them from the transport of dicarboxylic acids on the dicarboxylic acid carrier. The specificity they exhibit towards the dicarboxylic acid structure indicates the exchange-diffusions to be mediated by two different carriers, one for tricarboxylic acids and another for α-oxoglutarate.

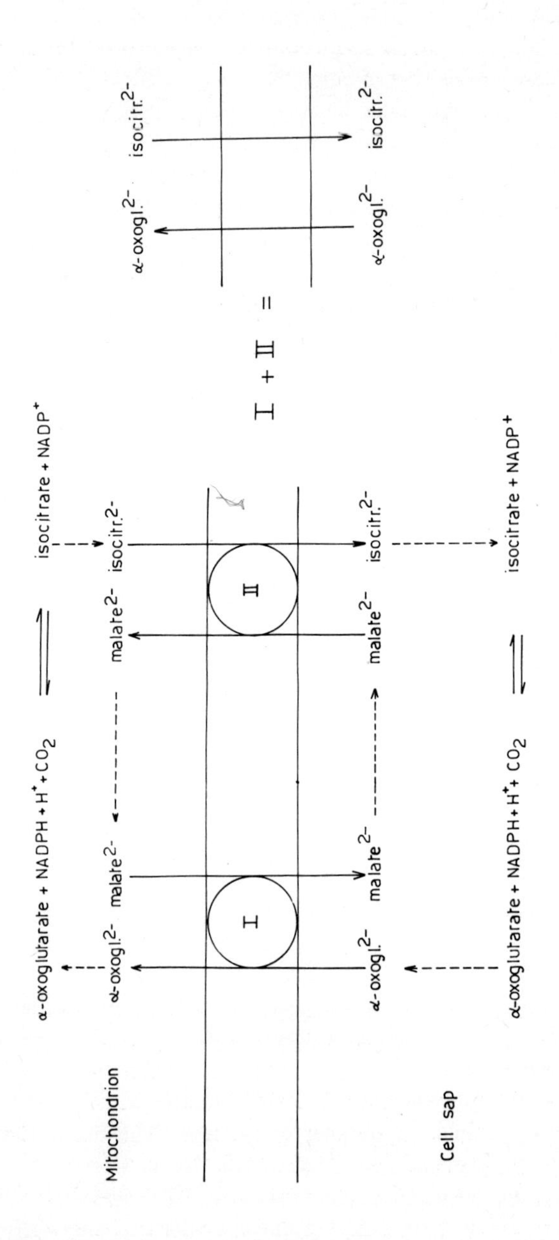

Figure 5. Scheme.

The study of the transport of α-oxoglutarate and tricarboxylic acids in mitochondria is still in a preliminary stage. Further study of the exchange-diffusion reactions described might prove useful in the understanding of the basic mechanisms of the catalysis of solute translocation through natural membranes.

Examples have been presented on how transport of dicarboxylic acids by the dicarboxylic acid carrier may act in conjunction with exchange-diffusion reactions on the tricarboxylic acid or α-oxoglutarate carrier to control utilization of isocitrate in the aconitate hydratase reaction and of α-oxoglutarate in the Krebs-Cohen dismutation.

Chappell [1] has already presented schemes on the way in which the carriers for tricarboxylic acid cycle intermediates may take part in the transfer of reducing equivalents between the cell sap and mitochondria in rat liver. We conclude by showing a scheme (Fig. 5) on how a direct coupling between antiport reactions on the tricarboxylic acid and the α-oxoglutarate carrier can mediate the transport of reducing equivalents between NADP in the mitochondrial matrix and that in the cell sap. The mitochondrial NADP-linked isocitrate dehydrogenase can make use of mitochondrial NADPH to synthesize isocitrate. Isocitrate goes out of the mitochondria in exchange-diffusion with malate. In the cell sap isocitrate can be oxidized by the soluble NADP-linked isocitrate dehydrogenase. NADPH is now available for extramitochondrial reductive synthesis. α-Oxoglutarate goes into the mitochondria in exchange-diffusion with malate. According to this mechanism, malate will catalyse a net 1:1 exchange of isocitrate against α-oxoglutarate. This, in combination with the two NADP-linked isocitrate dehydrogenases, will function as a hydrogen transport shuttle system between the two compartments [27]. The rate of the hydrogen transfer will be dependent on the level of malate in the cell, and its direction, on the energy level of the two compartments.

REFERENCES

1. Chappell, J. B., *Br. med. Bull.* **24** (1968) 150.
2. Slater, E. C., Quagliariello, E., Papa, S. and Tager, J. M., *in* "The Energy Level and Metabolic Control in Mitochondria" (edited by S. Papa, J. M. Tager, E. Quagliariello and E. C. Slater), Adriatica Editrice, Bari, 1969, p. 1.
3. Chappell, J. B., *Abstr. 6th Int. Congr. Biochem.* New York, 1964, Vol. 8, p. 625.
4. Chappell, J. B., *Biochem. J.* **100** (1966) 43P.
5. Chappell, J. B. and Haarhoff, K. N., *in* "Biochemistry of Mitochondria" (edited by E. C. Slater, Z. Kaniuga and L. Wojtczak), Academic Press and Polish Scientific Publishers, London and Warsaw, 1967, p. 75.

6. Chappell, J. B., Henderson, P. J. F., McGivan, J. D. and Robinson, B. H., *in* "The Interaction of Drugs and Subcellular Components in Animal Cells" (edited by P. N. Campbell), Churchill, London, 1968, p. 71.
7. Robinson, B. H. and Chappell, J. B., *Biochem. biophys. Res. Commun.* **28** (1967) 249.
8. de Haan, E. J. and Tager, J. M., "Abstracts, 3rd Meeting of the Federation of European Biochemical Societies", Warsaw, 1966, Academic Press and Polish Scientific Publishers, London and Warsaw, 1966, p. 159.
9. Chappell, J. B., *in* "Biological Structure and Function," Vol. 2, (edited by T. W. Goodwin and O. Lindberg), Academic Press, New York, 1961, p. 71.
10. Ferguson, S. M. F. and Williams, G. R., *J. biol. Chem.* **241** (1966) 3696.
11. Meijer, A. J. and Tager, J. M., in preparation.
12. Werkheiser, W. C. and Bartley, W., *Biochem. J.* **66** (1957) 79.
13. Harris, E. J. and van Dam, K., *Biochem. J.* **106** (1968) 759.
14. Davis, E. J., personal communication.
15. Ochoa, S., Stern, I. R. and Schneider, W. C., *J. biol. Chem.* **193** (1951) 691.
16. Peters, R. A., *Proc. R. Soc.* Ser. B, **139** (1952) 143.
17. Schneider, W. C., Striebich, M. J. and Hogeboom, G. H., *J. biol. Chem.* **222** (1956) 969.
18. Bellamy, D., *Biochem. J.* **82** (1962) 218.
19. Pfaff, E., Ph.D. Thesis, Marburg, 1965.
20. Palmieri, F. and Quagliariello, E., *Abstr. 5th Meeting Fed. Eur. Biochem. Soc.* Prague, 1968, p. 133.
21. Meijer, A. J., Tager, J. M. and van Dam, K., *in* "The Energy Level and Metabolic Control in Mitochondria" (edited by S. Papa, J. M. Tager, E. Quagliariello and E. C. Slater), Adriatica Editrice, Bari, 1969, p. 147.
22. Palmieri, F. and Quagliariello, E., *in* "The Energy Level and Metabolic Control in Mitochondria" (edited by S. Papa, J. M. Tager, E. Quagliariello and E. C. Slater), Adriatica Editrice, Bari, 1969, p. 172.
23. Papa, S., Tager, J. M., Francavilla, A., de Haan E. J. and Quagliariello, E., *Biochim. biophys. Acta* **131** (1967) 14.
24. de Haan, E. J. and Tager, J. M., *Biochim. biophys. Acta* **143** (1968) 98.
25. Papa, S., d'Aloya, R., Meijer, A. J. and Quagliariello, E., *in* "The Energy Level and Metabolic Control in Mitochondria" (edited by S. Papa, J. M. Tager, E. Quagliariello and E. C. Slater), Adriatica Editrice, Bari, 1969, p. 159.
26. Mitchell, P., *Adv. Enzymol.* **29** (1967) 33.
27. Lowenstein, J. M., *in* "The Control of Lipid Metabolism (edited by J. K. Grant), Academic Press, London, 1963, p. 57.

FEBS Symposium, Volume 17, 1969, pp. 347-352

Mitochondrial Anion Uptakes in Relation to Metabolism

E. J. HARRIS

Department of Biophysics, University College London, London, England

The maximum rates of respiration and phosphorylation obtainable from a suspension of rat liver mitochondria using not more than 3 mM substrate can, with several substrates, be increased by inducing an uptake of K [5, 9]. The cation gain was obtained by addition of one of the ionophorous antibiotics (e.g. valinomycin, dinactin) and was accompanied by a net gain of substrate [7]. A similar beneficial effect on maximum metabolic rate can be obtained with the same substrates by substituting a saline medium for the sucrose medium, and is presumably also attributable to improved penetration of the mitochondrial membrane. It is notable that the rate of oxidation of hydroxybutyrate is insensitive to the changes in medium, suggesting that its metabolism is not limited in the same way (Table 1).

These results led to further study of the metabolic control which could be exerted by a combination of the permeability of mitochondrial compartments and by the tendency for substrates to be accumulated. The following factors have to be taken into account:

(a) there is a requirement for residual, stored energy if accumulation is to occur to its full extent. It is uncertain if mitochondria in the cytoplasm can ever become so depleted that this requirement is exposed.

(b) there are interactions between the anions, both competitive and facilitatory. With normal levels of metabolite present only the competitions are likely to be important.

(c) the rate of permeation through the mitochondrial membrane in relation to the rate of substrate consumption will affect the apparent "K_m" of the system.

The uptake of a substrate by untreated mitochondria proceeds even in the presence of a respiratory inhibitor. The quantities found when the applied concentration is mM depend on the charge and not on the nature of the anion [4]. This suggests some electrical interaction, either with a positive interior or with positive sites. When the mitochondria are de-energized, either by exposure to oligomycin in the absence of added oxidizible substrate, or to uncoupler, the

Table 1. Stimulation of oxidation of substrates other than hydroxybutyrate by valinomycin in a sucrose-KCl medium and by saline media

Substrate at 10 mM	Maximum rate of DNP-uncoupled respiration μatom oxygen per g protein per min		
	In 0·25M sucrose with or without 10 mM KCl	In 0·25M sucrose, 10 mM KCl and valinomycin 1·6 μg/g	In 120 mM KCl or NaCl
Oxoglutarate	13	61	42
Glutamate	33	103	100
Glutamate + malate	120	135	160
Succinate (rotenone)	59	190	190
Hydroxybutyrate	32	28	32

All media included 20 mM tris chloride pH 7·4.

ability to take up anions falls [3, 4]. Minimum values are likely to result from adsorption because 1-2 μmole anion/g has been found to associate with sonicated particles. The failure to accumulate is presumably related to low or negligible rates of metabolism [4, 7]; Haslam and Krebs [8] have described an energy requirement for malate or oxalacetate metabolism which could have the same origin.

Uptake of a substrate is presumably accompanied by displacement of other anions (compare Gamble [2]), but in some examples more citrate is taken up than the measured displacement of malate and phosphate. That there is some unidentified anion in mitochondria is also implied in the work of Chappell [1].

The quantities of substrate taken up can be related to the applied concentration by adsorption isotherms like the well-known Michaelis enzyme kinetic equation. Some substances tend to saturation uptakes approaching, though less than, the K^+ content of the particles, while under the conditions used here others tend to much lower maximum uptakes [4].

Competition for accumulation must play a part in metabolic control and it is relevant to compare effects of third substances on accumulation and on oxidation by intact and ruptured particles. The rationale for this is that accumulation, with metabolism blocked, allows evaluation of competition for occupation of the mitochondrial compartment. Oxidation by particles should be independent of accumulation. There should be a combination of the effect of accumulation and of enzyme interaction when intact particles are used. This approach leads to two mechanisms for metabolic inhibition: there can either be simple competition for entry or a combination of this with competition at the enzyme. As one example, phosphate competes with succinate for accumulation

Table 2. Effect of phosphate and malonate on accumulation and oxidation of succinate

PHOSPHATE		
Accumulation	Oxidation by particles	Oxidation by intact mitochondria
Competes K_i 3·1 to 4·4 mM	No effect	Inhibits K_i 3·2 mM
MALONATE		
Accumulation	Oxidation by sonicate	Oxidation by intact mitochondria
Competes K_i 1·7 mM	Inhibits K_i 0·8 mM	Inhibits K_i 0·15 mM

and also slows its oxidation in intact particles, but is without effect when added to disrupt particles so it seems to act at the membrane only (Table 2). Malonate is more effective, having a lower K_m when used to inhibit succinate oxidation by intact mitochondria than when applied to a sonicate. This can be explained by its being concentrated internally by a bigger factor than the succinate which is being consumed. One mechanism for stimulation of metabolism by an added anion is by its displacing or combining with inhibitory products. This may be the origin of the stimulation of malate oxidation in intact mitochondria by phosphate, which is without effect on the sonicated preparation.

Competitions can occur between specific substances while others are unaffected. This points to the existence of compartments for some at least of the substances. For example, although malate or succinate accumulations are reduced by simultaneous presence of phosphate, citrate or malonate, the latter have no obvious effect on hydroxybutyrate accumulation. Support for the special compartmentation of hydroxybutyrate is also derived from the observation that, unlike most other substrates, the content of this compound is quite unaffected when K^+ is gained under the influence of valinomycin (Table 3). In this respect hydroxybutyrate behaves like ATP and presumably the reason is that a permeability barrier prevents its net movement. For other substrates the K_m's are diminished in the presence of valinomycin (see Table 4 and Ref. 6) which suggests that the induction of high K permeability increases the force acting on the substrate without providing more carriers. One consequence of additional entry of substrate, when this does occur, is to keep up a better supply

Table 3. Gains of anions with K under influence of valinomycin

Anion	Ratio: equivalents anion gained/K^+ gained	Inhibitor and energy source
Citrate	0·9	Antimycin, TMPD ascorbate
Malate	0·6	Rotenone succinate
Malonate	1·2	Glutamate
Oxoglutarate	0·26	Arsenite hydroxybutyrate
Glutamate	0·02	Rotenone TMPD ascorbate
Succinate	0·32	Antimycin TMPD ascorbate
Hydroxybutyrate	0·02	Rotenone TMPD ascorbate
ATP (2 mM)	0·00	Glutamate & malate

Anion at 1 mM, valinomycin at 10 μg/g protein; medium with 10 mM KCl, 250 mM sucrose. Inhibitors of oxidation of the measured compound were present.

Table 4. Effects of valinomycin at 10 μg/g protein on K_m's

	Substrate	K_m no Val. mM	K_m with Val. mM
A. Accumulation	Oxoglutarate	1·7	1·0
	Pyruvate	1·4	0·9
B. Oxidation	Succinate	2·6	0·25
C. Inhibition of succinate oxidation by malate		K_i	K_i
	Malate	2·6	0·45

Medium: 250 mM sucrose, 10 mM KCl, 20 mM tris Chloride, pH 7·4.

Table 5. Computed example to show variation of C_i with C_e

C_e mM	C_i mM	Metabolic rate
0·1	0·02	5·0
0·3	0·07	15·5
0·5	0·14	24·7
0·6	0·19	29·2
1·0	0·75	47·5
2·0	10·0	59·0
3·0	19·0	59·5

The values assumed were $V_m/a = 60$, $b/a = 1{\cdot}8$ mM, K_m for enzyme = 0·2 mM, n = 2 and X = 100 μequiv/ml.

to the enzymes so that the apparent K_m is changed from that set by the penetration process towards that characteristic of the enzyme proper. Equally, access of a competitor is improved: for example, valinomycin at 10 μg/g reduced the K_m for succinate oxidation and the K_i for malate inhibition of the oxidation (Table 4, B & C).

These results led to the examination of the effect of metabolic consumption on the internal level. It was assumed that the classical Langmuir treatment for occupation of sites could be used for the accumulation process; this leads to the saturable uptake found experimentally. To this was added a Michaelis rate equation describing the rate of use of the internal substrate.

The scheme for combined control of internal level by entry and metabolism is as follows:

Rate entry = Rate leak + Rate oxidation

$$C_e \cdot a \cdot (X/n - C_i) = b \cdot C_i + \frac{V_m C_i}{V_m + C_i}$$

a—rate constant
C_e—applied conc.
X—internal capacity in equivs.
n—charge on anion
C_i—internal conc.

b—rate constant

V_m—max. velocity of enzyme reaction
K_m—Michaelis constant for enzyme reaction

With no metabolism the second term on the right vanishes and we get

$$C_i = \frac{X/n \cdot C_e}{b/a + C_e} \quad \text{or, rewriting: } C_i = \frac{X/n \cdot C_e}{K'_m + C_e} \quad (\text{with } K'_m \equiv b/a)$$

For substrates which can attain an extrapolated maximum content of 100 μequiv/g, the X/n term is taken as 100/net charge, V_m and K_m are obtained from the enzyme kinetics. For a hypothetical case for a dicarboxylic acid (n = 2) and taking $K'_m \equiv b/a = 1{\cdot}8$ mM and $V_m/a = 60$ mM the values for internal concentration and rate of oxidation were calculated over a range of applied concentrations, with $K_m = 0{\cdot}2$ mM. The numbers obtained (Table 5) show that the system has an apparent K_m for substrate about 0·6 mM which is less than the K_m for penetration (1·8 mM) but more than that for the enzyme (0·2 mM). Increase of a and b relative to V_m shifts the apparent K_m down towards value characteristic of the enzyme. An inert accumulated competitor for entry would inhibit the overall process because more external substrate would be required to reach a given internal concentration.

To justify and further test this approach we need to measure "a" and "b" and to relate X to the energy state or reserves of the mitochondria.

ACKNOWLEDGEMENTS

Supported by grants from the Medical Research Council, the Muscular Dystrophy Association of America and the Wellcome Trust. Thanks are due to Dr. B. C. Pressman for helpful discussions.

REFERENCES

1. Chappell, B., *in* "Round Table Discussion on Energy Level and Metabolic Control in Mitochondria", Adriatica Editrice, Bari, 1968.
2. Gamble, J. L., *J. biol. Chem.* **240** (1965) 2668.
3. Harris, E. J., *Biochem. J.* **109** (1968) 247.
4. Harris, E. J., *in* "Round Table Discussion on Energy Level and Metabolic Control in Mitochondria", Adriatica Editrice, Bari, 1968.
5. Harris, E. J., Höfer, M. P. and Pressman, B. C., *Biochemistry* **6** (1967) 1348.
6. Harris, E. J. and Manger, J. R., *Biochem. J.* **109** (1968) 239.
7. Harris, E. J., van Dam, K. and Pressman, B. C., *Nature, Lond.* **213** (1967) 1126.
8. Haslam, J. M. and Krebs, H. A., *Biochem. J.* **107** (1968) 659.
9. Höfer, M. P. and Pressman, B. C., *Biochemistry* **5** (1966) 3919.

FEBS Symposium, Volume 17, 1969, pp. 353-361

The Effects of Uncouplers and Detergents on the Permeability of the Mitochondrial Membrane

E. CARAFOLI

Institute of General Pathology, University of Modena, Modena, Italy, and

C. S. ROSSI

Institute of Biological Chemistry, University of Padova, Padova, Italy

According to the chemiosmotic hypothesis of oxidative phosphorylation [1], electron transport along the respiratory chain generates a pH or electrochemical gradient across the mitochondrial membrane: the gradient is discharged by a membrane ATPase operating in reverse, thus becoming the driving force for the synthesis of ATP. As for the phenomenon of uncoupling, the chemiosmotic hypothesis proposes that uncouplers carry protons across the normally impermeable mitochondrial membrane, thus eliminating the pH or electrochemical gradient formed during the transport of electrons. As a result of considerable conceptual and experimental activity, evidence both for and against the chemiosmotic hypothesis has been produced [2-5]. In particular, recent studies have shown that uncouplers increase the specific electric conductance of phospholipid bilayer membrane models by facilitating the transport of H^+ or OH^- across the model membrane [6-8]. On the other hand, Chappell and Haarhoff [9] have shown that dinitrophenol greatly stimulates the valinomycin-induced efflux of K^+ from another membrane model, the so-called "banghasomes".

The experiments described in the present paper have been designed to establish whether uncoupling agents can in fact abolish the normal impermeability of the membrane of intact mitochondria to H^+, as required by the chemiosmotic hypothesis. Preliminary accounts of these results have been communicated elsewhere [10, 11]. Rat liver mitochondria were prepared in 0·25M sucrose according to the conventional procedure of Schneider [12]. The movements of H^+ and K^+ were followed in vessels especially designed to minimize the back-diffusion of O_2 from the air, and monitored with Beckman specific glass electrodes, connected with Beckman Expandomatic pH meters. Simultaneous recordings were obtained on a Texas Servo/Riter strip chart recorder. Other technical details are found in the legends to the figures.

When a suspension of rat liver mitochondria in 0·25M sucrose is added to lightly buffered isotonic media, a very rapid pH shift ensues. Gear and Lehninger [13] have shown that the shift is due to the rapid exchange of monovalent

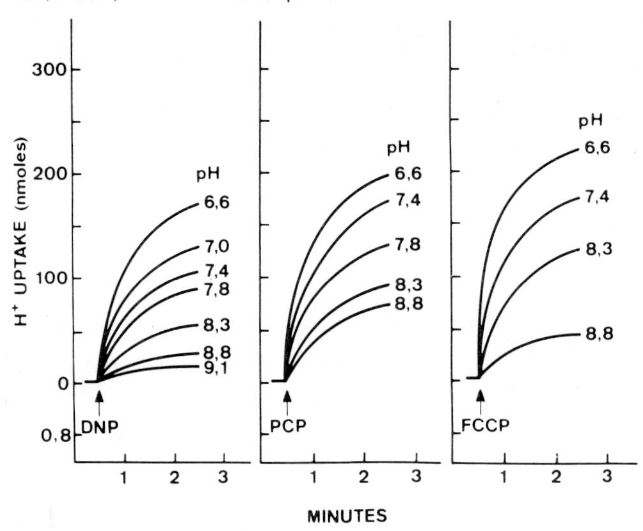

Figure 1. Effect of DNP, PCP, and FCCP on the movements of H+ between mitochondria and the extramitochondrial medium. The incubation medium contained 100 mM choline-Cl (buffered at the pH shown in the figure with 3 mM imidazole-Cl), and 8 mg mitochondrial protein. The final volume was 3·8 ml, inclusive of 0·15 ml of mitochondrial suspension in 0·25M sucrose, and the temperature was 25°C. The addition of mitochondria caused a rapid pH shift, which reached a steady state in less than 1 min. The pH's reported in the figure are those actually measured 40 s after addition of mitochondria. At the point indicated by the arrows, i.e. 40 s after addition of mitochondria, 10^{-4}M DNP, or 10^{-6}M PCP, or 5 x 10^{-5}M FCCP were added. The pH shifts induced by them have been normalized for the different buffering capacity of imidazole-Cl at the pH's shown in the figures. The curves were redrawn accordingly.

cations of the medium with protons from dissociating groups on the mitochondrial surface. Under the conditions used in the present paper, the size and the direction of the shift have been found to depend on the monovalent cation present in the medium, and on the initial pH: in lightly buffered media containing 80 mM NaCl, or KCl, or choline-Cl, the addition of mitochondria induced a shift towards alkalinity when the pH of the medium was below 6·2, and towards acidity if it was above 6·2. In either case, after a few seconds the pH trace reached a steady state: evidently an equilibrium had been reached between the concentration of H^+ in the incubation medium and protons bound to anionic groups on the surface of the mitochondria. The effect of a series of uncouplers on this equilibrium has been the subject of the present study; they were compared with the non-ionic detergents Lubrol and Triton X-100, which are known

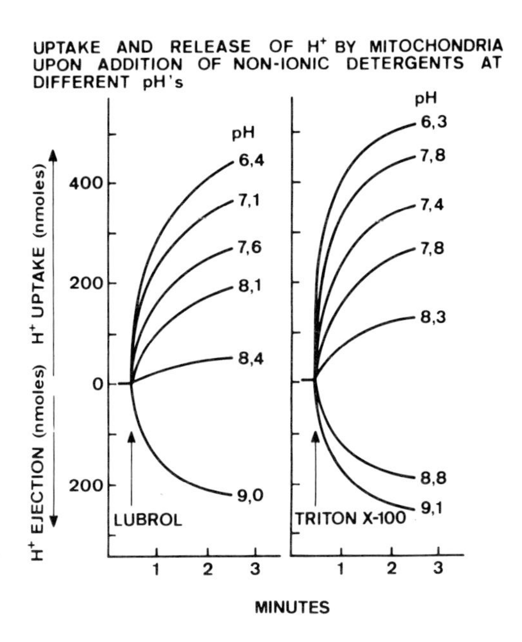

Figure 2. Effect of the non-ionic detergents Lubrol and Triton X-100 on the movements of H^+ between mitochondria and the extramitochondrial medium. Conditions were as in Fig. 1. Mitochondrial concentration was 8·9 mg in the experiment with Lubrol, and 9·2 mg in the experiment with Triton X-100. Final concentration of Lubrol was 0·025%, and of Triton X-100, 0·027%.

to abolish the permeability barriers in the mitochondrial membrane. Two experimental approaches have been used: in the first, uncoupling agents were added to mitochondria already equilibrated with salt-containing media, and the changes in pH induced by them were recorded. In the second, uncouplers were present in the medium before the addition of mitochondria. In this second case, their influence on the pH shift induced by the addition of the mitochondrial suspension was studied.

In the experiment shown in Fig. 1, the first approach was used. Addition of dinitrophenol (DNP), pentachlorophenol (PCP), or carbonyl cyanide-*p*-trifluoromethoxyphenylhydrazone (FCCP), induced a rapid absorption

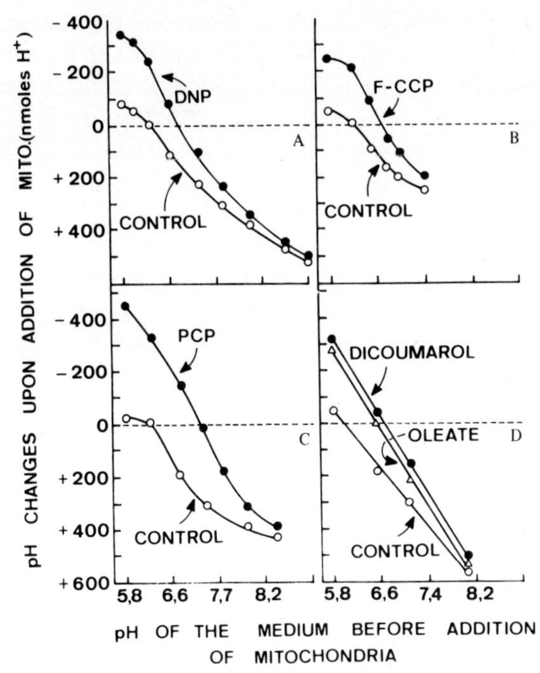

Figure 3. Movements of H^+ between mitochondria and the external medium, and effect of uncouplers. Addition of mitochondria to media of different pH, already containing the uncouplers. The reaction medium contained 80 mM choline-Cl (buffered with 5 mM imidazole-Cl), 4·5 mg of mitochondrial protein in A, B and C and 6·2 mg in D. Final volume was 1·9 ml, inclusive of 0·1 ml of mitochondrial suspension in 0·25M sucrose. The following concentrations of inhibitors were present, as shown in the figure: 10^{-4}M DNP, 10^{-6}M FCCP, 5 x 10^{-5}M PCP, 10^{-5}M dicoumarol, and 3 x 10^{-4}M oleate. The pH shifts reported in the figure were measured 2 min after the addition of mitochondria, after the glass electrode trace had reached a stable steady state.

of H^+ by mitochondria: the total amount of H^+ absorbed decreased as the initial pH of the incubation medium was increased from 6·6 to 9·1; however, at no pH did the uncouplers induce a deflection of the pH trace in the direction of the ejection of H^+ from mitochondria. By contrast, ejection of protons from mitochondria at alkaline pH was observed upon addition of the non-ionic detergents Lubrol or Triton X-100, which are known to abolish the permeability barriers in the mitochondrial membrane. Fig. 2 shows that when the initial pH of the incubation medium was 8·8 or more, the addition of the non-ionic detergents Lubrol or Triton X-100 did induce ejection of H^+ from mitochondria.

In Fig. 3, the effects of a larger group of uncoupling agents were tested by using the second experimental approach, i.e. by adding mitochondria to incubation media already containing the uncoupling agents. It is seen in the

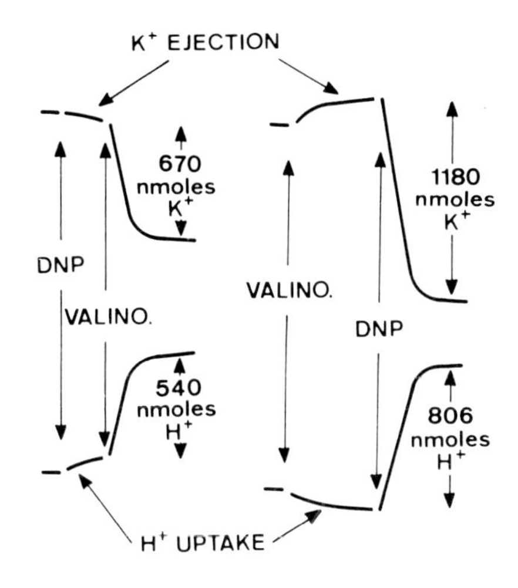

Figure 4. Effect of the sequential addition of DNP and valinomycin on H^+ and K^+ movements in isolated mitochondria at alkaline pH. The incubation medium contained 100 mM choline-Cl (buffered with 5 mM imidazole-Cl). After the addition of 10·3 mg rat liver mitochondrial protein (0·15 ml in 0·25M sucrose), the pH of the medium reached a steady state at 8·02 in about 60 s. The K^+ electrode trace also reached a steady state in about the same time. The final volume was 3·8 ml. The uptake of H^+ and the ejection of K^+ were followed as described in the methods, and the additions indicated in the figure were made at the times shown. The final concentration of DNP was 10^{-4}M, of valinomycin, $7{\cdot}6 \times 10^{-8}$M.

figure that, when the uncouplers were present in the medium, the output of protons from mitochondria at an initial pH of the medium higher than 6·2 was decreased; the alkalization of the incubation medium upon addition of mitochondrial suspensions at an initial pH below 6·2 was, on the other hand, considerably increased in the presence of all uncoupling agents tested. At this stage, it therefore seemed permissible to draw the conclusion that uncoupling agents and non-ionic detergents had the same effect on the permeability of the mitochondrial membrane to protons at a low pH of the incubation medium; however, the effect of uncouplers was lost almost completely at neutral and alkaline pH.

The experiments of Chappell and Haarhoff [9] mentioned above, and also experiments by Mitchell [1] on the collapse of pH gradients induced artificially across the membrane of rat liver mitochondria by the addition of HCl to the incubation medium, had suggested that DNP could act in cooperation with

EFFECT OF THE COMBINED ADDITION OF DNP AND VALINOMYCIN ON THE MOVEMENTS OF H^+ AND K^+ IN MITOCHONDRIA AT ACID pH

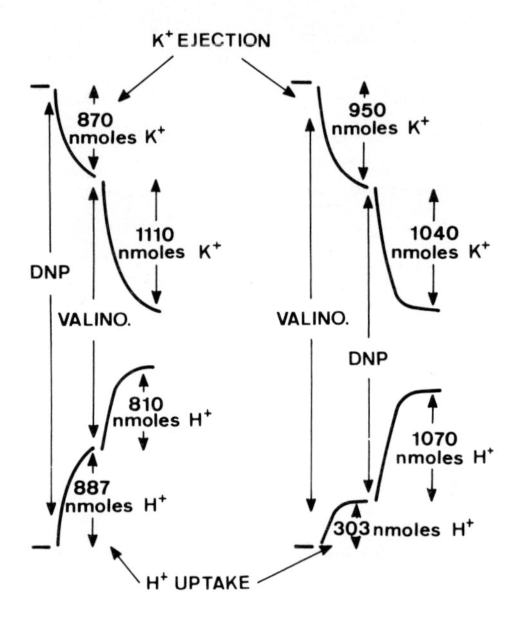

pH OF THE MEDIUM AFTER ADDITION OF MITOCHONDRIA = 6,19

Figure 5. Effect of the sequential addition of DNP and valinomycin on H^+ and K^+ movements in isolated mitochondria at acid pH. Conditions as in Fig. 4. At the end of the pH shift induced by the addition of mitochondria the pH of the medium was 6·19.

valinomycin, and that the movements of H^+ and K^+ could be dependent on each other. Therefore, DNP and valinomycin were added sequentially to mitochondria which had been equilibrated with a medium of alkaline pH (8·02), where no pH shifts were induced by either DNP or valinomycin added alone. Experiments not reported here had previously shown that valinomycin was also capable of inducing absorption of H^+ by mitochondria at acid pH, losing its effect more or less completely as the initial pH of the medium was raised to neutrality and above. It is seen in Fig. 4 that if DNP was added to valinomycin-treated mitochondria, or if valinomycin was added to DNP-treated mitochondria, a very rapid and large absorption of H^+ by mitochondria was observed. Simultaneous recordings of the K^+ movements showed that the movements of H^+ were accompanied by opposite movements of K^+. This experiment thus demonstrated that the penetration of H^+ into mitochondria depended on the permeabilization of the membrane to K^+; H^+ could only pass

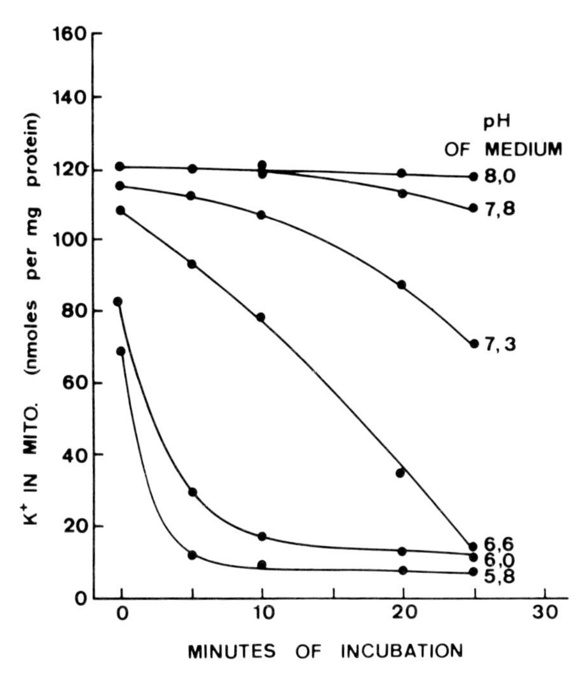

Figure 6. K^+ release from mitochondria as a function of the pH of external medium. The incubation medium contained 0·25M sucrose, 5 mM imidazole-Cl and 98·5 mg mitochondrial protein. Final volume, 40 ml. The incubation was carried out in air, at 25°C, with gentle shaking. At the times indicated in the figure, aliquots of the reaction medium were withdrawn, put in centrifuge tubes, immediately cooled to zero in ice-cold water, and centrifuged at 20,000 x g for 4 min to separate mitochondria. The pellets were resuspended in 5 ml 5% TCA, containing 14·42 meq. Li_2SO_4 per litre. The denaturated proteins were discarded by a low-speed centrifugation, and aliquots of the supernatants, diluted with the Li_2SO_4-TCA solution, were analyzed for potassium in a Beckman B spectrophotometer equipped with a Beckman flame attachment.

into mitochondria if K^+ left them in exchange. This is the reason why no penetration of H^+ into DNP-treated mitochondria took place at this alkaline pH, despite the fact that DNP had abolished the impermeability of the mitochondrial membrane to H^+. Similarly, the exit of K^+ from mitochondria seems to depend on the permeability of the membrane to H^+: indeed, at this alkaline pH, and despite the presence of valinomycin, no K^+ left mitochondria unless H^+ was carried into them following the addition of DNP. The same experiment on sequential additions of DNP and valinomycin was then repeated at pH 6·19, i.e. at a pH where either DNP or valinomycin added alone to mitochondria induced loss of K^+ and uptake of H^+; it is seen in Fig. 5 that at this low pH no "extra" effect on the movements of H^+ or K^+ was induced by the combined addition of

DNP and valinomycin. The experiment suggested that at acid pH the mitochondrial membrane was apparently freely permeable to K^+, since no valinomycin was needed for DNP to induce penetration of H^+ into mitochondria. Direct demonstration of the validity of this suggestion was provided by the experiment shown in Fig. 6. Freshly isolated mitochondria were incubated at 25°C in 0·25M sucrose lightly buffered at different pH values with imidazole-Cl; at different time intervals their K^+ content was analyzed by flame photometry. The figure shows that K^+ was rapidly lost by mitochondria when the pH of the medium was 6·0 or lower, and was maintained for at least 20 min at pH 7·8. At pH 7·3, only negligible amounts of K^+ were lost by mitochondria in 10 min. Therefore, it can be concluded that the mitochondrial membrane was in fact permeable to K^+ at acid pH values. It can also be concluded that the

EFFECT OF THE COMBINED ADDITION OF VARIOUS UNCOUPLERS AND VALINOMYCIN ON THE MOVEMENTS OF H^+ AND K^+ IN MITOCHONDRIA

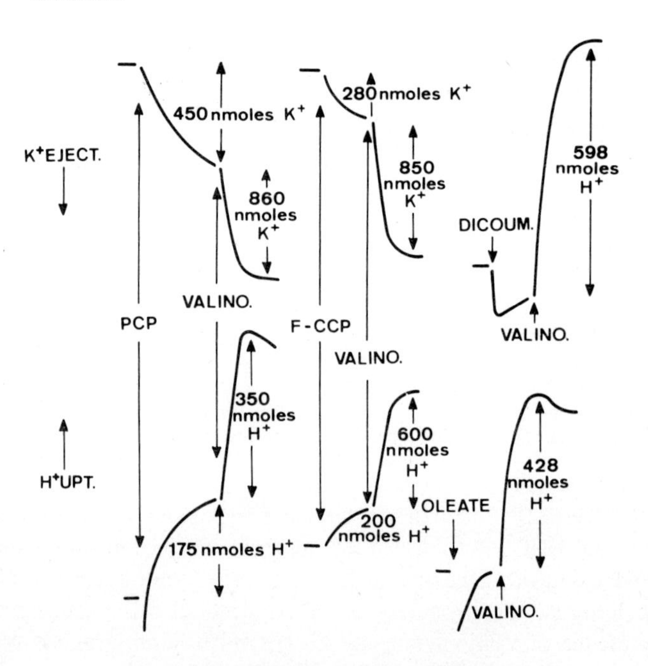

pH OF THE MEDIUM AFTER ADDITION OF MITOCHONDRIA = 7,30

Figure 7. Effect of the sequential addition of valinomycin and several uncoupling agents on K^+ and H^+ movements between mitochondria and the medium. Conditions as in Fig. 4. At the end of the pH shift induced by addition of 12·4 mg mitochondrial protein the pH of the medium was 7·3. Concentrations of uncoupling agents as in Fig. 3.

entrance of H^+ into mitochondria induced by DNP was obligatorily linked to the opposite translocation of K^+.

The effect of the combined additions of valinomycin and a series of uncouplers different from DNP on the movements of H^+ and K^+ between mitochondria and a medium of pH 7·3 are illustrated in Fig. 7. It is seen that all uncouplers behaved similarly, i.e. they all required the previous permeabilization of the membrane to K^+ induced by valinomycin to cause extensive and rapid penetration of H^+ into mitochondria. However, in contrast to DNP, PCP and FCCP added alone were capable of inducing a modest loss of K^+ from mitochondria, and some absorption of H^+ by them, at pH 7·3. They were thus apparently capable of inducing a partial permeabilization of the mitochondrial membrane to K^+, in addition to inducing the expected permeabilization to H^+.

ACKNOWLEDGMENTS

The investigation has been supported by grants from the U.S. Public Health Service (Grant No. RO5 TW 00216 to E. Carafoli, and Grant No. RO5 TW 00115 to C. S. Rossi), and from the National Research Council of Italy. The excellent collaboration of Mr. Paolo Gazzotti is gratefully acknowledged.

REFERENCES

1. Mitchell, P., "Chemiosmotic Coupling in Oxidative and Photosynthetic Phosphorylation", Glynn Research Ltd., Bodmin, Cornwall, (1966).
2. Chance, B. and Mela, L., *Nature, Lond.* **212** (1966) 369.
3. Chance, B. and Mela, L., *Nature, Lond.* **212** (1966) 372.
4. Tager, J. M., Veldsema-Currie, R. D. and Slater, E. C., *Nature, Lond.* **212** (1966) 376.
5. Slater, E. C., *Eur. J. Biochem.* **1** (1967) 317.
6. Bielawski, J., Thompson, T. E. and Lehninger, A. L., *Biochem. biophys. Res. Commun.* **24** (1966) 948.
7. Bielawski, J., Thompson, T. E. and Lehninger, A. L., *in* "Mitochondrial Structure and Compartmentation" (edited by E. Quagliariello, S. Papa, J. M. Tager and E. C. Slater), Adriatica Editrice, Bari, 1967, p. 181.
8. Hopfer, U., Lehninger, A. L. and Thompson, T. E., *Proc. natn. Acad. Sci. U.S.A.* **50** (1968) 484.
9. Chappell, J. B. and Haarhoff, K., *in* "Biochemistry of Mitochondria" (edited by E. C. Slater, Z. Kaniuga, and L. Wojtczack) Academic Press and Polish Scientific Publishers, London and Warsaw, 1966.
10. Carafoli, E. and Rossi, C. S., *Biochem. biophys. Res. Commun.* **29** (1967) 153.
11. Rossi, C. S. and Carafoli, E., *in* "Proceedings of International Conference on Biological Membranes" (edited by L. Bolis and B. A. Pethica), North Holland, Amsterdam, 1968, p. 264.
12. Schneider, W. C., *in* "Manometric Techniques" (edited by W. W. Umbreit, R. Burris and J. F. Stauffer) Burgess, Minneapolis, 1967.
13. Gear, A. R. L. and Lehninger, A. L., *J. biol. Chem.* **243** (1968) 3953.

FEBS Symposium, Volume 17, 1969, pp. 363-368.

The Effects of Oligoamines on Cation Binding in Mitochondria

N.-E. L. SARIS, M. F. WIKSTRÖM and A. J. SEPPÄLÄ

Department of Clinical Chemistry, University of Helsinki, Helsinki, Finland

The divalent cations Ca^{++}, Sr^{++}, and Mn^{++} are bound or taken up by mitochondria in an energy-dependent reaction, which is well known, but little understood at present. The main features of the reactions are stimulation of respiration or ATPase activity, depending on the available energy-generating system, ejection and apparent alkalinization of the mitochondrial inner compartment observable as an increased absorbance of BTB* at 618 nm [1-3], and, finally, accumulation of phosphate and other anions with accompanying swelling phenomena [4-8]. The aim of this study was to find out whether various divalent amines are accumulated by mitochondria and/or whether the accumulation of divalent cations is affected. The programme was enlarged to encompass the oligoamines spermidine and spermine, which are physiological substances that in some systems may replace divalent cations [9].

The oligoamines have been found to stabilize various membrane systems, including mitochondria [9-12]. Similar, but weaker, effects are seen with the higher homologues of the diamine series, 1,5-diaminopentane (cadaverine) and 1,4-diaminobutane (putrescine). Thus, these substances inhibit various types of swelling, including succinate or Ca^{++}-induced swelling [10-12]. In tissue homogenates, they are found associated mainly with the nuclear and microsomal fractions in proportion to their content of RNA, but they are partly recovered from the mitochondrial fraction [13]. These amines have been found to form complexes with acidic substances other than nucleic acids, such as phospholipids [9, 14, 15]. It seems likely that the binding is ionic, since it is inhibited by increasing the salt concentration of the medium [13-15]. In mitochondria aggregation is brought about by high concentrations indicating a neutralization of the repelling negative charge of the outer surface [9, 10]. These agents may thus be used to study the effect of decreasing the amount of fixed negative charges in the membranes.

In a paper published in 1963 it was shown that the cyclic pH changes produced by the addition of Ca^{++} can be used to evaluate membrane-stabilizing

*Non-standard abbreviations: Bromthymol blue, BTB; Rat liver mitochondria, RLM.

action [16]. Stabilizing agents, like Mg^{++}, delayed the pH-rising phase, or greater amounts of Ca^{++} had to be added in order to provoke it. The stabilizing effects of oligoamines are confirmed by the experiments shown in Fig. 1. The signals recorded on the addition of oligoamines are artefactual and are seen also in the absence of RLM. No effects were seen in the presence of the lower homologues of diamines, 1,2-diaminoethane to 1,4-diaminobutane.

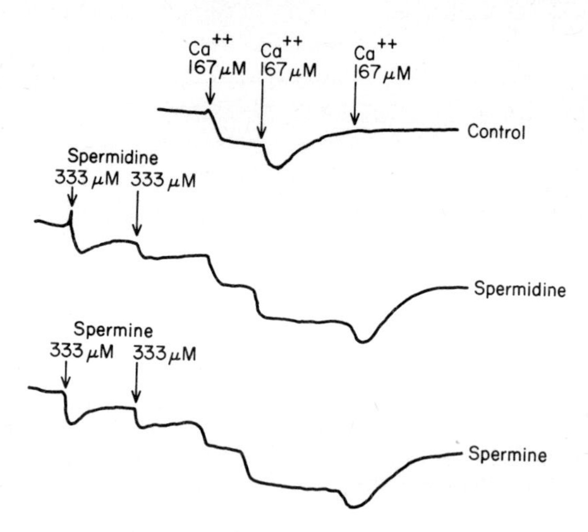

Figure 1. Ca++-induced pH cycles in the presence and absence of oligoamines. Experimental conditions: 225 mM mannitol, 75 mM sucrose, 20 mM KCl, 10 mM Tris-HCl, 7 mM tris-succinate, 5 mM Pi, 7μM rotenone, pH 7·4, RLM 1·3 mg/ml. pH rise: upward deflection.

The stabilizing effect of the oligoamines means that greater total amounts of Ca^{++} can be accumulated by RLM in their presence. This has recently been found also for the massive accumulation of Ca^{++} by microsomes derived from the sarcoplasmic reticulum of skeletal muscle [17]. In this system, the Ca^{++}-stimulated extra ATPase activity was partially inhibited; the same is true also for actomyosin ATPase activity [18]. We have found no effects of di- and oligoamines in RLM either on the latent ATPase, the Ca^{++}-induced ATPase activity associated with the uptake process, or the Mg^{++}-stimulated ATPase activity of aged mitochondria.

The pH electrode responds too slowly to give reliable information on the initial rate of Ca^{++} and Sr^{++} accumulation; the same is true of methods based on the separation of RLM from the incubation medium by filtration or centrifugation. We preferred the use of the BTB response obtained on the addition of a fairly large concentration of the cationic species (0·33 mM). Table 1 shows that the initial rates are slightly lower in the presence of about 1 mM spermine, while at 2 mM the rate is back at control values or maybe even

Table 1. Effect of spermine on the rate of change of absorbance of BTB, induced by Ca^{++} (0·33 mM)

Concentration (mM)	Relative rate
—	100
0·67	74
1·33	78
2·0	110

Experimental conditions: 200 mM Mannitol, 70 mM sucrose, 20 mM KCl, 30 mM Tris-HCl, 7 mM Tris-succinate, 8 μM rotenone, RLM 1·15 mg protein/ml, pH 7·4. An Aminco-Chance double-beam photometer was used at 633 nm with reference at 680 nm.

increased. The overall picture is the same with spermidine, and when Ca^{++} is replaced by Sr^{++}. Slight as these effects may be, they are still outside the experimental error, even though a rapid flow attachment was not available for the measurements.

The BTB response to the addition of Mn^{++} is smaller, slower, and occurs only after a lag (Fig. 2). In the presence of spermine, the lag is shortened or abolished, the plateau reached is higher, and the rates substantially increased (see Fig. 2 and Table 2). The increase in rate is linear over the range of concentrations used (Table 2). When tested at the same concentrations, the effect of spermidine is slightly smaller, that of diaminopentane clearly smaller, and with diaminobutane the rate approaches that of the control (Table 3). A corresponding increase in the respiration rate is observed. Addition of Mn^{++} slightly increases the respiration rate; in the presence of 2 mM spermine there is a clear respiration "jump" (Fig. 3) followed by a slowing respiration. Addition of spermine does not cause any change in the respiration, nor do any of the amines cause a change in the BTB signal.

Table 2. Effect of spermine on the rate of change in absorbance of BTB, induced by Mn^{++} (0·33 mM)

Concentration (mM)	Relative initial rate
	100
0·17	106
0·33	112
0·67	118
1·33	138
2·00	156

Experimental conditions as in Table 1.

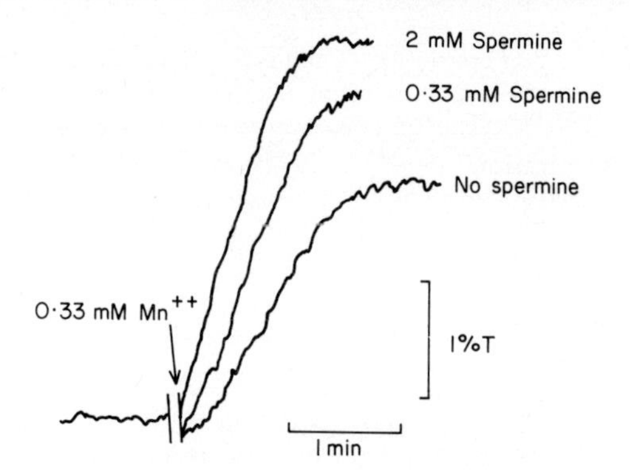

Figure 2. The effect of spermine on the Mn++-induced BTB change 633-680 nm. Experimental conditions as in Table 1.

The stimulation of Mn^{++} uptake indicated by the data above is similar to the enhancing action of Ca^{++} studied by Chance and Mela [19] and Ernster and Nordenbrand [20]. There are some differences, however; Ca^{++} has no effect if added before Mn^{++}, and is potent in much smaller concentrations, while the amines can be added before Mn^{++}. This difference is probably due to the rapid accumulation of Ca^{++} into RLM, while we have no evidence for any such reaction in the case of the amines. The stabilizing action of oligoamines on RLM accumulating Ca^{++} is again similar to the effect of Mn^{++} on Ca^{++}-binding [20].

Both the membrane stabilization and the effects on the uptake rates by the various inorganic and organic cations are most readily understood as resulting from binding to the phospholipids of the membranes. Inorganic divalent cations with high enough affinities could form bridges between the negatively charged phosphate groups of phospholipids. Organic oligoamines with suitably spaced positive charges separated by lipophilic carbon chains would be even more effective "cementing" agents.

Table 3. Effect of amines on the rate of change of absorbance of BTB, induced by Mn^{++} (0·33 mM)

Amine (2 mM)	Relative rate
Control	100
Diaminobutane (n)	115
Diaminopentane (n)	125
Spermidine	175
Spermine	190

Experimental conditions as in Table 1; RLM 0·59 mg protein/ml.

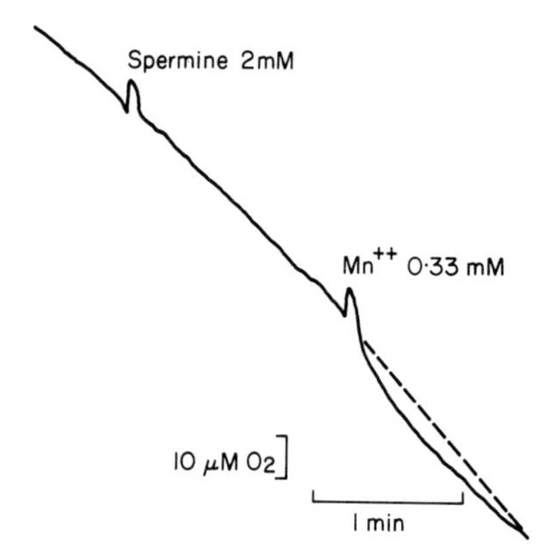

Figure 3. The effect of spermine on the Mn^{++}-stimulated respiration rate. Experimental conditions as in Table 1; RLM 0·59 mg protein/ml.

The effects on the rates of penetration are compatible with a mechanism in which fixed negative charges in the membrane retard the flow of cations by ion binding. When they are neutralized, the diffusion rates are increased. This mechanism would imply that inorganic divalent cations enhance their own penetration which would result in sigmoidal uptake kinetics. In the case of Mn^{++} the uptake rates are comparatively slow and it might be possible to demonstrate such kinetics. The sigmoidal form of the BTB response is compatible with this although it may be due to the buffering capacity of the intramitochondrial space [19] or to the pH dependence of the BTB signal [21]. The lag reported for the stimulation of ATPase activity by Mn^{++} [16] could also be cited in this connection.

It may be of interest to compare the effects of the oligoamines with those of other agents capable of reacting with phospholipids. Mela has reported a strong inhibition of divalent cation uptake by minute quantities of La^{+++} [22, 23] and other lanthanides [24], the mechanism probably being an interaction with cation carriers [23]. However, with small amounts of La^{+++}, a stimulation of M^{++}-induced responses were seen, while larger amounts inhibited these as well as the Ca^{++}-induced responses. Butacaine and some other local anaesthetic drugs increased the BTB response without affecting respiration [23], indicating a decrease in the buffering capacity of the membrane [23] or in the extent of binding of BTB. The oligoamines, on the other hand, increase the Mn^{++}-induced BTB response with a stimulation of respiration, while the effects on Ca^{++}-induced responses were slightly inhibitory at low oligoamine concentrations. By binding to phospholipids, the oligoamines seem to change the permeability of divalent cations, possibly by affecting the access to a carrier. A direct effect on the carrier [20] seems unlikely in view of the relative bulkiness of the oligoamines.

ACKNOWLEDGMENTS

This study was aided by a grant by the Sigrid Jusélius Foundation. The technical assistance of Mrs. A. Sarasjoki is gratefully acknowledged.

REFERENCES

1. Mela, L., *Fedn Proc. Fedn Am. Socs exp. Biol.* **25** (1966) 414.
2. Chance, B. and Mela, L., *J. biol. Chem.* **241** (1966) 4588.
3. Chance, B. and Mela, L., *J. biol. Chem.* **242** (1967) 830.
4. DeLuca, H. F. and Engström, G. W., *Proc. natn. Acad. Sci, U.S.A.* **47** (1961) 1744.
5. Rossi, C., Azzone, G. F. and Azzi, A., *Eur. J. Biochem.* **1** (1967) 141.
6. Lynn, W. S. and Brown, R. H., *Archs Biochem. Biophys.* **114** (1966) 260.
7. Ogata, E. and Rasmussen, H., *Biochemistry* **5** (1966) 57.
8. Rossi, E., Scarpa, A. and Azzone, G. F., *Biochemistry,* **6** (1967) 3902.
9. Tabor, H. and Tabor, C. W., *Pharmac. Rev.* **16** (1964) 245.
10. Tabor, C. W., *Biochem. biophys. Res. Commun.* **2** (1960) 117.
11. Herbst, E. J. and Witherspoon, B. H., *Fedn Proc. Fedn Am. Socs exp. Biol.* **19** (1960) 138.
12. Tabor, C. W., *Fedn Proc. Fedn Am. Socs exp. Biol.* **19** (1960) 139.
13. Raina, A. and Telaranta, T., *Biochim. biophys. Acta* **138** (1967) 200.
14. Razin, S. and Rozansky, R., *Archs Biochem. Biophys.* **81** (1959) 36.
15. Siekevitz, P. and Palade, G. E., *J. Cell Biol.* **13** (1962) 217.
16. Saris, N.-E. L., *Soc. Sci. Fenn. Comment. Phys.-Math.* **28 n:r 11** (1963) 1.
17. de Meis, L., *J. biol. Chem.* **243** (1968) 1174.
18. de Meis, L. and de Paula, H. J., *Archs Biochem. Biophys.* **119** (1967) 16.
19. Chance, B. and Mela, L., *Biochemistry,* (1966) 3220.
20. Ernster, L. and Nordenbrand, K., Abstr. 4th FEBS Meeting, Oslo, 1967, p. 108.
21. Saris, N.-E. L. and Seppälä, A. J., *Eur. J. Biochem.* in press.
22. Mela, L., *Fedn Proc. Fedn Am. Socs exp. Biol.* **26** (1967) 456.
23. Mela, L., *Archs Biochem. Biophys.* **123** (1968) 286.
24. Mela, L., *Biochemistry* **8** (1969) 2481.

FEBS Symposium, Volume 17, 1969, pp. 369-377

Energy-independent and Energy-dependent Interactions of Cations with Mitochondria

A. L. LEHNINGER, C. S. ROSSI*, E. CARAFOLI† and B. REYNAFARJE

Department of Physiological Chemistry, Johns Hopkins University School of Medicine Baltimore, Maryland, U.S.A.

It is now well known that the divalent cations Ca^{++}, Mn^{++}, and Sr^{++} are readily accumulated by mitochondria in energy-linked reactions, in contrast to Mg^{++}, K^{+}, and Na^{+} which are accumulated only after the membrane is made permeable by appropriate pretreatment (cf. [1]). Two alternative explanations have been offered. One holds that Ca^{++} is freely diffusible through the membrane and enters mitochondria without intervention of specific transport mechanisms, in exchange for H^{+} generated by electron transport [1, 2]. The other postulates that Ca^{++} must first undergo energy-independent binding to a specific membrane component before it interacts with a respiration-energized structure, a postulated high-energy chemical intermediate, and is then actively transported into the mitochondria [3]. That mitochondria can bind Ca^{++} ions from the medium in an energy-independent reaction is now clear from the work of several investigators [4-6]. However, it seems doubtful whether the type of Ca^{++} binding described actually represents an intermediate step in respiration-dependent accumulation of this cation, since the reported affinity of the mitochondrial membrane for binding Ca^{++} in the absence of electron transport [6] is relatively low in comparison with the very high affinity for Ca^{++} shown by respiring mitochondria [1, 3]. These uncertainties have prompted a more detailed investigation of energy-independent Ca^{++} binding by rat liver mitochondria in relation to net energy-linked accumulation [7, 8].

The first point examined was the behaviour of the endogenous-bound Ca^{++} of rat liver mitochondria. When isolated from sucrose homogenates they were found to contain between 12–16 μmoles Ca^{++} per g mitochondrial protein. Following isolation from sucrose fortified with 1 mM EDTA, they contained approximately 8–9 μmoles Ca^{++} per g protein. When they were incubated

*Present address: Institute of Biological Chemistry, University of Padova, Padova, Italy.
†Present address: Institute of General Pathology, University of Modena, Modena, Italy.

Table 1. Effect of inhibitors on endogenous Ca^{++}

	Loss of Ca^{++} nmoles mg^{-1}
None	1·9
DNP (30 μM)	6·1
Gramicidin (40 μg)	1·1
Oligomycin (15 μg)	1·0
Rotenone (0·5 μM) + antimycin A (0·125 μM)	0·0
Cl-CCP	6·8
Dicumarol	5·8

Endogenous Ca^{++} at zero time = 15·8 nmoles per mg protein. Incubation 1 min at 0°C.

anaerobically or in the presence of respiratory inhibitors for 15 min at 25°C, mitochondria isolated from pure sucrose slowly lost approximately 10–20 per cent of their endogenous Ca^{++}. On the other hand, the addition of 2,4-dinitrophenol and other true uncoupling agents caused extremely rapid loss of about 40 per cent of the endogenous Ca^{++}, which was complete within 10 s at 0°C. The remainder of the endogenous Ca^{++} was completely stable to dinitrophenol (Table 1). If mitochondrial Ca^{++} is labeled *in vivo* by injection of $^{45}Ca^{++}$ into rats prior to isolation, about two-thirds of the radioactivity is discharged upon brief incubation in the presence of DNP or other true uncouplers. The remaining one-third of the mitochondrial $^{45}Ca^{++}$ was not discharged by uncouplers, even upon prolonged incubation (Fig. 1).

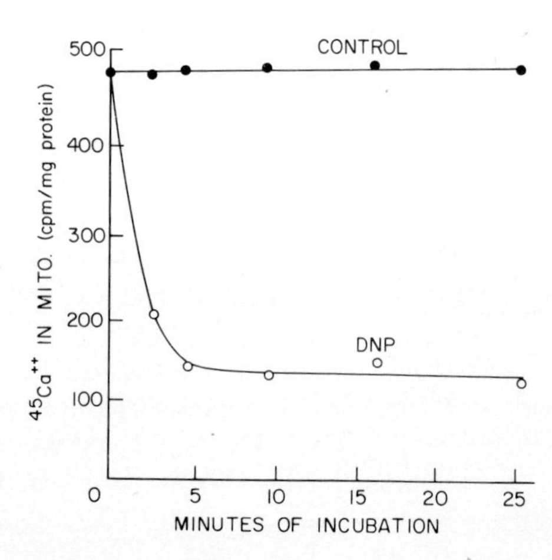

Figure 1. Discharge of endogenous $^{45}Ca^{++}$ from *in vivo*-labelled mitochondria. Rats were injected with 10 μc $^{45}CaCl_2$ 5 min prior to death.

Presumably, injected $^{45}Ca^{++}$ equilibrates very rapidly with one pool of endogenous mitochondrial Ca^{++} which is labile and discharged by uncouplers. A second pool of the mitochondrial Ca^{++} is probably in more sluggish equilibrium with injected $^{45}Ca^{++}$, and is not accessible to uncouplers. Then there is still another fraction or pool (about 5 μmoles per g protein), which does not equilibrate with the injected $^{45}Ca^{++}$, and is not discharged by uncouplers. Isotopic tracer experiments, carried out both *in vivo* and *in vitro* [20], have indicated that the labile pools turn over rapidly, while the stable pool does not. Neither form of endogenous Ca^{++} was discharged by gramicidin, valinomycin, or oligomycin.

Detailed study was then carried out of the affinity of freshly prepared rat liver mitochondria for binding exogenous Ca^{++} in the presence of antimycin A and rotenone to inhibit respiration [7]. In confirmation of the work of Rossi *et al.* [6] it was found that rat liver mitochondria contain a large number (~40-50 μmoles Ca^{++} per g protein) of respiration-independent binding sites for Ca^{++}, having a relatively low affinity (half-saturated at ~200 μM Ca^{++}). However, a more detailed study of the Ca^{++} affinity of mitochondria, measured down to extremely low concentrations of added Ca^{++}, revealed a second class of Ca^{++} binding sites, very few in number, but extremely high in affinity. This

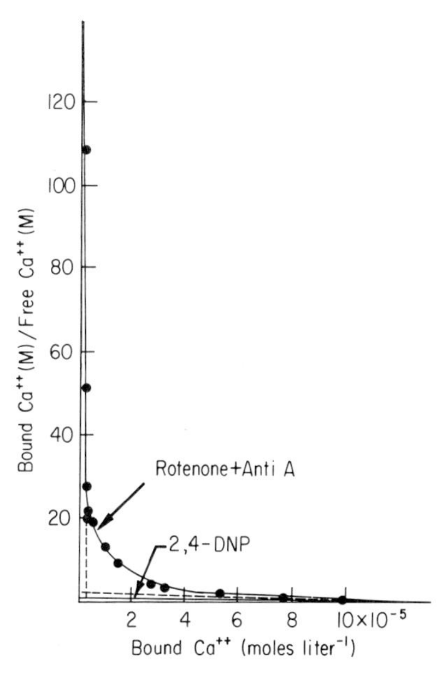

Figure 2. Scatchard plots of Ca^{++} binding. In the absence of dinitrophenol the biphasic curve indicates the presence of two classes of binding sites. In the presence of dinitrophenol the high affinity sites are abolished.

relationship was revealed by means of Scatchard plots of Ca^{++} binding data, employing either $^{40}Ca^{++}$ or $^{45}Ca^{++}$. Extrapolation from the Scatchard plots (Fig. 2) showed that the high affinity sites could bind only about 0·6 μmoles added Ca^{++} per g of protein and that they are half-saturated at about 0·01 μM Ca^{++}, which corresponds to an extraordinarily high affinity, higher than that of most enzymes for their substrates. All these measurements were made in the presence of sufficient antimycin A + rotenone to block net electron transport completely [7].

The addition of 2,4-dinitrophenol and other uncoupling agents caused not only discharge of some of the endogenous Ca^{++} but also inhibition of Ca^{++} binding at the high affinity sites, without affecting Ca^{++} binding at the low affinity sites significantly. Other true uncoupling agents, such as FCCP and dicumarol, showed precisely identical effects. On the other hand, inducers of alkali metal cation permeability, such as valinomycin and gramicidin, did not discharge endogenous Ca^{++} nor did they prevent high-affinity Ca^{++} binding (Table 2). Neither oligomycin nor atractyloside alone or in combination with

Table 2. Effect of inhibitors on high-affinity Ca^{++} binding

Inhibitor	Concentration	Ca^{++} bound nmoles per mg protein	Per cent inhibition
Antimycin A	0·225 μM	1·996	0
Rotenone	0·250 μM	1·991	0
Antimycin A + rotenone	—	1·980	0
KCN	50 μM	1·960	2·5
Dicoumarol	5 μM	1·598	54
2,4-Dinitrophenol	50 μM	1·264	90
Cl-CCP	0·25 μM	1·176	100
Aurovertin	10 μg	1·996	0
Oligomycin	10 μg	1·980	0
Oligomycin + antimycin A + rotenone	—	1·965	<2
Gramicidin	10 μg	1·976	<1
Gramicidin + NaCl	—	1·974	<1
Valinomycin	0·4 μg	1·976	<1
Valinomycin + KCl	—	1·976	<1

The test system (2·0 ml) consisted of Sucrose-Tris medium (pH 7·4) and the inhibitors or combinations thereof as indicated. When the inhibitors were used in combination, their concentrations were the same as when used alone. KCl and NaCl were added at 2·0 mM where shown. Rat liver mitochondria (5 mg protein) were added at 0°C, followed by addition of 2·0 nmoles $^{45}CaCl_2$ per mg protein. The tubes were centrifuged at 0°C immediately after mixing and the $^{45}Ca^{++}$ remaining in the supernatant medium was counted. The tube containing antimycin A + rotenone bound 1·98 nmoles $^{45}Ca^{++}$ per mg protein; the tube containing Cl-CCP bound 1·18 nmoles $^{45}Ca^{++}$.

antimycin A + rotenone prevented high-affinity Ca^{++} binding, even when oligomycin was added in huge excess [7]. These results showed that the high-affinity Ca^{++} binding is not due to traces of endogenous ATP.

Because some 4-6 μmoles endogenous Ca^{++} per g protein can be rapidly discharged by dinitrophenol, it appears possible it may also be bound to high-affinity sites. Tentatively it may be concluded that the latter may number anywhere from about 0·6 to 6 μmoles Ca^{++} per g protein.

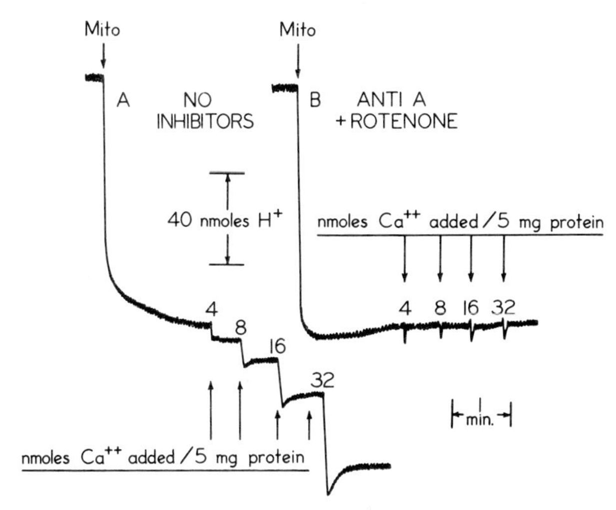

Figure 3. Effect of electron transport on H^+ ejection in the presence of Ca^{++}.

Another indication that high-affinity Ca^{++} binding is respiration-independent under the conditions examined was provided by comparison of hydrogen ion exchanges during Ca^{++} binding in the absence and presence of active electron transport. Although Rossi *et al.* [6] had earlier concluded that respiration-independent binding of Ca^{++} by the low-affinity sites of rat liver mitochondria is accompanied by a nearly equivalent discharge of protons into the suspending medium, we could find no evidence whatsoever for discharge of protons during high-affinity Ca^{++} binding when respiration was inhibited, as is shown in Fig. 3. Although the glass electrode system was found to be capable of measuring acid production equivalent to the amounts of Ca^{++} bound at the high-affinity sites, no measurable H^+ ejection occurred during Ca^{++} binding in the absence of respiration. On the other hand, when electron transport was allowed to take place, normal amounts of H^+ ejection occurred on addition of small amounts of Ca^{++}. It must therefore be concluded that binding of Ca^{++} to high-affinity groups in the mitochondria may occur without displacement of H^+; ejection of H^+ ions occurs only in a later stage when electron transport is initiated. Further investigation of low-affinity respiration-independent Ca^{++} binding revealed that it also is unaccompanied by H^+ ejection, contrary to earlier reports [6], but in agreement with Wenner and Hackney [9]. The acid production observed by Rossi *et al.* [6] can probably be rationalized on the basis of the experimental conditions that were employed, which permitted acid-base changes, such as described by Gear and Lehninger [10], to occur.

Further proof that high-affinity Ca^{++} binding can be dissociated into a respiration-independent phase, unaccompanied by H^+ ejection, and a respiration-dependent phase, accompanied by H^+ ejection, was provided by use of other inhibitors, namely ethanol and the La^{3+} ion [8, 11, 21]. Ethanol in concentrations of 100-200 mM completely blocked respiration-dependent Ca^{++} accumulation by rat liver mitochondria, even when Ca^{++} is added in large excess. It also blocks Ca^{++}-induced jumps in oxygen uptake. On the other hand, ethanol

Table 3. Inhibition of high affinity Ca^{++} binding by La^{3+}

La^{3+} μM	Per cent inhibition
0·1	0
0·3	0
0·1	7
0·3	16
1·0	26
3·0	48
10·0	78

in these concentrations does not block high-affinity Ca^{++} binding by mitochondria. This finding therefore confirms the results with the specific respiratory inhibitors antimycin A + rotenone. A second line of evidence came from an investigation of the capacity of other di- and trivalent metal ions to interfere with high-affinity Ca^{++} binding [8, 11]. Although Na^+, K^+, and Mg^{++} did not interfere with high-affinity energy-independent Ca^{++} binding, both Sr^{++} and Mn^{++} inhibited in a competitive manner. They were bound by mitochondria, independently of respiration, and showed biphasic Scatchard binding plots very similar to those observed with Ca^{++}. Mn^{++} and Sr^{++} have somewhat lower affinity than Ca^{++} for the high-affinity sites [7]. A number of other di- and trivalent cations compete with Ca^{++} at the high-affinity sites. The most striking effect was provided by La^{3+} ions, which inhibited Ca^{++} binding at very low concentrations (Table 3). La^{3+} in concentrations sufficient to block non-energized high-affinity Ca^{++} binding completely, also blocked respiration-energized net Ca^{++} accumulation [8]. La^{3+} alone did not stimulate electron transport in State 4 [8] and completely prevented the normal stimulation given by Ca^{++}. These findings are therefore consistent with the conclusion that high-affinity Ca^{++} binding is a necessary step preceding energized Ca^{++} uptake. Preliminary experiments indicate that La^{3+} is itself bound by mitochondria. Ca^{++} uptake in the presence of phosphate was not influenced by La^{3+}, nor was oxidative phosphorylation of ADP. However, La^{3+} forms an extremely insoluble salt with phosphate.

The high-affinity binding sites for Ca^{++} are extremely labile, whereas the low-affinity sites are rather stable to a number of treatments [7]. Ageing of mitochondria for short periods at 30°C caused complete loss of the capacity for high-affinity binding. Water lysis of mitochondria, which removes the outer membrane and the matrix to produce ghosts, also caused loss of capacity for high-affinity Ca^{++} binding, although such preparations can still phosphorylate, but without acceptor control [12]. Such ghosts do not accumulate Ca^{++} unless ADP is present [12]. Inner membrane preparations made by the digitonin procedure of Schnaitman *et al.* [13] retain the high-affinity Ca^{++} binding sites, as do sonic particles. On the other hand, phosphorylating particles prepared with the non-ionic detergent Lubrol-WX do not; the latter particles are "inside out" and show knobs on the outer surface of the membrane under negative contrast electron microscopy. None of these procedures caused loss of low-affinity binding but some increased the amount of Ca^{++} so bound, such as the water lysis treatment. Heating of mitochondria at 100°C for 5 min completely abolished high-affinity Ca^{++} binding, but caused only a slight diminution of low-affinity Ca^{++} binding [8].

Some attempts were made to identify the chemical nature of the low-affinity Ca^{++} binding sites. Since no net H^+ is ejected during non-energized Ca^{++} binding, a mechanism in which Ca^{++} exchanges for H^+ at some protonated group on the mitochondria may be excluded. Significantly, some phospholipids, such as phosphatidyl serine and cardiolipin, will readily bind Ca^{++} at pH 7·0, which is far above the pK′ of their dissociating secondary phosphate groups (pK′ $\simeq$ 2·0); Ca^{++} binding by such groups would not lead to proton displacement [14]. Phosphatidyl serine and cardiolipin exist in more than sufficient amounts in the mitochondrial membrane to account for both high-affinity and low-affinity Ca^{++} binding. The fact that boiled mitochondria are almost as effective as fresh mitochondria in low-affinity Ca^{++} binding suggests that a heat-stable moiety, such as the phosphoglycerides, could well account for low-affinity binding. Binding of Ca^{++} by proteins is also well known; the actomyosin system can bind Ca^{++} very strongly (see ref. [15]). Ca^{++} binding is also known to occur at sialic acid residues of polysialogangliosides and of some glyco-proteins. We have established the presence of hexosamine in both protein and lipid fractions of the mitochondrial membranes. Other experiments on low-affinity Ca^{++} binding suggest that it may take place inside mitochondria which have been opened by valinomycin and that K^+ may compete with Ca^{++}, in accordance with an earlier report [6].

The preceding experiments showed that the high-affinity sites differ greatly in affinity and number from the low-affinity sites and therefore are probably chemically different. At least two working hypotheses for the nature of the high-affinity Ca^{++} binding sites may be entertained. One is that they are themselves the widely postulated high-energy intermediates of oxidative phosphorylation, since their number, affinity (K_m = 0·01 μM), and sensitivity to dinitrophenol are those expected for such intermediates and the enzymes that can generate them. However, there are some problems in accepting this hypothesis, in addition to the fact that such intermediates have never been found to exist. High-affinity Ca^{++} binding is shown only by preparations still capable of oxidative phosphorylation, but not all such preparations show high-affinity Ca^{++} binding when respiration is inhibited. This is an inconsistency, since Ca^{++} normally has a higher affinity for the respiratory chain than ADP [1, 3]. It is possible, however, that high-affinity Ca^{++} binding in the presence of a respiratory block occurs only with *preformed* high-energy intermediates, which have been postulated to be the energy-source for the fast bursts of ATP formation observed earlier in both intact and submitochondrial preparations [16, 17]. Attempts to titrate away such preformed intermediates with ADP or DNP are under way.

An alternative presents itself, namely, that the high-affinity Ca^{++} binding sites of mitochondria represent the action of a membrane-linked carrier molecule similar to that operating for energy-independent transport of ATP, and those invoked for other specific anions. The number and affinity of the Ca^{++} binding sites are in the range of those of the ATP carrier reported earlier [18, 19]. Whether such a Ca^{++} carrier is an exchange diffusion carrier, capable of exchanging Ca^{++} for K^+ or Ca^{++} for H^+, is now under study. So far, our experiments suggest the possibility that Ca^{++} may be bound to such a carrier without H^+ exchange, but can be transported to the inside only if H^+ generated by electron transport is transported out.

Whatever the mechanism involved, our findings indicate strongly that Ca^{++} binds to some specific membrane element with an affinity that greatly exceeds the affinity of most enzymes for their substrates. The entry of Ca^{++} into mitochondria is therefore not to be regarded as a simple concentration-dependent physical diffusion, but as the result of the action of a molecular component that shows saturation characteristics and specificity, as is to be expected in view of the exceedingly great importance of Ca^{++} in intracellular excitation phenomena. In our opinion, Ca^{++} accumulation by mitochondria is not merely an accidental feature of mitochondrial activity or the casual result of its ability to cause collapse of a membrane gradient. In addition to providing an approach to crucial experiments on the mechanism of proton movements during respiration, Ca^{++} movements in mitochondria may represent an important process in cell physiology [20].

REFERENCES

1. Lehninger, A. L., Carafoli, E. and Rossi, C. S., *Adv. Enzymol.* **29** (1967) 259.
2. Mitchell, P., *Biol. Rev.* **41** (1966) 445.
3. Chance, B., *J. biol. Chem.* **240** (1965) 2729.
4. Chappell, J. B., Cohn, M. and Greville, G. D., *in* "Energy-linked Functions of Mitochondria" (edited by B. Chance), Academic Press, New York, 1963, p. 219.
5. Rasmussen, H., Chance, B. and Ogata, E., *Proc. natn. Acad. Sci. U.S.A.* **53** (1965) 1069.
6. Rossi, C., Azzi, A. and Azzone, G. F., *J. biol. Chem.* **242** (1967) 951.
7. Reynafarje, B. and Lehninger, A. L., *J. biol. Chem.* **244** (1969) 584.
8. Lehninger, A. L., *J. biol. Chem.* to be submitted.
9. Wenner, C. E. and Hackney, J. H., *J. biol. chem.* **242** (1967) 5053.
10. Gear A. R. L., and Lehninger, A. L., *Biochem. biophys. Res. Commun.* 28, (1967) 840.
11. Lehninger, A. L., *Ann. N.Y. Acad. Sci.* in press.
12. Caplan A. I. and Greenawalt, J. W., *J. Cell Biol.* **32** (1967) 719.
13. Schnaitman, C., Erwin, V. G. and Greenawalt, J. W., *J. Cell Biol.* **32** (1967) 719.
14. Hauser, H. and Dawson, R. M. C., *Eur. J. Biochem.* **1** (1967) 61.
15. Weber, A. and Herz, R., *J. biol. Chem.* **238** (1963) 599.
16. Eisenhardt, R. H. and Rosenthal, O., *Fedn Proc. Fedn Am. Socs exp. Biol.* **21** (1962) 56.
17. Vignais, P. V., *Biochim. biophys. Acta* **78** (1963) 404.
18. Winkler, H. H., Bygrave, F. L. and Lehninger, A. L., *J. biol. Chem.* **243** (1968) 20.
19. Winkler, H. H. and Lehninger, A. L., *J. biol. Chem.* **243** (1968) 3000.
20. Carafoli, E., *J. gen. Physiol.* **50** (1967) 1849.
21. Mela, L., *Archs Biochem. Biophys.* **123** (1968) 286.

Author Index

Numbers in parentheses are reference numbers and are included to assist in locating references. Numbers followed by an asterisk refer to the page on which the reference is listed.

C

D

E

F

I

J

M

Q

R

S

Y

Z

THE LIBRARY
UNIVERSITY OF CALIFORNIA
San Francisco
THIS BOOK IS DUE ON THE LAST DATE STAMPED BELOW

Books not returned on time are subject to fines according to the Library Lending Code. A renewal may be made on certain materials. For details consult Lending Code.

14 DAY
MAY 22 1975
RETURNED
MAY 21 1975
14 DAY
DEC 15 1975
RETURNED
DEC 15 1975
14 DAY
JUL 21 1976
RETURNED
JUL 14 1976
14 DAY
OCT 20 1976
31
RETURNED
OCT 29 1976
14 DAY
JAN 24 1977
RETURNED
JAN 17 1977
14 DAY
JUN 21 1977
RETURNED
JUN 20 1977

15m-7,'72 (Q3550s4) 4128–A33-9

MAY 13 1974